Praise for

Continental Philosophy of Science

"Continental philosophers in Britain and the United States have for the most part ignored the enormous contribution of continental philosophy to the philosophy of science, just as philosophers of science in Britain and the United States have done. Gary Gutting has long been a leading exponent of the importance of this contribution and his superb collection, with its many new translations, should go a long way toward turning the tide."

ROBERT BERNASCONI, UNIVERSITY OF MEMPHIS

"This masterful collection of original texts and expert commentary demonstrates Continental philosophers' rich and diverse engagement with science, dispelling the notion that significant philosophical thinking about science is the sole prerogative of 'analytic' philosophers."

DANIEL DAHLSTROM, BOSTON UNIVERSITY

"This book makes a welcome contribution to the secondary literature on the history and philosophy of modern science. Gary Gutting has assembled an impressive gallery of essays, which collectively advance a powerful, if relatively negected, interpretation of the development of scientific method and practice. The pairing of influential historical figures with leading contemporary commentators is especially valuable."

DANIEL W. CONWAY, THE PENNSYLVANIA STATE UNIVERSITY

Blackwell Readings in Continental Philosophy

Series Editor: Simon Critchley, University of Essex

Each volume in this series provides a detailed introduction to and overview of a central philosophical topic in the continental tradition. In contrast to the author-based model that has hitherto dominated the reception of the continental philosophical tradition in the English-speaking world, this series presents the central issues of that tradition, topics that should be of interest to anyone concerned with philosophy. Cutting across the stagnant ideological boundaries that mark the analytic/continental divide, the series will initiate discussions that reflect the growing dissatisfaction with the organization of the English-speaking philosophical world. Edited by a distinguished international forum of philosophers, each volume provides a critical overview of a distinct topic in continental philosophy through a mix of both classic and newly commissioned essays from both philosophical traditions.

Continental Philosophy of Science

Edited by

Gary Gutting

Blackwell
Publishing

© 2005 by Blackwell Publishing Ltd

BLACKWELL PUBLISHING
350 Main Street, Malden, MA 02148-5020, USA
108 Cowley Road, Oxford OX4 1JF, UK
550 Swanston Street, Carlton, Victoria 3053, Australia

The right of Gary Gutting to be identified as the Author of the Editorial Material
in this Work has been asserted in accordance with the UK Copyright, Designs,
and Patents Act 1988.

First published 2005 by Blackwell Publishing Ltd.

Library of Congress Cataloging-in-Publication Data

Continental philosophy of science / [edited by] Gary Gutting.
p. cm. — (Blackwell readings in Continental philosophy)
Includes bibliographical references and index.
ISBN 0-631-23609-0 (alk. paper) — ISBN 0-631-23610-4 (pbk. : alk. paper)
1. Science—Philosophy. 2. Philosophy, European—20th century. I. Gutting,
Gary. II. Series.

Q175.C6936 2005
501—dc22
2004016924

A catalogue record for this title is available from the British Library.

Set in 10.5/12.5pt Bembo
by Kolam Information Services Pvt. Ltd, Pondicherry, India
Printed and bound in the United Kingdom
by TJ International, Padstow, Cornwall

The publisher's policy is to use permanent paper from mills that
operate a sustainable forestry policy, and which has been manufactured from pulp
processed using acid-free and elementary chlorine-free practices. Furthermore,
the publisher ensures that the text paper and cover board used have met acceptable
environmental accreditation standards.

For further information on
Blackwell Publishing, visit our website:
www.blackwellpublishing.com

CONTENTS

NOTES ON CONTRIBUTORS

Linda Martín Alcoff is Professor of Philosophy, Political Science, and Women's Studies at Syracuse University. Her books include *Feminist Epistemologies* (1993), co-edited with Elizabeth Potter; *Real Knowing: New Versions of the Coherence Theory of Knowledge*; *Epistemology: The Big Questions* (Blackwell, 1996); *Thinking From the Underside of History* (2000), co-edited with Eduardo Mendieta; *Identities: Race, Class, Gender, and Nationality* (Blackwell, 2003), co-edited with Eduardo Mendieta; and *Singing in the Fire: Stories of Women in Philosophy* (2003). *Visible Identities: Race, Gender and the Self* is forthcoming.

Penelope Deutscher is Associate Professor in the Department of Philosophy, Northwestern University. She is the author of *Yielding Gender: Feminism, Deconstruction and the History of Philosophy* (1997) and *A Politics of Impossible Difference: The Later Work of Luce Irigaray* (2002), and co-editor (with Kelly Oliver) of *Enigmas: Essays on Sarah Kofman* (1999) and (with Françoise Collin) *Repenser la politique: l'apport féministe*.

Michael Friedman is currently Frederick P. Rehmus Family Professor of Humanities at Stanford University. His publications include *Foundations of Space-Time Theories: Relativistic Physics and Philosophy of Science* (1983), *Kant and the Exact Sciences* (1992), *Reconsidering Logical Positivism* (1999), *A Parting of the Ways: Carnap, Cassirer, and Heidegger* (2000), and *Dynamics of Reason* (2001).

Jean Gayon is Professor of History and Philosophy of Science at the University of Paris 1, Panthéon Sorbonne, and a member of the Deutsche Akademie der Naturforscher Leopoldina. He has published 12 books (11 as editor) and 140 articles, including *Darwinism's Struggle for Survival: Heredity and the Hypothesis of Natural Selection* (1998); *Buffon 1988, Actes du colloque international du bicentenaire de la mort de Buffon, Paris-Montbard-Dijon* (ed., 1992); and *Bachelard dans le monde* (ed. with J. J. Wunenburger, 2000).

Gary Gutting holds the Notre Dame Chair in Philosophy at the University of Notre Dame. His books include *Religious Faith and Religious Skepticism* (1982), *Michel Foucault's Archaeology of Scientific Reason* (1989), *Pragmatic Liberalism and the Critique of Modernity* (1999), and *French Philosophy in the Twentieth Century* (2001). His edited volumes include *Paradigms and Revolutions: Applications and Appraisals of Thomas Kuhn's Philosophy of Science* (1980) and *The Cambridge Companion to Foucault* (2nd ed. 2005). He is the founder and editor of the electronic journal, *Notre Dame Philosophical Reviews* (http://ndpr.icaap.org).

Axel Honneth, born 1949 in Essen, studied Philosophy, Sociology, and German at Bonn, Bochum, and Berlin. At present he is Professor of Social Philosophy at the Johann Wolfgang Goethe-University and Director of the Institute for Social Research in Frankfurt/Main. Publications include *Social Action and Human Nature* (with Hans Joas, 1988), *Critique of Power* (1990), *Struggle for Recognition* (1994), *The Fragmented World of the Social: Essays in Social and Political Philosophy* (1995), *Suffering from Indeterminacy: Spinoza Lectures* (1999), and *Redistribution or Recognition? A Political-Philosophical Exchange* (with Nancy Fraser, 2003).

Todd May is a Professor of Philosophy at Clemson University. He has published on the work of contemporary French theorists, especially Michel Foucault, Gilles Deleuze, and Jean-François Lyotard. Most recently he is author of the forthcoming book, *Gilles Deleuze: An Introduction*, for Cambridge University Press.

Terry Pinkard is Professor of Philosophy at Northwestern University. His most recent books include *Hegel's Phenomenology: The Sociality of Reason* (1994), *Hegel: A Biography* (2000), and *German Philosophy 1760–1860: The Legacy of Idealism* (2002). He has had fellowships from the German Academic Exchange Service and the Alexander von Humboldt Foundation. In 1998, he was made an "Honorary Professor" (Ehrenprofessor) and "Honorary Teacher" (Ehrenlehrbeauftragte) at Tübingen University, Germany. He serves as Adviser to the Editorial Board (Mitwirker zur Redaktionsbeirat) for the *Zeitschrift für philosophische Forschung*. Currently, he is working on a translation of Hegel's *Phänomenologie des Geistes* for Cambridge University Press.

Hans-Jörg Rheinberger is Director at the Max Planck Institute for the History of Science in Berlin. He has published numerous articles on molecular biology and the history of science. He has written *Experiment, Differenz, Schrift* (1992), *Toward a History of Epistemic Things* (1997), and *Experimentalsysteme und epistemische Dinge* (2001). Among his edited books are *Die Experimentalisierung des Lebens* (together with M. Hagner, 1993), *Räume des Wissens* (together with M. Hagner and B. Wahrig-Schmidt, 1997), *The Concept of the Gene in Development and Evolution* (together with P. Beurton and R. Falk, 2000), and *Reworking the Bench: Research Notebooks in the History of Science* (with F. Holmes and J. Renn, 2003).

Joseph Rouse is the Hedding Professor of Moral Science and Chair of the Science in Society Program at Wesleyan University. He is the author of *How Scientific Practices Matter: Reclaiming Philosophical Naturalism* (2002), *Engaging Science: How to Understand*

Its Practices Philosophically (1996), and *Knowledge and Power: Toward a Political Philosophy of Science* (1987).

Richard Tieszen is Professor of Philosophy at San Jose State University, located in California's Silicon Valley. He has been a visiting professor at Stanford University and the University of Utrecht in the Netherlands, and is the author of *Mathematical Intuition: Phenomenology and Mathematical Knowledge* (1989) and *Phenomenology, Logic, and the Philosophy of Mathematics* (forthcoming). In addition, he is co-editor of *Between Logic and Intuition: Essays in Honor of Charles Parsons* (2000), and has been guest editor of several special issues of the journal *Philosophia Mathematica*. Professor Tieszen is the author of numerous articles and he has lectured widely in Europe and the United States.

Mary Tiles is Professor of Philosophy at the University of Hawaii at Manoa. Her major publications include *Living in a Technological Culture: Human Tools and Human Values* (co-authored with Hans Obderdiek, 1995); *The Authority of Knowledge: An Introduction to Historical Epistemology* (co-authored with Jim Tiles, 1993), *Mathematics and the Image of Reason* (1991), *The Philosophy of Set Theory: An Introduction to Cantor's Paradise* (Blackwell, 1989), and *Bachelard: Science and Objectivity* (1984).

ACKNOWLEDGMENTS

The editor and publisher gratefully acknowledge the permission granted to reproduce the copyright material in this book:

2. From G. W. F. Hegel, "Naturphilosophie," §246 in *Werke in zwanzig Bänden*, vol. 9 (Part Two of Hegel's *Encyclopedia of the Philosophical Sciences*), eds. Eva Moldenhauer and Karl Markus Michel. Frankfurt am Main: Suhrkamp Verlag, 1971. New English translation by Terry Pinkard.

4. Henri Bergson, "Psycho-physical Parallelism and Positive Metaphysics," pp. 33–4 and 43–57 from *Bulletin de la Société française de philosophie*, séance du 2 mai 1901. New English translation by Matthew Cobb.

6. Ernst Cassirer, selections from chapter 5, §III, and chapter 7 of *Substanzbegriff und Funktionsbegriff*. Berlin: Bruno Cassirer, 1910. New English translation by Michael Friedman.

8. (i) Edmund Husserl, pp. 28–31 from *Introduction to the Logical Investigations*, ed. Eugen Fink, trans. Philip J. Bossert and Curtis H. Peters. The Hague: Martinus Nijhoff, 1975. Reproduced by kind permission of Kluwer Academic Publishers; (ii) Edmund Husserl, pp. 48–57 (sections 9h and 9i) from *The Crisis of European Sciences and Transcendental Phenomenology*, ed. David Carr. Evanston, IL: Northwestern University Press, 1970. Reproduced by kind permission of Northwestern University Press.

10. Martin Heidegger, pp. 55–73 from *On "Time and Being"*, trans. Joan Stambaugh. New York: Harper & Row, 1972 (reissued 1977). English language © 1972 by Harper & Row Publishers Inc. Reprinted by permission of HarperCollins Publishers Inc.

12. Gaston Bachelard, pp. 156–67 (ch. IX) from *Essai sur la connaissance approchée*. Paris: J. Vrin, 1973 (first published 1928). French text © 1973 by Librarie Philosophique J. Vrin – Paris. Reproduced by permission of the publisher, http://www.vrin.fr. New English translation by Mary Tiles.

14. Georges Canguilhem, "The Object of the History of Sciences," from *Etudes d'histoire et de philosophie des sciences*. Paris: J. Vrin, 1983. French text © 1983 by Librarie Philosophique J. Vrin – Paris. Reproduced by permission of the publisher, http://www.vrin.fr. New English translation by Mary Tiles.

16. Michel Foucault, "Objectives" (pp. 81–91) and "Method" (pp. 92–102) from *History of Sexuality*, vol. I, trans. Robert Hurley. New York: Vintage, 1980. © 1978 by Random House Inc., New York. Originally published in French as *La Volonté du Savoir*. French text © 1976 by Editions Gallimard. Reprinted by permission of Georges Borchardt Inc., Éditions Gallimard, and Penguin Books Ltd.

18. Gilles Deleuze and Félix Guattari, pp. 117–20 and 125–29 from *What Is Philosophy?*, trans. H. Tomlinson and G. Burchell. New York: Columbia University Press, 1994. © 1994 by Columbia University Press. Reprinted with the permission of the publisher.

20. Luce Irigaray, "In Science, Is the Subject Sexed?" pp. 247–58 from *To Speak Is Never Neutral*, trans. Gail Schwab. New York: Routledge, 2002. English translation © 2002 by Continuum and reproduced by kind permission of Continuum International Publishing Group. Originally published in 1985 as *Parler n'est jamais neutre* by Les Editions de Minuit.

22. Jürgen Habermas, "*Knowledge and Human Interests*: A General Perspective," pp. 301–17 from *Knowledge and Human Interests*, trans. Jeremy J. Shapiro. Boston: Beacon Press, 1971. German text © 1968 by Suhrkamp Verlag, Frankfurt am Main. English translation © 1971 by Beacon Press. Reprinted by kind permission of Beacon Press and Suhrkamp Verlag.

Every effort has been made to trace copyright holders and to obtain their permission for the use of copyright material. The publisher apologizes for any errors or omissions in the above list and would be grateful if notified of any corrections that should be incorporated in future reprints or editions of this book.

INTRODUCTION: WHAT IS CONTINENTAL PHILOSOPHY OF SCIENCE?

Gary Gutting

Philosophy vs. Science, Continental vs. Analytic

The subdiscipline we call "philosophy of science" originated in the nineteenth century in the wake of Kant's critical philosophy. It derives from the challenge posed by modern science to the very idea of a distinctively philosophical enterprise. The "scientific" achievements of Galileo, Descartes, and Newton realized long-sought philosophical goals of answering fundamental questions about the nature of planetary and terrestrial motions. Over the next two centuries, however, it became apparent that the empirical methods that produced the seventeenth-century revolution could and should be separated from the a priori methods of traditional philosophy; and the question gradually arose of what, if anything, there remained for philosophy to do. This question became entirely explicit with Kant and has continued to be at the center of the philosophical enterprise ever since.

As a rough but useful categorization of philosophies of science I propose distinguishing three basic attitudes to scientific knowledge. The first, which I will call empiricist or positivist, regards science as the only knowledge worthy of the name. Philosophy is at best a metareflection that makes explicit the conclusions of science and the methods whereby it has produced them. The second, Kantian or critical, attitude is that science provides the only first-order knowledge, while philosophy reveals a distinctive domain of truth by deriving the necessary conditions for the possibility of scientific knowledge. The justification of philosophical claims requires the assumption of the validity of science, but the claims themselves (unlike those of positivist philosophy of science) constitute a domain of "transcendental" truth that is of a different order than that of science. The third, ontological or metaphysical, attitude claims access to a domain of philosophical truth that is entirely independent of (and, indeed, in some sense superior to) science. This autonomous philosophical truth provides a more general, more fundamental, or more concrete vision of reality, of which science is just one subordinate part and in terms of which it must be understood.

The positivist attitude is typically found among reflective scientists and philosophers deeply involved in science. The most famous proponents were (in Germany) Ernst

Mach and (in France) Poincaré and Duhem.[1] During the first two decades of the twentieth-century, positivism was overshadowed by a revival of Kantian thinking (neo-Kantianism): in France by Lachelier, Boutroux, Brunschvicg, and Bachelard; in Germany by the rival Marburg (Cohen, Natorp, Cassirer) and Southwest (Windel-band, Rickert, Lask) schools. Later the Frankfurt School produced what can be regarded as a version of neo-Kantian philosophy of science (Habermas). The onto-logical attitude arose first through *Lebensphilosopie* (e.g., Bergson and Dilthey) and, later, through phenomenology (Husserl) and existentialism (Heidegger and Merleau-Ponty). It continued in France through varieties of poststructuralism, particularly the philosophers of "difference," Deleuze and Irigaray.

This quick review of philosophy of science on the European continent covers much more than "continental" philosophy of science. The reason is that the split between what we call continental and analytic philosophy emerged from the decline of the neo-Kantianism that dominated French and German universities in the late nineteenth and early twentieth centuries. Before that, even the deepest philosophical divisions (say between a Bergson and a Poincaré or between the early Husserl and the early Carnap) did not prevent informed and fruitful discussion. For all its manifest inadequacies, the continental–analytic divide is grounded in the undeniable fact that, sometime around the end of the 1920s, philosophers split into two camps that, in short order, had nothing to say to one another.

We still do not entirely understand how the division arose, but, as Michael Fried-man has suggested, its root is in two opposing views of the role of logic in philosoph-ical thought. On the one hand, there was the idea that logic, particularly the new mathematical logic of *Principia Mathematica*, was the privileged tool for formulating and resolving philosophical problems. On this view, most fully and powerfully de-veloped by Carnap, philosophical questions could be resolved (or dissolved) by insisting on the highest standards of logical clarity and argument. On the other hand, there was the idea that logical categories and techniques are themselves abstractions from the fullness of lived experience and therefore are severely limited for the pur-pose of understanding concrete existence. This, for example, was the view of Heideg-ger in *Being and Time*, where he deployed Husserl's phenomenological method to describe aspects of the human situation regarded as inaccessible to merely logical analysis. Adapting some of Derrida's terminology, we might formulate the analytic–continental division as one between logocentric and nonlogocentric philosophy. It is, however, important to emphasize that the continental rejection of logical analysis as the privileged instrument of philosophical understanding is not equivalent – as some analytic philosophers seem to think – to a rejection of logical principles (e.g., non-contradiction) as a necessary condition on the intelligibility of discourse. Nor is it – as some continental philosophers seem to think – an abrogation of the philosopher's duty to be as logically clear and rigorous as the subject at hand permits.

It is not surprising that philosophers committed to the analytic approach have often been sympathetic to positivist philosophy of science. The analytic ideal is modeled on a commonly accepted ideal of scientific thought, so that those who hold to the analytic ideal may well privilege scientific knowledge, and those who privilege scientific know-ledge are likely to prefer the analytic model of philosophy. Similarly, we might expect that continental philosophers will embrace the centrality of nonscientific modes of knowing and so reject positivist philosophy of science in favor of the ontological attitude. But none of this is logically entailed. Analytic philosophers (for example, in the ordinary-language movement) can and have contested the positivist assertion of science's cognitive privilege. Correspondingly, a continental philosopher (I will suggest Foucault as an example) may hold that it is less logically rigorous sciences, such as history, that offer the best philosophical perspectives on human existence and, accordingly, endorse a distinctively continental version of positivist philosophy of science. Moreover, some important but now often neglected strands of continental philosophy are based on what I have called the critical stance toward science.

There is also an important, if less emphasized, split within the domain of continental philosophy: that between philosophy in France and philosophy in Germany. The great and obviously important exception has been the significance, from the 1930s on, of Husserl and Heidegger for French philosophy (though even here it is important to appreciate the large extent to which the French did not simply import phenomenology but appropriated it for purposes arising from their own distinctive philosophical tradition). But other important German developments, for example, Marburg neo-Kantianism and the Frankfurt School, had very little impact in France, even on philosophers with parallel interests and approaches. And, until a late twentieth-century interest in French poststructuralism (mostly, however, for the sake of refuting it), German philosophy has on the whole been indifferent to most French developments.

In what follows, I offer a survey of the major treatments of science by French and German philosophers of the twentieth century. The discussion will, of course, be very schematic, but I hope it provides a useful background for the more detailed essays that follow.

France: Neo-Kantians and Bergson

For nearly the first third of the twentieth century, French philosophy, the philosophy of the Third Republic, was dominated by a distinctive version of neo-Kantian idealism, which combined a particular reading of Kant's critical philosophy with the French "spiritualist" tradition going back to Descartes and Maine de Biran. Spiritualism was sympathetic to the Kantian idea that the mind constituted its objects of knowledge but strongly resisted idealistic extensions of Kant that undermined the metaphysical and moral autonomy of the individual human agent. There was, for example, never any serious French sympathy for the Hegel of absolute idealism. Nor was there much interest in romantic versions of idealism that challenged science as the paradigm of knowing. Like Kant himself, the French neo-Kantians took the cognitive authority of science as a given and developed their philosophical systems by deducing the conditions necessary for this authority. Jules Lachelier and Émile Boutroux were important early representatives of this approach, with Lachelier offering an elegant transcendental derivation of the principles of induction and Boutroux developing a revised Kantianism that allowed for freedom (indeterminacy) in the phenomenal world. But the most important figure, both for French neo-Kantianism in general and for French philosophy of science, was Léon Brunschvicg.

Brunschvicg combined a general neo-Kantian philosophical perspective with a strong emphasis on the importance of the history of science. In this latter emphasis, he was continuing the strong French tradition, beginning with Comte and continuing with Duhem, Poincaré, and Meyerson, that insisted on understanding science through its historical development. His own distinctive contribution was to join this historical approach to a critical philosophy of science, in contrast to the earlier thinkers' predominantly empiricist viewpoint. While rejecting a naive empiricism that sees knowledge as the result of what the mind passively receives from a predetermined world, he likewise denies that knowledge arises simply from the mind's reflection on itself. Truth is expressed in "mixed judgments" that combine what is given in experience with intellectual frameworks developed, through scientific investigation, over the course of human history. In a sense combining positivism and idealism, Brunschvicg sees our knowledge of the world as the outcome of the mind's historical reflection on scientists' continually more successful interpretations of experience. He rejected Kant's assumption that, from a particular stage of science (the Newtonian), he could deduce final truths that would regulate all subsequent accounts of the world, and saw Einstein's theory of relativity as a clear refutation of Kant's "dogmatism" on this point.

Brunschvicg's approach was continued, although in a much less idealistic manner, by Gaston Bachelard. This is reflected, first, in his insistence, contrary to Brunschvicg, on radical discontinuities in the history of science. Over 30 years before Kuhn, Bachelard read the history of physics as a series of epistemic "breaks" whereby one conception of a natural domain is replaced by a radically different conception. He also emphasized an initial "break" that introduces a scientific vision of the world in opposition to the common-sense categories of ordinary experience. Second, Bachelard insisted that philosophy, which always has to "go to the school of the sciences," must develop new conceptions corresponding to each new historical stage of science. The philosophy of an age of relativity and quantum physics has to be essentially different from a philosophy of the Newtonian era, since Newtonian concepts are now "epistemological obstacles" to an adequate understanding of nature. Bachelard accordingly worked to develop a philosophical standpoint (a non-Cartesian and, in some ways, non-Kantian epistemology) that would mirror the radically new conceptions of physics. He also offered striking insights into the power of the images through which common-sense and outdated scientific views maintain their attraction, even after they have lost their scientific value. He also pursued the positive role of such images in the nonscientific contexts of poetry and art, and he developed what he called a "psychoanalysis" of the attraction of primordial images such as earth, fire, air, and water.

Bachelard's position remains broadly rationalist (indeed Kantian) in that it emphasizes an active role of the mind in knowledge and sees an irreducible role for philosophy in reflecting on the epistemological significance of scientific results. But his view is, in his terminology, an "applied rationalism" in two senses. First, as we have seen, the categories the mind constructs are relative to the historical situation. Second, Bachelard sees the mind's "constitution" of its objects as mediated through scientific instruments, which are "theories materialized." (For more on this topic, see Mary Tiles's essay on Bachelard below.) Given the priority of the scientific accounts that correct and replace the categories of common-sense experience, what we need is not Husserl's phenomenological descriptions of the constitution of everyday objects but a "phenomeno-technics" describing how instrumental technology constitutes scientific objects.

Despite the dominance of neo-Kantian idealism, the greatest philosopher of the Third Republic, Henri Bergson, did not share its privileging of science as the unique source of our knowledge of nature. Kant, on Bergson's reading, starts from the early modern rationalist vision of a world made intelligible by the relational power of mind, but asks why this cannot be the human rather than the divine mind. Even more important, Kant goes on to make a distinction between the forms and the material of knowledge, a distinction no doubt tied to the fact that the human mind does not have the creative power of the divine mind. The crucial question for Bergson concerns the status of this "matter" from which the objects of knowledge are constituted. For Kant, it is merely the vehicle for the mind's structuring of the world by the imposition of its forms. But, according to Bergson, this neglects the possibility, opened up in principle by Kant's approach, that this matter of knowledge is something with significance in its own right, beyond what it is given by the forms of the intellect. Kant, unfortunately, uncritically assumed that knowledge could be only scientific knowledge; given this, since the realm of science is defined by intellectual forms, there could be no knowledge beyond these forms (no "extra-intellectual" knowledge).

But, according to Bergson, this assumption ignores the obvious limitations (incompleteness) of scientific knowledge, particularly as we move from the inanimate through the vital to the psychological. If we avoid Kant's mistake, we will recognize "a supra-intellectual intuition" of reality that gives us knowledge of reality in itself, not just the phenomenal constructions of the intellect. For Bergson, of course, the object of this intuition is the duration (lived time) that science excludes from its purview but which is in fact the "richer" whole from which scientific objects are abstracted. Kant's idealistic successors (Fichte, Hegel) recognized the need to find intuitive knowledge beyond the forms of the intellect that would put us in contact with reality in itself. But they wrongly sought this in a *nontemporal* intuition, which is really just a reformulation of the pre-Kantian mechanism (Leibniz, Spinoza) in mentalistic terms. Abandoning these intellectual constructions for the concreteness of experience brings us back to duration.

Science's abstraction from the concreteness of duration, results in what Bergson calls its "cinematographical method," whereby science views reality not as a continuous flux (the duration that it in fact is) but as a series of instantaneous "snapshots" extracted from this flux. In terms of a simple but fundamental example, science derives from the mindset that makes Zeno's paradoxes both inevitable and unsolvable. Such a view is essential for science, given that its goal is control of nature and therefore more effective action in the world. For, Bergson maintains, action is always directed from a starting-point to an end-point and therefore has no concern with whatever comes between the two. The practical (instrumental) nature of science leads to its abstraction from the reality of duration, and a full philosophical account of the world *in concreto* must restore what science omits. Indeed, the heart of Bergson's philosophical effort was to show, for a succession of key philosophical questions (concerning freedom, the mind–body relation, the nature of existence, the truth of religion) how answering them requires supplementing the abstractions of science with the intuition of duration.

Germany: Neo-Kantians and Phenomenology

Twentieth-century German philosophy through the 1920s runs roughly parallel to the course of French philosophy. But French neo-Kantianism was a general spirit informing a group of thinkers who, despite disagreements, saw themselves as part of a common enterprise, as illustrated by the remarkable collaborative venture of the *Vocabulaire critique and technique de philosophie*, coordinated by André Lalande. By contrast the German neo-Kantians were divided into fiercely competitive schools that thrived on controversy with one another. (The difference may correspond to the centralization of French philosophical education in the related Parisian institutions of the Ecole Normale and the Sorbonne, in contrast to the separate university centers of German philosophical education.) Also, far more than the French discussions, the German debates were rooted in close textual disputes over the meaning of Kantian texts.

There were two dominant neo-Kantian schools, one associated with the University of Marburg and the other with the University of Heidelberg (or, more generally, the southwest region). Both schools adopted the critical (Kantian) attitude toward science, accepting it as the primary instance of knowledge and developing a distinctive realm of philosophical knowledge through reflection on the conditions of possibility of science. They also accepted Kant's basic idea that knowledge of an object requires the structuring of the "matter" of pure sensation by the conceptual "forms" of the understanding. The classic Kantian problem, of course, is how this structuring is achieved. According to Kant himself the structuring is possible only because there is an intermediate epistemic domain, the a priori forms of sensibility (space and time), that allows the application of pure logical concepts to preconceptual sensibility. Both neo-Kantian schools, however, rejected such intermediate forms. There is, according to them, no intermediary between the pure logical forms of the understanding and the preconceptual matter of sensation. How, then, are the pure conceptual forms able to structure the preconceptual matter?

This is the key point over which the two schools disagreed. The Marburg school in effect denied Kant's sharp distinction of epistemic form and matter; or, rather, it maintained that the distinction is merely an abstraction from the concrete reality of objects of knowledge that have both formal (conceptual) and material (sensible) aspects. By contrast, the southwestern school maintained the distinction and offered new ways of bridging the gap between the two extremes.

What may seem to be merely technical disputes within the Kantian tradition in fact turned out to have major significance for the understanding of science. This becomes especially clear in the work of the Marburg school, which, particularly in the area of philosophy of science, was brought to its fullest development by Ernst Cassirer. For one thing, the rejection of Kant's forms of sensibility avoided the objection that Kantianism was refuted by the development of non-Euclidean geometry and the theory of relativity. For it was only these forms that committed Kant to Euclidean geometry and absolute time. Further, denying the sharp distinction of epistemic form and matter led the Marburg school to the idea that the constitution of empirical objects was something carried out in the course of the history of science, with each stage of development corresponding to a new articulation by scientists of the precise formal stuctures required to understand the world. This genetic view led to the position, similar to that of Brunschvicg and Bachelard, that science can be understood only through its history. Finally, the Marburg refusal to isolate pure formal structure allowed Cassirer to argue that mathematics has a synthetic character that prevents it from being reduced to pure logic, which is itself only an abstraction from the concrete generative process whereby the mathematical methods of science constitute the objects of the world.

The Neo-Kantian schools were eventually defeated by challenges from three directions. The first, which lies outside our concerns here, was that of logical positivism, particularly the work of Schlick and Carnap. Recent historical scholarship has shown how the founders of logical positivism began working from within neo-Kantianism and only gradually developed a distinctively different standpoint. This shows that, contrary to Ayer's account in *Language, Truth, and Logic*, logical positivism was not a simple return to Hume combined with the tools of the new logic. It – and therefore the analytic philosophy it engendered – needs to be understood in terms of its neo-Kantian origins.

The other two challenges came from phenomenology, first from Husserl's original version and second from Heidegger's radical transformation of Husserl's project. Both Husserl and Heidegger rejected critical philosophy's privileging of empirical science on the grounds that its objectivizing methods could not take adequate account of what we actually encounter in experience. Beyond science, there was need for phenomenology, a rigorous and complete description of "the things themselves"; that is, of what we find in experience prior to the objectifying abstractions and idealizations of science. In Husserl's case, the appeal to experience was primarily for the sake of certainty. He saw phenomenology as a source of absolute certainty in its pure intuitions of essential meanings. As such, phenomenology would provide an unshakeable foundation for all other human knowledge, including science. According to Husserl, the alternative to such a phenomenological foundation is collapse into self-refuting relativism or historicism.

Husserl's claim is that (empirical) science must be grounded in a philosophical project that is itself scientific: with the highest standards of clarity, rigor, and objectivity. But the standards of "philosophy as rigorous science" (i.e., phenomenology) are quite different from those of empirical science. This is because the object of this science is not the natural world of material, sensible things – about which absolute certainty is not possible – but a realm of ideal essences, not existing as independent Platonic Forms but as the intentional objects of acts of consciousness and therefore exhaustively knowable through self-reflection.

Husserlian phenomenology has important similarities to both the conceptual analysis of logical positivism and the transcendental deductions of neo-Kantianism. Like the positivists, Husserl sees philosophy as reaching non-empirical, necessary truths through the analysis of meanings. But for the logical positivists, "analysis" is a matter of applying the categories and techniques of mathematical logic to common-sense and scientific concepts. For Husserl, this is not sufficient, since both our logic and our concepts are based on unexamined presuppositions, which can only be uncovered through a phenomenological return to the immediate experience from which logic, science, and common sense are all abstractions. Similarly, like the neo-Kantians, Husserl wants to determine the necessary conditions of experience (eidetic truths implicit in experience). But he rejects the neo-Kantian project of deducing such truths from the (uncritical) assumption that empirical science is a valid body of knowledge. Instead, Husserl insists, these truths must be given in direct phenomenological intuition.

Heidegger shared Husserl's commitment to the primacy of the everyday world, but objected to Husserl's (and Dilthey's or Scheler's) attempts to express that in terms of experience or consciousness. He also objected to Husserl's aspiration to certainty as an epistemic ideal. There is a complex story to be told about his reasons for parting with Husserl on both these points, a story involving his critique of Husserl's subject–object distinction, his insistence on the need for our understanding of beings to be rooted in a fundamental understanding of Being, and his development of a hermeneutical method that aims at interpretation rather than pure description. The details of this story need not concern us here, but its outcome is that Heidegger replaces Husserl's eidetic analysis of ideal essences with an "existential analysis" of human beings (*Dasein*) as they exist in the everyday world. Science is then understood in its relation to this world.

Of course, Husserl too, especially in the *Crisis*, emphasized the need to understand science in its relation to the everyday world (the *Lebenswelt*). Moreover, his analysis of science in these terms is in many ways similar to Heidegger's; both see science as an abstraction, for the sake of prediction and control, from the lifeworld, and both warn against the cultural dangers of substituting scientific abstractions for the fullness of human reality. For Husserl as much as for Heidegger, relating science to the lifeworld allows us to situate science in its historical context. Husserl, of course, continues to insist, even in the *Crisis*, on the need (for the sake of foundational certainty) to ground our historical experience of the lifeworld in an eidetic analysis of the ideal, ahistorical essences that define its ultimate meaning. But Heidegger's rejection of this further level of analysis does not alter his substantial general agreement with Husserl on the historical perils of scientistic misunderstandings of our world.

On the other hand, Heidegger's existential phenomenology of human life in the world (what he calls his *Daseinanalysis*) reveals dimensions of science that Husserl either ignores or denies. Whereas Husserl regards science as primarily a theoretical account of nature, developed by the scientist as a disengaged spectator, Heidegger sees the lifeworld in terms of our practical engagement with it, and so, in particular, sees science as fundamentally a set of practices rather than a theoretical vision. This, in turn, leads to Heidegger's emphasis on and critique of technology.

France: From Existentialism to Foucault

The French reaction against neo-Kantianism was less complex, and not only because French philosophers were not so heavily invested in scholastic disputes about the meanings of Kant's texts. There were also no developments parallel to the rise of logical positivism and no strong interest in anything like the Husserlian program of foundational certainty through eidetic analysis. The former point has two main explanations. First, in France the spirit of positivism (which, after all, had been born there with Comte) had for a long time been channeled out of philosophy and into the social sciences. Second, most of the promising French philosophers of logic and mathematics – Louis Courturat, Jean Nicod, Jacques Herbrand, and Jean Cavaillès – who might well have developed along something like positivist lines, died at an early age.

There was, of course, considerable French interest in Husserl. But this interest arose from the fascination with concrete experience that characterized French existentialism. Contrary to a common opinion, philosophical existentialism did not first develop in France from Sartre's and Merleau-Ponty's readings of Husserl and Heidegger, but rather from Jean Wahl's existential interpretations of Hegel (1929) and Kierkegaard (1938).[2] Husserl and Heidegger were read with an eye to what they had to offer philosophers attuned to the need for a concrete immersion in the world, but with little interest in Husserl's foundational project or Heidegger's problem of Being. Husserl's strongly foundationalist *Cartesian Meditations* (given as lectures at the Sorbonne in 1929) were an unfortunate choice and not well received.

Lacking engagement with the issues raised by logical positivism and Husserlian "rigorous science," French existential phenomenology not surprisingly had little to say about the philosophy of the natural sciences (which then, as for so long, defined the main concerns of the philosophical study of science). The same, however, was not true of psychology and the social sciences, which were a major concern, particularly in the work of Maurice Merleau-Ponty.

His first important publication was *The Structure of Behavior* (1942; hereafter *S*), which uses Gestalt psychology to construct a scientifically detailed argument against behaviorist models and then goes on to show the deficiencies of even the Gestalt account. Phenomenology is explicitly mentioned only in the last chapter, where Merleau-Ponty suggests that it provides a superior standpoint for an adequate understanding of consciousness and its relation to the natural world.

Subsequently (particularly in *The Phenomenology of Perception*; hereafter *PP*), Merleau-Ponty develops in detail his claim of phenomenology's superiority to scientific explanation. The basic problem with a scientific approach is, he maintains, that the deployment of its rigorously empirical and quantitative methodology requires regarding the contents of our lived experience as fully determinate and totally objective (that is, in no way dependent on our experience of them). Science must conceive of its objects in a way that allows them to be understood entirely in terms of ideal mathematical constructs. This means that science understands everything, including living, feeling, and thinking bodies, as nothing more than a set of physical elements connected by causal relations. As a result, even the human body becomes pure exteriority, a mere collection of parts outside of parts, interacting with one another according to scientific laws. On this view, genuine subjectivity is eliminated – an obvious travesty of our experience. This is the motivation behind Merleau-Ponty's dramatic statement that phenomenology's "return to the 'things themselves' . . . is from the start a rejection of science" (*PP*, viii).

Subsequently, however, Merleau-Ponty came to maintain that phenomenology could avoid idealism only by accepting the fact that the domain of lived experience was itself essentially tied to the world of scientific objectivity. His line of thought was as follows: his analysis of lived experience led him to the conclusion that there was an "ultimate truth" of idealism in the fact that all phenomenological description took place from the standpoint of the "cogito" (perception, which, Merleau-Ponty always insisted is primary, implies a perceiver). To avoid subjective idealism (which is contradicted, moreover, by the givens of lived experience), this cogito must be understood as a an impersonal subject (a "tacit cogito"), other than my personal self. But then, to avoid absolute idealism (or at least an ahistorical transcendental idealism), this tacit cogito had to be viewed as having a real content; that is, a content that made it in at least some respects not constituted by consciousness. Specifically, Merleau-Ponty suggested that this objective content could be introduced through the phenomenological "recognition" and "appreciation" of the structures revealed by the social sciences, especially the anthropology of his good friend Claude Lévi-Strauss and the linguistics of Ferdinand de Saussure.

Both Lévi-Strauss and Saussure give accounts of social realities (e.g., language, kinship relations) in terms of structures. These structures are meanings (that is, they "organize [their] constituent parts according to an internal principle" [*Signs*, 117]) and are therefore not reducible to causal relations among objects. At the same time, they are not the idealist's "crystallized ideas," since the subjects who live in accord with the meanings typically have no conscious grasp of them. People "make use of [structure] as a matter of course," but "rather than their having got it, it has, if we may put it this way, 'got them'" (*Signs*, 117).

Because structures are both objective realities, independent of any mind, and meanings informing the lives of individuals, they are the vehicle of the concrete unity of man-in-the-world. The problem, of course, is how to join objective structural analysis to lived experience. Part of the answer is available from phenomenology, which describes our lived experience of structural meanings. But our particular consciousness of such meanings is just one perspective on them. There is also a need for "ethnological experience," which results from inserting ourselves into another culture through anthropological fieldwork and provides an "experience" that is more comprehensive than what phenomenology has access to. Merleau-Ponty's phenomenology has revealed its own need to be complemented by social-scientific knowledge.

Although existential phenomenology dominated French philosophy for the 15 years after the Second World War, there was another line of thought, centered on science, that was a major force, particularly in university philosophical training. This was the French, broadly positivist, tradition of history and philosophy of science, ultimately rooted in Comte's positivism, classically developed by Duhem, Poincaré, and Meyerson, and brought to fruition in the work of Bachelard. From the 1940s, this approach was primarily represented by Georges Canguilhem, Bachelard's successor as director of the Sorbonne's Institut d'Histoire des Sciences et des Techniques. Canguilhem trained a large number of historians and philosophers of science, and even nonspecialists frequently followed his courses.

Canguilhem was more a historian than a philosopher, although his historical work cannot be sharply separated from his generally Bachelardian philosophical viewpoint. Moreover, his specialty was biology, rather than the natural sciences on which Bachelard focused. The Bachelard–Canguilhem approach provided a distinct alternative to existential phenomenolgy: it accepted the cognitive priority of science and regarded the domain of lived experience as merely a first approximation to the truth about the world, a truth toward which science moved by revising and even rejecting the concepts of everyday experience. As Foucault put it, Canguilhem offered not a philosophy of experience but a philosophy of (scientific) concepts.

Indeed, Canguilhem's major contribution to the philosophy of science is his analysis of the relation between scientific theories and the concepts in terms of which they are formulated. In much twentieth-century philosophy of science, concepts are functions of theories, deriving their meaning from the roles they play in theoretical accounts of phenomena. Newtonian and Einsteinian mass, for example, are regarded as fundamentally different concepts because they are embedded in fundamentally different physical theories. This subordination of concept to theory derives from the view that the interpretation of phenomena (that is, their subsumption under a given set of concepts) is a matter of explaining them on the basis of a particular theoretical framework. For Canguilhem, by contrast, there is a crucial distinction between the interpretation of phenomena (via concepts) and their theoretical explanation. According to him, a given set of concepts provides the preliminary descriptions of a phenomenon that allow the formulation of questions about how to explain it. Different theories (all, however, formulated in terms of the same set of basic concepts) will provide competing answers to these questions. Galileo, for example, introduced a new conception of the motion of falling bodies to replace the Aristotelian conception. Galileo, Descartes, and Newton all employed this new conception in their description of the motion of falling bodies and in the theories they developed to explain this motion. Although the basic concept of motion was the same, the explanatory theories were very different. This shows, according to Canguilhem, the "theoretical polyvalence" of concepts: their ability to function in the context of widely differing theories. His own historical studies (for example, of reflex movement) are typically histories of concepts that persist through a series of theoretical formulations.

Canguilhem supervised Michel Foucault's doctoral thesis (on the history of madness), and his history of concepts was a model for what Foucault called his "archaeological" histories of knowledge. Foucault's primary focus was the social sciences, and his *The Birth of the Clinic* and much of his *The Order of Things* can be read as history of concepts, à la Canguilhem.

A good case can be made for thinking of Foucault's attitude toward science as broadly positivist, in the sense defined above of recognizing no cognitive authority beyond that of science. Here a first point to note is that, although both critics and supporters often classify him as a epistemological skeptic or relativist, he never questions the objective validity of mathematics and the natural sciences. He does show how the social sciences (and the medicalized biological sciences) are essentially implicated in social power structures, but does not see such implication as automatically destroying the objective validity of a discipline's claims. Sometimes a discipline's role in a power regime is in part due precisely to its objective validity (if, for example, objectivity is a social value). Further, Foucault does not, like the neo-Kantians and even Bachelard, recognize any body of truth achieved by philosophical theorizing. He spins out the occasional philosophical theory (e.g., of language or of power), most often of

Nietzschean or Heideggerian inspiration. But this is for the *ad hoc* purpose of understanding a particular historical phenomenon and has no pretensions to universal validity.
The only general epistemic standard to which Foucault holds his own work is that of
historical accuracy.[3] If we count history as a broadly scientific enterprise, then Foucault
recognizes no knowledge outside the scientific domain and so counts as a positivist.

Unlike mainstream positivists, however, Foucault has little interest in questions
about the methodology or ontology of science. This is no doubt because his focus
was almost entirely on "dubious" scientific disciplines, such as psychiatry or crimin-
ology, or, at best, on the dubious aspects of more respectable disciplines, such as
economics or anthropology. Here a discipline is "dubious" to the extent that what it
presents as unquestionable objective truths about a certain domain (say, the mad or
criminals or homosexuals) are rather (or also) part of an eminently questionable
system of social power. So, for example, Foucault argues in his *History of Madness* that
the modern conception of madness as "mental illness" is grounded much more in the
effort of bourgeois morality to control the mad than in any scientific truth about the
nature of madness. Foucault's concern with the cognitive limitations of disciplines
implicated in the power network left little room for standard discussions of the
positive (methodological and ontological) achievements of science.

On the other hand, Foucault's critical historiography was very fertile in developing
new ways of viewing science, ways that would reveal aspects not available to the self-
understanding of a discipline. Here his two great innovations were the archaeology and
the genealogy of thought. Archaeology is a synchronic technique of unearthing and
comparing the deep structures (the epistemic "unconscious") of historical bodies of
thought. Foucault's assumption was that there are rules of "discursive formations" (the
bodies of discourse that express the scientific and would-be-scientific disciplines),
beyond those of grammar and logic. These rules materially constrain the possibilities of
what can be said and define a limited conceptual domain in which the thought of a
certain period about a given subject-matter must operate. Genealogy is a complementary diachronic technique for understanding the emergence of new disciplines and the
discursive formations that structure them. Its two main postulates are that systems of
knowledge develop in symbiotic relations with systems of social power and that social
power consists of a diffuse network of many microcenters of power, with no central-
ized, hierarchical structure. As a result, a genealogical history of knowledge avoids
unitary teleological narratives of domination (such as Marxism) while still allowing us
to question alleged cognitive necessities that mask techniques of disciplinary control.

The Bachelard–Canguilhem approach to science has produced a number of other
important contemporary French philosophers/historians of science. In his early work,
Michel Serres emphasized (as did Bachelard) the dispersed, regional character of scientific work. Each domain is like a Leibnizian monad, with a life and intelligibility of
its own. But here, unlike Bachelard but like Foucault, Serres sees a structural unity
that connects independent scientific domains. He explicates this unity in terms of
the concept of communication, which he expresses through both the metaphor of the
Greek god Hermes and the formalism of modern communication theory. Serres also
offers disconcertingly flamboyant interpretations designed to show how domains conventionally regarded as nonscientific, such as art and literature, share the structures of
scientific disciplines and must be regarded as their epistemic peers. So, for example,

he claims that Emile Zola expressed thermodynamics in his novels before it was explicitly formulated by physicists, tries to show the structural identity of Descartes's *Meditations* and La Fontaine's fables, argues that "Turner translates Carnot," and presents Lucretius's *De Rerum Natura* as a contribution to twentieth-century physical theory. He later developed, in a series of academic bestsellers, a poetico-philosophical cosmology that presents a metaphysics inspired by chaos theory and fractal geometry.

In quite a different vein, Michèle Le Doeuff has continued the tradition of Bachelard by articulating the images that dominate systems of scientific and philosophical thought (for example, in the works of Francis Bacon). She has also employed the Bachelardian notion of epistemological breaks to develop feminist analyses highlighting fissures and discontinuities in the history of scientific reason that, she argues, reveal the complexity and ultimate incoherence of the sexist attitudes implicit in much scientific thought.

Germany: Habermas and the Frankfurt School

Foucault's approach to science is interestingly similar to that of the critical theory of the Frankfurt School, which in many ways anticipates his social critique of science. (However, according to Foucault, he learned virtually nothing of the Frankfurt School during his philosophical formation.) The Frankfurt School began with Max Horkheimer, who developed a very important neo-Marxist approach to the problem of reason at just the time that Husserl was writing his *Crisis*. A major obstacle to Horkheimer's approach was the subtle and very persuasive social analysis of rationality recently put forward by Max Weber, who maintained that the very application of reason to the practical sphere (the key idea of critical theory) was the primary form of social control in modern society and the destroyer of any hope of objective values. Horkheimer and other members of the Frankfurt School (here influenced by Lukács) tried to show that Weber's analysis applied not to reason as such but only to the form it inevitably took in capitalist societies. But their own analyses (as well as the reality of Marxist totalitarianism) eventually led them to the conclusion, especially in Horkheimer and Adorno's *Dialectic of Enlightenment*, that the fault did indeed lie in reason itself. Horkheimer and Adorno argue, for example, that social oppression inevitably follows from the "identity logic" at the core of scientific rationality; this is the root drive to eliminate otherness and reduce everything to a single identity. This line of thought culminated in Herbert Marcuse's rejection of technology as such (i.e., any practical applications of reason) as a form of domination.

Since the 1960s, Jürgen Habermas has developed a new approach to critical theory. He agrees that modern deployments of reason have in fact undermined values and curtailed human freedom. But he maintains that this is not due to the nature of reason as such but to a one-sided modern conception of reason. Modern accounts, he maintains, have viewed rationality as limited to the techniques of "instrumental reason": the means/end reasoning characteristic of empirical science. There are, however, other forms of reason (e.g., the understanding of hermeneutics) that provide a key to unlock Weber's iron cage. As Habermas sees it, the goal of philosophy should be to offer a fully comprehensive account of rationality in all its aspects and, on the basis of it, provide a foundation for human values and restore reason as the avant-garde of human liberation.

In *Knowledge and Human Interests* (1968) Habermas offered a neo-Kantian account (though in a Marxist vein) of knowledge as constituted by various human interests. Natural objects are knowable by scientific methods precisely because this is the only way that we can fulfill our interest in technical control of nature. In this sense, the technical interest constitutes the natural world as an object of our scientific know-ledge, just as Kant thought the forms of sensibility and categories of the understanding did. However, our survival and development as a species requires not only the control of nature but also our forming social groups. Further, society is not possible without effective communication, which requires mutual understanding through intersubjec-tively shared symbols. This reveals the second of Habermas's cognitive interests: the communicative interest whereby humans understand one another in social contexts. Habermas sees the fundamental flaw of positivism as its failure to recognize the role of the communicative interest (and the knowledge as understanding correlated with it). Further, he thinks that it is the same failure that led philosophers of the Frankfurt School to see reason as destructive of human values and freedom. If our only cogni-tive interest were the empirical-analytic interest in technical control, then natural science would be the only form of knowledge and reason would be nothing but instrumental reason and could construct nothing but Weber's iron cage. If we ignore the communicative interest and think of reason in purely instrumental terms, we will find no alternative to postmodern dystopias. But a recognition of the understanding of human beings as a distinct realm of objective knowledge opens the way to the grounding of human values in a liberating practical reason (which, he argued, corres-ponded to a third, emancipative, interest).

In subsequent work Habermas brought his entire discussion of knowledge and interests under the heading of communication, arguing that the interest in technical control itself required a particular form of communication ("discourse") among scien-tists and that an orientation toward emancipation is implicit in the norms of effective communication (which involve, for example, the right of all to equal participation). Developing this view, his treatment begins to sound less like a continuation (in a more practical and historical mode) of Kant's transcendental reflection and more like a social-scientific construction of models for social practices. The detailed analyses of Habermas's theory of communicative action can be read as a kind of higher positiv-ism, in which "reconstructive social science" (like history for Foucault) replaces tran-scendental reason. However, unlike Foucault, Habermas insists on the irreducibly normative character of such social science and so maintains a stronger link with traditional conceptions of philosophy.

France: Poststructuralism and the Abuse of Science?

As we have discussed it so far, continental philosophy, whatever its reservations and critiques, clearly takes the cognitive enterprise seriously and works from a responsible understanding of its methods and results. According to some recent commentators, this is not true of another group of continental thinkers, often called poststructuralists or postmodernists, whose views we have not yet discussed. Some of these (e.g., Lacan, Kristeva, Baudrillard) seem better classified with those working in disciplines other than philosophy. But others, especially Luce Irigaray and Gilles Deleuze, seem philosophers by any reasonably inclusive definition, and have in fact been included in this volume's roster of continental philosophers with views on science worthy of consideration. By way of conclusion, I want to address the concerns of those who find their treatment of science irresponsible.

The negative case has been most thoroughly and forthrightly stated by Alan Sokal and Jean Bricmont in their *Fashionable Nonsense: Postmodern Intellectuals' Abuse of Science* (1999). Their basic procedure is to quote from their targets' passages dealing with science and comment on their scientific intelligibility and accuracy. Because Sokal and Bricmont rightly focus on specific points rather than vague accusations, I will look at two particular examples. I think, however, that these are typical of their discussions.

Consider first their treatment of Deleuze on calculus. They cite several pages on the topic from Deleuze's *Difference and Repetition*, pointing out specific inadequacies in footnotes to Deleuze's text and also making more general comments about its deficiencies. The latter consist of two main claims: that Deleuze is discussing "classical problems in the conceptual foundations of differential and integral calculus" that "were solved by the work of d'Alembert around 1760 and Cauchy around 1820" (p. 160), and that most of the sentences of the cited passages simply don't make sense ("these texts contain [only] a handful of intelligible sentences" (p. 165).

It is hard not to sympathize with Sokal and Bricmont's frustration at the aggravating obscurity of Deleuze's writing. But since they admit to not understanding the bulk of what Deleuze is saying, it is also hard to see how they can judge themselves to be in a position to pick out parts of his text as inaccurate or confused formulations of calculus or to conclude that Deleuze is offering a discussion of classical problems in the foundations of mathematics. His discussion does not neatly separate what he is saying (or implying) about mathematics from his formulation of his own philosophical standpoint; the two are inextricably intertwined. Sokal and Bricmont get critical purchase on Deleuze's text only by assuming that he is using terms such as "difference," "differential," "continuity," "power" in the technical mathematical sense. There are certainly allusions to these technical senses, but the terms are also part of Deleuze's distinctive philosophical discourse. We are, therefore, in no position to assess what he is saying without understanding the philosophical language he is using. Sokal and Bricmont, of course, assert that this language is simply meaningless. But they offer no philosophical analysis to justify this claim, nor do they claim to be competent to do so.

The same general point applies to Sokal and Bricmont's critique of Luce Irigaray's provocative comments about the sexist nature of the physics of fluid dynamics (although here their case is weaker, since the main "mistakes" they attribute to Irigaray concern not the content of physics but philosophical issues about the limits of formalization and the role of idealization). But, beyond this, Sokal and Bricmont seem oblivious to the humorous and ironic tone of Irigaray's discussion, which is as much a tease as a sober critique, as much a matter of trying to disconcert and stimulate as of trying to refute.

Of course, these are very puzzling texts and thorough analysis might reveal that they have no plausible sense, although the essays in this volume on Deleuze and Irigaray make a good case to the contrary. But my point here is that the sort of critique Sokal and Bricmont propose is not capable of supporting such a conclusion, which would have to be based on an informed awareness of the text's possible meanings and connections, not uninformed exclamations of incomprehension.

The essays and complementary primary texts that follow offer much more detail on topics either ignored or treated only schematically above. (In some cases, for example that of Merleau-Ponty, my comments cover material on which we were not able to include essays.) We begin with two essays on the historical background of continental philosophy of science. Terry Pinkard's discussion of Hegel gives a good sense of the issues that developed out of Kant and the idealistic turn taken by his followers and critics, while Jean Gayon shows the relation between Bergson's spiritualist metaphysics and his critique of science. Moving into the German neo-Kantian context out of which the continental–analytic division arose, Michael Friedman analyzes Ernst Cassirer's views on philosophy of science. Next, we look at the phenomenological approach: Richard Tieszen provides an overview of Husserl's conception of philosophy as a science and of his critique of empirical science, and Joseph Rouse discusses Heidegger's treatment of science with particular reference to his attitude toward naturalism. Returning to France, we begin with three essays on figures in the important French tradition of history and philosophy of science. Mary Tiles discusses Bachelard's early work on science and technology, Hans-Jörg Rheinberger treats Canguilhem's historical approach to epistemology, and Linda Alcoff explores Foucault's approach to science in the context of his genealogies of power. Next there are two essays on poststructuralist views of science, with Todd May surveying Gilles Deleuze's view of science and Penelope Deutscher discussing Luce Irigaray and French feminist approaches to science. Finally, we return to Germany for one last time, with Axel Honneth's analysis of the Frankfurt School's critique of science.

Notes

1 Duhem in fact thought there was a body of metaphysical knowledge about the world, roughly that expressed in Aristotle's metaphysics. But his own philosophy of science dealt with the realm of appearances, not the underlying realm of metaphysical truth.
2 There was also Wahl's book, *Vers le concret* (Paris: Vrin, 1932), which Sartre says was particularly important for him and his friends.
3 On this point, see my "Foucault and the History of Madness," in Gary Gutting (ed.), *The Cambridge Companion to Foucault* (Cambridge: Cambridge University Press, 1994), 47–70.

HEGEL

1

SPECULATIVE *NATURPHILOSOPHIE* AND THE DEVELOPMENT OF THE EMPIRICAL SCIENCES: HEGEL'S PERSPECTIVE

Terry Pinkard

As a possible source for ideas about the philosophy of science, Hegel might seem like an unlikely prospect. Many of his basic ideas about history have, after all, already been put to use (even if quite unconsciously and often in full ignorance of their source) by people in the history and philosophy of science. Hegel's shade appears throughout the post-Kuhnian picture of science that sees science as going through revolutions in which one scheme of thought (or "paradigm") replaces another such that the new scheme grows out of the very specific failures (or, as Hegel would say, the "determinate negations") of the previous scheme, setting itself up not merely as what just comes later but as the rational successor to what preceded it. Likewise, Hegel's refusal to comment virtually at all on the nature of scientific "method" or the structure of scientific theories, and his insistence instead on treating the individual sciences (mechanics, physics, meteorology, geology, biology) in detail, has at least a passing resemblance to the kind of close-grained contemporary philosophies of physics and biology that are very much the mode in contemporary philosophy of science, but what Hegel actually has to say about those sciences hardly seems to have any contemporary resonance to it.

Curiously enough, however, Hegel, who took a historical approach to almost everything he did, did not himself take such a historical approach to science. Instead, his writings and extensive lecture series on the topic were titled "*Naturphilosophie*," and, contrary to what one might have expected, in his *Naturphilosophie* he did not offer a Collingwood-style treatment of the history of the "Idea" of nature but instead a reconstruction of the picture of nature that was emerging from the sciences of his time, and how that picture related to his conception of agency, of *Geist*. Even worse, although Hegel himself cut a rather impressive figure as a reader and commentator on the scientific literature of his time, his status as a prognosticator about which developments in science were going to be the winners and the losers turned out not to be nearly as imposing. In almost all cases, he simply placed his bets on the wrong horses – most famously in siding with Goethe's delightful but wrong-headed theory of colors against the Newtonian tradition.

To be sure, many of Hegel's own failures in this regard cannot be laid entirely at his feet. After all, he lived and wrote before the advent of the twentieth-century revolution in physics; in his time geology was dominated by the debate between vulcanists and neptunists — that is, by the debate over whether the earth's formations originate in internal fiery volcanoes or in more watery origins. Post-Euclidean geometries were barely even dreamed of in his time, and the fledgling efforts at creating them were for the most part unknown. Chemistry was still in its early infancy — Lavoisier's recognition of oxygen and banishment of phlogiston had not yet been fully accepted, and organic chemistry had not yet even been born. Modern biology was still several years off — Darwin's *Origin of the Species* was published in 1859, and Hegel died in 1831. It would be unfair to fault Hegel for failing to predict the upcoming "second" scientific revolution.

It is nonetheless worth attending to what Hegel took himself to be doing in offering a piece of what he called a *speculative* philosophy of nature in order to see whether there still is anything left to find in his lectures and writings on the topic other than matters now only of antiquarian interest.[1]

To get a grip on that, we need to understand what Hegel means by a "speculative philosophy." Hegel's use of the term originates in the post-Kantian predicament of how to use Kant to get beyond Kant, especially when the Kantian resolutions of certain key problems seemed so problematic.[2] Key to this was Kant's "third antinomy," which to his successors seemed to say that the problem of freedom in the modern world was not only theoretically irresolvable but was, literally speaking, theoretically unintelligible, and few seemed convinced by Kant's own solution to save freedom by appeal to the phenomenal/noumenal distinction. However, because so much of Kant seemed right, it also seemed especially important to the post-Kantians either to put the Kantian house in order (such as Reinhold and, at first, Fichte more radically tried to do) or to use Kant to get out of Kant into something appropriately post-Kantian.

The post-Kantian rejection of both Kant's hard-and-fast distinction between two separate faculties of knowledge — intuitions and concepts — and his language of an "imposition" of conceptual form onto intuitive content also put the issue of saving Kant from Kant high on the agenda. Hegel in particular joined in the arguments against intuition as an *independent* source of knowledge uninformed by concepts, arguing that Kant's own arguments to the effect that we could never be conscious of "unsynthesized intuitions" showed that intuitions could only play their epistemic, normative role as part of (or as a "moment" of) some larger normative "whole," that is, that classifying part of our experience as an intuition (as a representation) amounted to ascribing a normative status to it, an ascription which itself had to come from "reason." Likewise, Kant's own concern that concepts without intuitions were devoid of content showed that any attempt to completely unchain concepts from sense-experience was doomed to repeat the failures of previous metaphysics that Kant had so devastatingly diagnosed. Hegel's own leading idea, articulated partially in his first published monograph in 1801, *The Difference Between Fichte's and Schelling's Systems of Philosophy*, and then made more explicit in his long journal article the following year, "Faith and Knowledge," was that concepts and intuitions should be understood as having normative statuses within a larger "whole," that their epistemic roles and contributions could be separated only in light of understanding their place in that whole, which he identified as "reason," the capacity to draw inferences, which he then developed into a more social conception of the *practice* of giving and asking for reasons. To use Hegelian language: we must begin from the *unity* of intuitions and concepts, not from their separation, which is rightfully done only within the larger whole in which they play their roles.[3]

Rejecting pure intuitions as a source of epistemic content independent of all conceptual shaping put all those post-Kantians making that move into a predicament that Kant himself had grasped (even if somewhat inchoately) with regard to his practical philosophy. On Kant's view, the moral law and its bindingness on us were, of course, independent of intuition, representing only the full, unfettered spontaneity of reason (expressed as autonomy in the practical sphere), and thus, as Kant put it in an often-cited passage in the 1785 *Groundwork*, the will can be subject only to those laws of which it can regard itself as the author.[4] However, since a lawless will cannot bind an agent, the will needs a law to guide it in authoring whatever law it institutes, which implies that such a prior law cannot itself be self-chosen, but the law, paradoxically, can obligate the agent only if it is self-chosen. This "Kantian paradox" – that the will must have a self-chosen law that is not self-chosen – found its expression in Kant's "fact of reason" in the 1788 *Critique of Practical Reason*, which in some ways just restates the "paradox" as a "fact," namely, that in undertaking any commitments at all, we cannot get "outside of" or "beyond" the claims of reason *even while* we regard them as self-authored – that we are committed to the absolute normative priority of reason as a "fact" that we ourselves have "made."[5]

Moving this "Kantian paradox" to the forefront informs the problem that animates virtually all post-Kantian conceptions of normative authority.[6] Hegel's own position develops in part out of the implications of dropping intuition as a separate, independent faculty that must then be combined with a conceptual faculty – the implications, that is, of dropping intuition as a separate source of "content" which must then be organized in terms of some "scheme." This paradox – about how I can be both author of the law and subject to the law – was for Hegel simply *the* speculative problem, the great "speculative truth" that post-Kantian philosophy was called upon to articulate and explain.[7] The problems surrounding the bindingness of the claims of reason (and of what even counts as "internal" and "external" to reason) is the pulse of the Hegelian dialectic, which, for example, in the *Phenomenology* narrative moves through various shapes of "consciousness" as those "shapes" try to hold fast to some type of external reason only to find it "dissolving," which in turn motivates "consciousness" to "return into itself" after having originally taken its standards to have been "external to itself." The *Logic* in turn traces the progress of thought's finding that it is, in Hegel's speculative language, the "other of itself" as it comes to grips with how it, as autonomous thought, can be the author of the norms to which it is subject.[8] On the Hegelian understanding, the "Kantian way out of Kant" thus has to take Kantian idealism not to consist in a contrast between the mental (the ideal) and the real (or the "inner" and the "outer"); it instead rests on the contrast between the normative order versus some kind of comprehensive naturalism (in a way very similar to Wilfrid Sellars's conception of the contrast between the "space of reasons" and the causal order).[9]

Turning either to a purely "externalist" or a purely "internalist" account of reason would only be one-sided and would, as Hegel stressed in his *Difference* book and the monograph, "Faith and Knowledge," only lead to the endless seesaw between, for example, realism and subjective idealism so typical of modern philosophy.[10] The "Kantian way out of Kant" had to preserve the paradox while at the same time superseding it, and Hegel's solution involved a move both to *sociality* and a historical conception of *socialization* as a response to these problems.[11] Whereas Kant's monadic conception of agency effectively split the agent in two, with one side authoring the law, the other side being subjected to it – of which perhaps the clearest expression is Kant's discussion of conscience in *The Metaphysics of Ethics* in 1797 – Hegel argued that the "Kantian paradox" had to be resolved by a nonmonadic, *social* conception of agency and had to involve at least two agents, each of whom authors the law to which both they and the other are subject (or contends with the other to see who is to be the author and who is to be the subject, exemplified in the famous dialectic of mastery and servitude that shows that such a one-sided, nonmutual assumption of authorship and subjection fails). This kind of *sociality* is essential to the structure of agency; we assume authorship of such norms by virtue of our mutually subjecting ourselves and others to them, and the very idea of subjecting ourselves to a norm (or "following a rule") involves us in social practices. Subjecting ourselves to norms, that is, fundamentally involves us in the social practice of giving and asking for reasons.

Such sociality of agency, however, only takes the "Kantian paradox" and raises it to a higher level, since it now places the issue of subjecting ourselves to norms at a social instead of individual level. The "reflective" issue – in Kant's terms, of needing the law to determine the law, or in Wittgenstein's terms, needing an interpretation of the rule to determine what counts as the correct interpretation of following the rule – remains, and for that reason, we also require a *developmental* conception of agency as involving a kind of socialization, a way in which we find ourselves contingently thrown into a social world in which the determinate space of reasons that makes up that world is something to which we are initially subjected. We are *initiated* into the space of reasons, and our grasp of this space must be itself tacit, prereflective in order for us to be able to move about in that space. We could never learn to "follow the rule" by offering ever new interpretations of that rule (since we would need ever more interpretations to grasp the meaning of each interpretation); instead we require a kind of orientation that precedes all more reflective orientations, including the split between "subjectivity" and "objectivity" itself – a prereflective sense of what it means to have the world in view as an objective and public world which is already there for us as we participate in the practices of giving and asking for reasons.

Hegel rejects Kant's transcendental account of such agency (including, obviously, the account of freedom as involving noumenal causation) in favor of such a developmental and historical account of agency. (This is, of course, obviously a weighty topic that deserves much more space than can be given to it here; here I will have to content myself with just stating the rejection and looking at some of its consequences.[12]) For Hegel, agency itself is a kind of norm, something that is socially and historically instituted, not some metaphysical or natural fact. Our independence from nature, that is, is a normative historical and social *achievement*, not a fact (metaphysical or natural) about ourselves that we have only recently discovered. We are not agents who are constitutively or metaphysically independent of nature (either by virtue of being made of nonnatural "stuff" or by being noumenally free from nature's causation); rather, we establish or *institute* our freedom from nature by virtue of a complex historical process in which we have come to see nature as inadequate to agency's (that is, *Geist*'s) interests, not because of some kind of fundamental metaphysical mismatch between the two, but because by taking "nature" to be something to which we had to keep faith – by *making* nature normative for ourselves – we put ourselves in situations where the giving and asking for reasons in that kind of context turned out to be impossible and in which we therefore, as being *who we were* by virtue of holding fast to that conception of normativity, turned out not to be able to sustain our collective lives as those kinds of people – we could no longer *be* those kinds of agents. Our freedom, as the ability to understand our actions as coming from us and not as any kind of "agent-causation," is itself an achievement, not a transcendental condition of our agency, and it is bound up with the achievement of our normative independence from nature.

Nature, Hegel says, has "turned out to be the Idea in its otherness [*Anderssein*]."[13] Although this has quite traditionally been taken as evidence of Hegel's neoplatonist approach to philosophy – as if nature were only some kind of "emanation" from the eternal logos or mind of the world – that view is hard to square with Hegel's "speculative" approach to the nature of agency.[14] In terms of that approach, nature is to be taken as the "other" of the space of reasons, as that which is to be regarded as nonnormative, devoid of spirit (*geistlos*), as Hegel sometimes describes nature. *Naturphilosophie* and physics are both, as he stresses, modes of "conceptually knowing" nature (*denkende Erkenntnis der Natur*).[15] As Hegel saw, his own speculative approach to nature, in terms of the "Kantian paradox," puts great strains on itself if it is to square itself with nature's sheer "otherness," its "recalcitrance" to a priori investigation.[16] To study nature, we must, in Hegel's words, "step back from natural things, leaving them as they are and directing ourselves to them"[17] – to do what Wilfrid Sellars characterizes as letting "the claim ... so to speak, [be] evoked or wrung from the perceiver by the object perceived."[18] Such an approach always suffers from the temptation to rely on the all-too-prevalent metaphors of the "world's telling us" what we need to know about it, or the "world's cooperating" with our endeavors to know it. The problem, though, for speculative philosophy – how exactly are we to make sense of speculative thought and the fact that we must take in empirical information about the world in order to speak sensibly of it – is another version of the post-Kantian problem of how we are to reconcile thought's spontaneity with the necessity for attending to given fact.

Strikingly, Hegel simply has no problem with giving modern natural science the credit for finding out about the way the world is. In an often-cited passage – which if it were taken out of context might have been attributed to Quine – Hegel remarks, "Not only must philosophy be in agreement with the experience of nature, but the *origin* and *formation* of philosophical science has empirical physics as its presupposition and condition."[19] He goes on to add in several places, however, that the "basis" of empirical science must lie in the "necessity of the concept," and that philosophy converts the findings of natural science into some kind of conceptual structure that exhibits a non-empirical necessity.[20] Hegel speaks metaphorically of this as a "translation" of what physics has "empirically prepared" into conceptual form, into the "diamond net" of metaphysics in which things are first "made intelligible."[21] This of courses suggests (and many have taken it to mean just this) that philosophy can simply unchain itself from empirical research and in its own unfettered spontaneity develop the concepts that science "presupposes" and "must use" in its otherwise empirical activities – that philosophy offers some kind of conceptual foundation for empirical science that empirical science cannot provide for itself.

That this is not Hegel's intent can be seen by other things he says, and, in any event, it contradicts his practice. He notes that Kant's attempt to provide a construction for matter in the *Metaphysical Foundations for Natural Science* has the "merit of having made a beginning towards a *concept* of matter and of having revived with this attempt the concept of a *Naturphilosophie*."[22] That is, he sees Kant's attempt at constructing what would count as the "conceptual paradigms" of natural science as orienting his own investigations, although, of course, Hegel accuses Kant of staying only at the level of the "understanding" and of not therefore adequately grasping the "speculative" nature of this enterprise. The Kantian philosophy, as Hegel often says, stays at the level of "consciousness," by which he means that it takes the opposition of subject and object as fundamental, as the "final dichotomy," together with its attendant and one-sided notion that it is the object that functions normatively as the "truth-maker" for our judgments about it. This attitude, which Hegel also identifies as the standpoint of "representational" thought (of *Vorstellung*), is on Hegel's account not so much false as it is one-sided: the subject/object split is not a metaphysical division already present at hand in the world but is itself normatively established as a moment in the space of the reasons – that is, the thinking subject, conceived as a locus of epistemic responsibilities, involves the concept of responsibility as responsibility *to* something independent of us that is nonetheless binding on us only insofar as we can regard it as authored by ourselves. The problem with the investigation of nature is to understand how it is that science tells us something true about nature but does not force us back into the pre-Kantian worries about whether our "representations" match up with an reality external to them. Or, as Hegel describes it, in investigating nature, we *set up* an opposition between "subject and object," in which our "intention is rather to grasp nature, to conceptually comprehend it, to make it our own ... [and] it is here that the difficulty enters onto the scene: how do we subjects cross over to those objects? If we set ourselves to working on how we would leap over this gulf, we are certainly letting ourselves be led astray in thinking of nature by making nature, which is an other to us, into something other than what it is."[23]

Hegel's answer to his own rhetorical question, of course, comes down to saying that this is the wrong question to ask. Instead, we are supposed to see that the "reflective" picture of our comparing two "things" – say, an internal mental representation and an external object – is at best only a moment of our (largely prereflective) comprehension of how it is that prior even to that picture of comparing two "things" (and prior even to the so-called subject/object gulf itself) are the socially instituted and sanctioned ways in which we make moves within the space of reasons – undertaking commitments, attributing entitlements, and so forth. Beginning our inquiries with the subject/object split already in place is beginning with too much conceptual baggage already in tow, since the split between subject and object is itself a normative distinction made within the "Idea" – the space of reasons as the basis of intelligibility itself – which must first be grasped prereflectively through socialization and training.

In his *Logic*, Hegel offers an analysis of what it would mean to think of the relation of mind and world as proceeding from this "Idea" instead of thinking of it by beginning with the distinction of subject and object as already presupposed – an analysis, that is, of what it would *mean* to speak of embodied subjects as having the world in view through the unity of concepts and intuitions, as operating within the whole that makes up the "Idea," the basis of intelligibility. Rather than looking to overcome the subject–object gulf, it would be more correct, Hegel says, to speak of a "*subject-object*," or an "intuiting understanding" rather than divide our thinking activities into distinct faculties that then have to be combined.[24] There nonetheless cannot be any purely *logical* or "analytical" transition from what it would *mean* for us to have the world in view to what that world *actually* looks like when subjected to empirical investigation; or, as Hegel says in the systematic transition from the *Science of Logic* to the *Naturphilosophie*, "in this freedom therefore no [logical] transition takes place."[25] Just as our agency itself is to be taken not as a metaphysical fact but as a normative, social, and historical achievement, this unity of concept and intuition, of having the world in view, is itself not something that is "given" but also something to be achieved. Or, to slip into Hegelian language again, the "certainty" that can be taken for the "truth" that we are grasping things as they are, not merely as they appear, does not come about naturally (as if the harmony between thought and reality were always already in place) but as a result of historical probing and error. As Hegel puts it, "This unity of intelligence and intuition, of the spirit's being-within-itself and its comportment to externality, must however not be the beginning but the goal, not be something immediate but rather a unity that is produced."[26]

The enterprise of natural science is part of that story, and clearly (at least to Hegel) the practice of science does not aim at overcoming the so-called gulf between subject and object, since the idea that there is a metaphysical "gulf" is one of the "pictures" that his *Phenomenology* and *Logic* were supposed to overcome. However, the space of reasons is already in play in the enterprise of natural science, even though, as self-legislated, it is, like all of our other norms, subject to challenge. The "a priori" of nature that *Naturphilosophie* has as its object is not unrevisable, as Kant had held, but is instead dynamic – as Hegel puts it: "All revolutions in the sciences no less than in world history only come about in that spirit, in order to understand and examine itself with a view to possessing itself, has changed its categories, grasping itself more truly, more deeply, more inwardly, and more at one with itself."[27] Moreover, as a good post-Kantian, Hegel does not hold that all our beliefs are of a piece (that is, are all empirical) since he rejects the view according to which it follows that all our beliefs are empirical from the idea that both empirical belief and the conceptual background of the sciences are revisable. Like Kant, Hegel holds that certain parts of, for example, Newtonianism (the mathematical parts, to be specific) are on a different order from the inductively and empirically established laws of Newtonian physics. Those non-empirical components *orient* the whole investigation and not simply more entrenched items of a "whole." (Hegel's thought is thus not to be identified with the kind of holism in vogue these days.[28])

Nonetheless, scientific investigation, as empirical investigation, begins with the observations of particular things and attempts to grasp what laws and principles are operative in them. *Empirical* research is necessary because we cannot through any kind of act of intellectual intuition simply read off the "universals" inherent in things, since, to cite Hegel again, "things are individual, and the lion as such does not exist."[29] We must instead *construct* our theories and test them, which implies more than merely inductively establishing laws, since such theory construction must take place against a background of an overall picture of nature. That we must construct our theories, however, does not mean that we should take an instrumental interpretation of the laws and principles established by science, as if, as Hegel puts it, they were not be taken as part of "objective actuality" but merely to serve as tools for "our convenience in order for us to make a note of things."[30] Like Kant, Hegel takes the construction of theories to be required in order for our experience to have any objective validity; but unlike Kant, Hegel does not hold that there can be a transcendental analysis of experience that will yield, for example, mathematics as constructed out of pure intuitions or the laws of inertia out of the constitutive rules of causality. Instead, we have only the way in which certain revolutions that have been achieved at the conceptual level have led to spirit's grasping itself more thoroughly. To that end, Hegel understands nature in the terms of his own jargon – itself constructed to express and explicate the "Kantian paradox" – as "externality" in general. Its basic laws and principles require an attention to that for which no purely "logical" (that is, a priori) reason can be given.

The philosophical issue therefore with which a *Naturphilosophie* is supposed to concern itself is the rationality, and therefore the truth, of those basic concepts themselves, which cannot be abstracted out of the findings of the natural sciences without begging all the questions at stake. To use the metaphor of levels: *Naturphilosophie* proceeds at a higher level than do the empirical sciences; its paradigms (or "shapes": "Gestalts," in Hegel's jargon) are reconstructions of the basic paradigms at work in the practices of the sciences themselves, and it is crucial to understand what the normative authority of those higher-level paradigms is – whether they can be seen to be norms of which we could regard ourselves as the authors by virtue of their rationality. This itself only points up what is Hegel's stated goal in his *Naturphilosophie*: to comprehend the normative authority of the modern scientific concept of nature without relying on "givens" in intuition and to understand how that concept fits into the true concept of our own agency as a normative status – that is, to get a better grip on who and what *we* are. Or, to put it a slightly different way, his concern is with those norms that we *bring to* empirical research and their status, even though he thinks those norms have historically developed *out of* the practice of empirical research. That we can bring certain concepts to empirical research means that they can be given an a priori status with regard to that research, even though that a priori is to be regarded as historically dynamic and not as some timeless set of conditions for possible experience.

Thus, like Kant, Hegel takes Newton's use of the concepts of absolute space and time to be paradigmatic for how proper science is to proceed, since these concepts function in just that kind of a priori way for Newton, although they were hardly there before Newton introduced them. Unlike Kant, of course, Hegel ascribes objective reality (not transcendental ideality) to both space and time. As part of nature's externality, (Newtonian) space and time are not purely logical conceptions but must be developed as the most abstract categories necessary to think of nature as an objective world, as that *to which* our claims are *responsible*, and not, for example, simply as a construct out of experience or out of different languages (as if in making claims about space and time we were only answering to each other and not to the world). In turn we get the right construction of the concepts of both space and time (that is, three-dimensionality and temporal direction) by thinking of what is unavoidable in conceiving the world as a whole of individual things coming-to-be and passing-away (which Hegel takes himself to have shown in his *Logic* to be unavoidable for us if we are to articulate the difference between "being" and "nothing" itself). Likewise, such abstractions require the determination of both place and movement, which in turn requires a conception of something that moves, namely, matter. Our experience of an objective world independent of us, to which we are responsible in making empirical claims, requires as a condition of our having it in view as "externality" itself space, time, place, motion, and matter. Those concepts are not *logical* concepts (in Hegel's extended use of "logical") but are taken from our having the world "in view" and from what is required for us to be able to exercise our abilities to speak about and measure that world as we have come to experience it in light of the development of the modern scientific revolutions such that our experience can be shown to have objective validity. It is not an analysis of our untutored experience (as if there were such a thing), nor is it an examination of the conditions of the possibility of ordinary experience (as if we all experienced the world as little Newtons); it is rather Hegel's own attempt to show that the basic concepts and norms that lay at the heart of the scientific, Newtonian picture themselves are rational, that is, they have the feature of being the way in which we must think of nature as the "other" of the modern, post-Kantian conception of agency. It is the analysis and reconstruction of the normative force of the prereflective notion of nature as a whole that "we moderns" bring with us to experience in light of our modern experience of ourselves as free beings.

Hegel therefore takes the concept of matter, in its post-Newtonian sense, to be constructed out of the forces of attraction and repulsion, as Kant did, but he criticizes Kant as having tried to derive the concept of matter both from experience and in terms of "the understanding," instead of seeing it "speculatively," that is, as a historically constructed concept (or "norm") that we can nonetheless "regard" as authored by ourselves while at the same time obligating us to be responsible *to it* in our investigations of nature. Taking nature as "externality," as consisting of units that can be counted, means taking nature, at least in terms of mechanics, to be the kind of thing for which mathematics is the appropriate language for the accounts of it.

For the "speculative" philosophy of nature, we thus need not stand pat with the opposition of invention *or* discovery. Our norms for the investigation of nature are indeed instituted by us, but when successful, they capture that part of nature that makes a difference to our understanding of it. The success of the disenchanted Newtonian account of the universe grew out of the ways in which earlier accounts themselves fail to make sense on their own terms, and which as a result were driven to seeing nature as "disenchanted," as a sphere of *geistlose* "externality" best described mathematically. Thus, Hegel is also concerned to show that Newton's concept of universal gravitation effectively responds to and answers the issues raised by the earlier division between celestial and sublunary motion that comes out of the more modern conception of nature as disenchanted "externality." (Hegel also criticizes Newton quite severely on some points, but since he draws on textbook accounts of his own day that themselves got Newton wrong, he is also off the mark in many places.[31])

Hegel tries to make a case for the irreducibility of other spheres of nature to the sphere of mechanics. The sphere of "physics" (heat, light, magnetism, electricity, sound, meteorology, and chemical processes) does not account for nature in terms of the pure externality of matter in motion; instead matter is said within those accounts to exhibit an affinity for other determinate matters (as in chemical processes), and, of course, he distinguishes organic life (and quasi-organic processes) from both of them. In each case, nature is said to become less and less "external," and in the case of organic animal life, as self-organizing, begins to prefigure the shape of subjectivity and self-direction themselves. In each case, mathematics comes to be less and less adequate to provide an account of that sphere.

The science of Hegel's own day was in all these regards undeveloped, and Hegel's own case for the irreducibility and uniqueness of each of these types of accounts cannot from a contemporary perspective be regarded as successful. In some respects, his regard for the working scientists of his own day in these fields was simply too high. This is not, however, the deepest failing in Hegel's own execution of his project. Like many people of his generation, Hegel was quite taken with the paradigm of comparative anatomy, and, in Hegel's own case, with the way in which Cuvier proposed that each species represents a fixed type that cannot be essentially changed. (Against Lamarck, Cuvier argued that evolution was an impossibility since it would require a species that is fully adapted in a fine-grained way to its environment as a whole to gradually alter itself, which would be impossible since any slight alteration would make it impossible for that species to survive at all.)

Hegel took that paradigm and expanded it to cover all of nature. Thus, Hegel thinks that there are three such "shapes" of nature (mechanical, physical, and organic), and within those shapes there are other, more determinate fixed shapes. Thus, by virtue of the requirements that thought imposes on itself to have "fixed shapes" for its thoughts, there is a requirement that we view nature as having certain natural kinds intrinsic to it, and that we see these natural kinds as being *in nature itself*, not merely in our own conceptual organization of the contingencies of nature. The division of nature into mechanical, physical, and organic systems is thus a division to be found *in nature* that is *required by thought* as it attempts to grasp nature as the "Idea in its otherness."

Following Cuvier's lead, Hegel argued against any kind of linear ordering of items in nature: in his words, "to seek to arrange in serial form the planets, the metals or chemical substances in general, plants and animals, and then ascertain the law of the series is a fruitless task, because nature does not arrange its shapes in such series and segments ... The concept differentiates things according to qualitative determinateness, and to that extent advances by leaps."[32] Moreover, as nature becomes organized into forms for which the category of "internal purposiveness" (or functional teleology) is necessary to render rational accounts of that sphere of nature, we are required to recognize an organism's intrinsic purposes in order for us to be able to sort out what counts as an aberration and what does not. In Hegel's own words, although nature "everywhere blurs the essential limits of species and genera by intermediate and defective forms, which continually furnish counter examples to every fixed distinction," the fact that we can recognize the existence of such transitional forms requires us to have already recognized the intrinsic purposes of various organisms; and "this type cannot be furnished by experience, for it is experience which also makes these so-called monstrosities, deformities, intermediate products, etc. available to us. Instead, the fixed type presupposes the independence and dignity of conceptual determination."[33] Thus, for example, Hegel thought that the concept of disease was teleological in that it demarcates some way in which an organism was not functioning (or functioning at odds with) its own internal purposes. (This focus on self-contained "shapes" of nature that cannot undergo any fundamental developmental alteration but which require leaps to new shapes also colors Hegel's own views about shapes of spirit or shapes of consciousness in his philosophy of history and in the Jena *Phenomenology*.)

The picture of the unity of nature that thereby emerges at the end of Hegel's treatment is not one that could be maintained today except by virtue of a rather weighty denial of a good bit of modern science itself. In particular, it rules out any Darwinian evolutionary account of life, and it rules out more broadly developmental accounts of the universe as a whole. This is, of course, all the more peculiar in that Hegel is the philosopher who champions developmental accounts above all. But Hegel's own interest in the philosophy of nature lies with showing how the scientific picture still emerging in his own time was compatible with a conception of subjectivity as a normative status, and that in fact the "disenchanted" view of nature that was part and parcel of the modern scientific image emerges together with the new conception of subjectivity at work in modern life, such that one cannot ascribe any priority to one or the other. (Neither is the "independent variable" explaining the other one.)

What strikes any contemporary reader as puzzling in Hegel's account of natural science (at least at first) are his assertions about the "untruth" of nature. This has to do with Hegel's own conception of truth as a kind of "primitive," in the sense that "truth" cannot be defined in terms of anything more fundamental than itself – it cannot be reduced to something else; thus, instead of offering something like a *theory* of truth (as correspondence, coherence, or warranted assertibility), Hegel – like Frege and Wittgenstein – offers instead an account of truth in terms of its *unfolding* (an *Entfaltung*) within the proper kind of reflective theory (such as his *Phenomenology* or *Logic*).[34] Hegel's conception of truth is thus closely linked to his general views about normativity and mindedness and to his implicit view, to use Robert Brandom's phrase, that the difference between the normative and the nonnormative (that is, the factual) is itself a normative distinction, a matter of how we ought to treat things (or to put it in Hegelian terms, that the difference between *Geist* and nature is a distinction, an achievement, made by *Geist* itself, and is not itself a natural distinction).[35] On Hegel's account, our *concepts* inevitably embody within them a conception of what it would mean for something to be the *best exemplification* of what it is. The problem with *nature* as it is (truly) conceived on the scientific model is that is it disenchanted: on its own, nature is incapable of organizing itself into "better and worse exemplifications," something Hegel nicely calls the "impotence [*Ohnmacht*] of nature."[36] Indeed, only when *life* appears in nature does it even make sense to speak of "better" and "worse," since only organisms display the kind of self-directing, functional teleological structure that makes the application of such terms meaningful. Only with organisms can one speak, for example, of disease; a planetary system, a mountain range, a piece of marble cannot be diseased. However, even "life," the stage at which "better" and "worse" exemplifications become meaningful, is itself still revelatory of the "impotence" of nature, since nature cannot organize itself even at the level of life into something like the "best" version of a lion, a rose, a trout. Nature *aims* at nothing. Only when human mindedness arrives on the scene does the issue arise of what it means for that kind of creature to be the best it can be, and that can only be formulated, as Hegel puts it, in terms of "self-consciousness," where we, as self-interpreting animals, have a historically developing conception of what it is to be the best exemplifications of the agents we are and actually aim at realizing those conceptions in our lives.

Hegel's ambition in constructing a speculative *Naturphilosophie* may be summarized succinctly: one can have science *and* subjectivity without having to sacrifice one or the other. Hegel's own philosophy of science would be opposed to any construal of science as a "mere" social construct − as if science were simply one more way of construing the world, no better, no worse than any other way − and it would be opposed to a comprehensively naturalized view of subjectivity, as if the causally oriented methods of natural science were adequate to grasp the meaning of things for human subjects and to provide any guidance for deliberation about what it means to lead a human life (even if *scientific* discoveries − as opposed to comprehensive natural-ist philosophies − about, for example, our genetic makeup offer valuable insight relevant to deliberations on that normative question). He therefore does not think that philosophy must subordinate itself to science, or naturalize itself and just become one part of a holistic (that is, empirical) account of mind and nature. Science seeks to explain nature, and both science and philosophy seek to grasp the unity of nature as a rational, intelligible whole. Philosophy's role in looking for the unity of nature is thus twofold: it seeks to understand the basic conceptual paradigms at work in the various natural sciences and to reconstruct them in terms of their rationality and intelligibility; and it does this with an eye to understanding agency itself, to grasp the role of subjectivity in the world. In reconstructing the paradigms of science, philosophy plays a different role than empirical science; it aims at the normative reconstruction of those paradigms, not at offering alternative empirical accounts or metaphysical views about how natural processes occur. Our own agency, spirit, is, as Hegel puts it, both "prior to and posterior to nature, and is not merely the metaphysical Idea of it," and agency (spirit) is to be understood as having "come forth out of nature, not, however, empirically but so that spirit is already always contained in nature, which spirit pre-supposes for itself."[37]

The revolution in modern science was an essential part of the modern revolution in "spirit," in our grasp of what it means to be human, just as the revolution in spirit's grasp of itself correspondingly called for a revolution in our theoretical stance to nature. To grasp the revolution in spirit required, so Hegel thought, grasping just what nature was so that it would become intelligible how it could be that spirit had to define itself as a *self-instituted liberation* from nature − or more metaphorically as "nature having become an other to itself in order again to cognize itself as the Idea and to reconcile itself with itself"[38] − not simply as the actualization of natural powers already dormant in human beings. It is not a priori given that all the ways in which natural science proceeds will cohere with each other, nor is it given that natural science, in its empirical procedures, will fit the concepts that we think in advance nature must have. Showing that there is a unity to nature by showing that the basic paradigms at work in the natural sciences all fit together – without their necessarily having to be reducible (in some sense of that word) to each other – is, on Hegel's view, something to be *achieved* in modern philosophy and science, not something that we can assume, and into which we can then shoehorn whatever results happen to turn up. The demand that our agency be subject only to a law that we can regard ourselves as authoring is correlative to the demand for the boundlessness, or "infinity," of the conceptual, a task made all the more daunting by the multifarious appearances of nature and the necessity of empirical research; but, in Hegel's words, "reason must nonetheless have confidence in itself, a confidence that in nature the concept speaks to the concept and that the genuine shape of the concept, which lies concealed beneath the externality of the infinitely many shapes [of nature], will show itself to reason."[39] In that rational faith, so Hegel claims, we find:

> Spirit thus *at first* grasps itself as emerging from that which is immediate but *then*, abstractly grasping itself, it wishes to liberate itself by developing nature out of itself; spirit's doing this is philosophy.[40]

Notes

1   There is a subtle difference between a "philosophy of nature" and a "nature philosophy" in the context of post-Kantian idealism. In Schelling's hands, it became an approach to philosophy itself, not an area within philosophy (like, for example, epistemology or metaphysics). I elaborate on this distinction in Terry Pinkard, *German Philosophy 1760–1860: The Legacy of Idealism* (Cambridge: Cambridge University Press, 2002).

2   On the notion of using Kant to get out of Kant, which played such a large role in the development of post-Kantian philosophy, see Rolf-Peter Horstmann, *Die Grenzen der Vernunft: Eine Untersuchung zu Zielen und Motiven des deutschen Idealismus* (Frankfurt a. M.: Anton Hain, 1991), and my own narrative of this movement in Pinkard, *German Philosophy 1760–1860*.

3   This is made all the more explicit in "Glauben und Wissen oder Reflexionsphilosophie der Subjektivität in der Vollständigkeit ihrer Formen als Kantische, Jacobische und Fichtesche Philosophie," in G. W. F. Hegel, *Werke in zwanzig Bänden*, eds. Eva Moldenhauer and Karl Markus Michel (Frankfurt a. M.: Suhrkamp Verlag, 1971), vol. 4, pp. 287–433. There Hegel says that "The original synthetic unity of apperception is also first recognized as the principle of the figurative synthesis, that is, of the forms of intuition: space and time are themselves conceived as synthetic unities, and the productive imagination, spontaneity, and the absolute synthetic activity is conceived as the principle of the sensibility which was previously characterized only as receptivity … One and the same synthetic unity … is the principle of intuition and of the understanding … And the imagination is nothing but reason itself" (pp. 305–8).

4   Immanuel Kant, *Groundwork of the Metaphysics of Morals*, trans. H. J. Paton (New York: Harper Torchbooks, 1964), p. 98 (AA 431). "The will is therefore not merely subject to the law, but is so subject that it must be considered as also *giving the law to itself* and precisely on this account as first of all subject to the law (of which it can regard itself as the author)." ["Der Wille wird also nicht lediglich dem Gesetze unterworfen, sondern so

unterworfen, dass er auch als selbstgesetzgebend, und eben um deswillen allererst dem Gesetze (davon er selbst sich als Urheber betrachten kann) unterworfen, angesehen werden muss."] Kant then later claims (p. 100 [A432]): "We need not now wonder, when we look back upon all the previous efforts that have been made to discover the principle of morality, why they have one and all been bound to fail. Their authors saw man as tied to laws by his duty, but it never occurred to them that he is subject only to laws which are made by himself and yet are universal and that he is bound only to act in conformity with a will which is his own but has as nature's purpose for it the function of making universal law."

5   Paul Franks, among others, has stressed this root of *Faktum* in "making," *facere*, an idea that Fichte tries to render by resurrecting the German term *Tathandlung*, to express how a subject binds itself only to laws of its own making. See Paul Franks, *All or Nothing: Skepticism, Transcendental Arguments, and Systematicity in German Idealism* (forthcoming)

6   I go into the role that what I have called the "Kantian paradox" plays in the development of post-Kantian philosophy in Pinkard, *German Philosophy 1760–1860*. This notion of the kind of paradox present in Kant's conception of autonomy and how post-Kantians generalized it into a problem concerning all issues of normative authority was first presented, to my knowledge, by Robert Pippin in his "The Actualization of Freedom," in Karl Ameriks (ed.), *Cambridge Companion to German Idealism* (Cambridge: Cambridge University Press, 2000). The issue of how such autonomy "binds itself" is also a leading theme in the writings of Christine Korsgaard. Her most recent and comprehensive statement is in Christine Korsgaard, *The Sources of Normativity* (Cambridge: Cambridge University Press, 1996). I do not see how Korsgaard provides a solution for how a will moves from a state of nonobligation to obligation in an act of will, except by restating the "fact of reason," that we always already in legislating these laws for ourselves assume that we are under the claims of reason. That seems to me to restate the paradox (as Kant himself did) rather than come to terms with it; but adequately making that point in a way that took Korsgaard's fine-grained reading of Kantian texts into proper account would require at least another essay.

7   See *Enzyklopädie der philosophischen Wissenschaften*, §436, *Zusatz*, where Hegel offers the relation between two subjects involving "universal self-consciousness as the affirmative knowing of oneself in an other self" as the paradigm of speculative truths. All references to the *Enzyklopädie*, unless otherwise noted, are to vol. 9 of G. W. F. Hegel, *Werke in zwanzig Bänden*, eds. Eva Moldenhauer and Karl Markus Michel.

8   The phrase, "*das Andere seiner selbst*," a phrase Hegel himself claimed to take from Plato, occurs in several places. See G. W. F. Hegel, *Wissenschaft der Logik*, (Hamburg: Felix Meiner, 1971), vol. 2, p. 494; *Science of Logic*, trans. A. V. Miller (Oxford: Oxford University Press, 1969), p. 834. In the *Enzyklopädie* see particularly §448, *Zusatz*. It also occurs in the *Enzyklopädie*, vol. 8, §81, *Zusatz*; vol. 8, §92, *Zusatz*; §389, *Zusatz*; §426, *Zusatz*.

9   Sellars makes this distinction in all of his works. The classical loci are Wilfrid Sellars, "Philosophy and the Scientific Image of Man," and "Empiricism and the Philosophy of Mind," in Wilfrid Sellars, *Science, Perception, and Reality* (London: Routledge and Kegan Paul, 1963), pp. 1–40, 127–96.

10 See G. W. F. Hegel, *Differenz des Fichteschen und Schellingschen Systems der Philosophie*, in *Werke*, eds. Moldenhauer and Michel, vol. 2, p. 48: "Die Einheit ist in einer nur relativen Identität erzwungen, die Identität, die eine absolute sein *soll*, *ist* eine unvollständige. Das System ist zu einem Dogmatismus – zu einem Realismus, der die Objektivität, oder zu einem Idealismus, der die Subjektivität absolut setzt – wider seine Philosophie geworden, wenn beide (was bei jenem zweideutiger ist als bei diesem) aus wahrer Spekulation

hervorgegangen sind." ["The unity is forced into an only relative identity, the identity that *ought* to be an absolute identity, *is* an incomplete identity. The system has changed into a dogmatism − into a realism that absolutely posits objectivity, or into an idealism that absolutely posits subjectivity − contra its philosophy, if both (which is more ambiguous with respect to the former than the latter) are to have emerged from pure speculation.]

11 On using "Kant to get out of Kant" in the move to Hegelianism, see Pinkard, "Virtues, Morality, and *Sittlichkeit*: From Maxims to Practices, Or: Using Kant to Get Out of Kant (and Using Hegel to Get Out of Hegel)," *European Journal of Philosophy* 7(2) (Aug. 1999).

12 I offer a fuller discussion of the issue in Pinkard, *German Philosophy 1760–1860*. On the issue of freedom in Hegel, see Robert Pippin's pathbreaking "Naturalness and Minded-ness: Hegel's Compatibilism," *Journal of European Philosophy* 7(2) (Aug. 1999): 194–212.

13 *Enzyklopädie* §247: "Die Natur hat sich als die Idee in der Form des Andersseins erge-ben." ["Nature has turned out to be the Idea in the form of otherness."]

14 The best statement in English of this way of reading Hegel (and of the implausibility of such an overall view) is Michael Rosen, *Hegel's Dialectic and Its Criticism* (Cambridge: Cambridge University Press, 1982).

15 *Enzyklopädie*, "Einleitung," p. 11: "Physik und Naturphilosophie unterscheiden sich also nicht wie Wahrnehmen und Denken voneinander, sondern nur *durch die Art und Weise des Denkens*; sie sind beide denkende Erkenntnis der Natur." ["Physics and *Naturphilosophie* are thus not differentiated as perception and thought are from each other, but rather *through the kind and mode of thought*; they are both a thoughtful cognition of nature."]

16 *Enzyklopädie* §376, *Zusatz*: "Die Schwierigkeit der Naturphilosophie liegt eben darin, einmal, daβ das Materielle so widerspenstig gegen die Einheit des Begriffes ist, und dann, daβ ein Detail den Geist in Anspruch nimmt, das sich immer mehr häuft" (p. 539). ["The difficulty of *Naturphilosophie* lies just in materiality being so recalcitrant to the unity of the concept, and then in its details continuing to pile up and making their claim on spirit."]

17 *Enzyklopädie* §246, *Zusatz*.

18 Wilfrid Sellars, "Empiricism and the Philosophy of Mind," para. 16, p. 144.

19 *Enzyklopädie* §246.

20 See *Enzyklopädie* §246 and *Zusatz*.

21 Ibid.

22 *Enzyklopädie* §262.

23 *Enzyklopädie* §246, *Zusatz*. "Unsere Absicht ist aber vielmehr, die Natur zu fassen, zu begreifen, zum Unsrigen zu machen, daβ sie uns nicht ein Fremdes, Jenseitiges sei. Hier also tritt die Schwierigkeit ein: Wie kommen wir Subjekte zu den Objekten hinüber? Lassen wir uns beigehen, diese Kluft zu überspringen, und wir lassen dazu uns allerdings verleiten, so denken wir diese Natur; wir machen sie, die ein Anderes ist als wir, zu einem Anderen, als sie ist."

24 "This identity has therefore been rightly determined as the *subject-object*, for it is as well the formal or subjective concept as it is the Object as such" and "having proceeded from the Idea, independent objectivity is immediate being only as the *predicate* of the judgment of the self-determination of the concept − a being that is indeed differentiated from the subject, but at the same time is essentially posited as a moment of the concept." Hegel, *Science of Logic*, p. 593; *Wissenschaft der Logik*, II, p. 232 in Moldenhauer and Michel (eds.), *Werke*, vol. 6, p. 266.; on the notion of the "intuiting understanding" see *Science of Logic*, pp. 758, 765; *Wissenschaft der Logik*, II, pp. 411, 418; *Werke*, 6, pp. 466, 475.

25 *Science of Logic*, p. 843; *Wissenschaft der Logik*, II, p. 505.

26 *Enzyklopädie* §246, *Zusatz*, p. 18.

27 *Enzyklopädie* §246, *Zusatz*: "Alle Revolutionen, in den Wissenschaften nicht weniger als in der Weltgeschichte, kommen nur daher, daβ der Geist jetzt zum Verstehen und

Vernehmen seiner, um sich zu besitzen, seine Kategorien geändert hat, sich wahrhafter, tiefer, sich inniger und einiger mit sich erfassend."

28 This obviously brings Hegel into conversation with the line of thought developed by Michael Friedman's attempt to construct a dynamic a priori in light of Kant's philosophy – Friedman's own version of "using Kant to get out of Kant," as it were. Friedman's own construction of a dynamic a priori in some kind of post-Kantian (or neo-Kantian) framework is indebted to his reading of Ernst Cassirer, who in turn was influenced by Hegel's own historicism. (Thus, the correspondences between a post-Kantian interpretation of Kant influenced by Cassirer and this kind of Hegelianism are not completely accidental.) Friedman's original interpretation of Kant is found in his now standard work: Michael Friedman, *Kant and the Exact Sciences* (Cambridge, MA: Harvard University Press, 1992). The development of his views on the dynamic a priori and Ernst Cassirer are to be found in Michael Friedman, *A Parting of the Ways: Carnap, Cassirer, and Heidegger* (Chicago: Open Court, 2000); his working out of a "dynamic" a priori in light of Kant and Cassirer is to be found in Michael Friedman, *Dynamics of Reason* (Stanford: CSLI Publications, 2001).

29 *Enzyklopädie* §246, *Zusatz*, p. 16.

30 *Enzyklopädie* §246, *Zusatz*, pp. 19–20.

31 See Wolfgang Neuser's helpful discussion in his contribution to Hermann Drüe, Annemarie Gethmann-Siefert, Christa Hackenesch, Walter Jaeschke, Wolfgang Neuser, and Herbert Schnädelbach (eds.), *Hegel's Enzyklopädie der philosophischen Wissenschaften (1830): Ein Kommentar zum Systemgrundriß* (Frankfurt a. M., 2000), pp. 139–205. The other comprehensive treatment of Hegel's *Naturphilosophie* that is virtually indispensable in this regard is Wolfgang Bonsiepen, *Die Begründung einer Naturphilosophie bei Kant, Schelling, Fries und Hegel: Mathematische versus spekulative Naturphilosophie* (Frankfurt a. M.: Vittorio Klostermann, 1997). In English, there is Michael Petry's indispensable and very learned commentary accompanied by his translation of the *Naturphilosophie*: see, *G. W. F. Hegel's Philosophy of Nature*, trans. with an Introduction by Michael John Petry (London: George Allen & Unwin, 1970).

32 *Enzyklopädie* §249, *Zusatz*.

33 *Enzyklopädie* §250.

34 Among the many passages that can be cited to illustrate this point, *Enzyklopädie der philosophischen Wissenschaften* §14 nicely sums up Hegel's point: "The science of this is essentially a system, because what is true exists as concrete only as unfolding itself and in both holding itself together and sustaining itself, i.e., as totality, and only through the differentiation and determination of its differences can the totality exist as the necessity of those differences and the freedom of the whole." ["Die Wissenschaft desselben ist wesentlich System, weil das Wahre als konkret nur als sich in sich entfaltend und in Einheit zusammennehmend und – haltend, d. i. als Totalität ist und nur durch Unterscheidung und Bestimmung seiner Unterschiede die Notwendigkeit derselben und die Freiheit des Ganzen sein kann."]

35 On this distinction between the normative and the nonnormative being itself normative, see Robert Brandom, "Freedom and Constraint by Norms," *American Philosophical Quarterly*, April 1977: 187–96.

36 *Enzyklopädie* §250.

37 *Enzyklopädie* §376, *Zusatz*, pp. 538–9.

38 Ibid., p. 538.

39 Ibid., p. 539.

40 Ibid.

2

NATURPHILOSOPHIE

G. W. F. Hegel

§246 What is now called *physics* was previously called *philosophy of nature*, and it is just as much a theoretical, indeed a *thinking* consideration of nature, which, on the one hand, does not set off from determinations external to nature ..., and, on the other hand, is aimed at a cognition (*Erkenntnis*) of what is universal in nature, in a manner so that the universal is at the same time determined within itself – it is aimed at the forces, laws and genera, whose content must also not be a mere aggregate but which is instead to be placed into arrangements and classes and which must have the appearance of organization. Because the philosophy of nature is a conceptually comprehending (*begreifende*) examination, it has as its object the same universal but taken for itself, and it considers this universal in its own immanent necessity in accordance with the concept's self-determination.

We have already spoken of the relationship between philosophy and what is empirical in the general introduction [to the philosophy of nature]. Not only must philosophy agree with our experience of nature, but philosophical science's unfolding and formation (*Bildung*) has empirical physics as its presupposition and condition. The path of this unfolding and the preliminaries of a science are, however, something different from the science itself; in the latter, the former can no longer appear as the foundation of the science, which here [in the philosophy of nature] should be the necessity of the concept. We have already been reminded that, in addition to specifying the object in terms of its conceptual determination in the course of its philosophical treatment, we must also take note of the empirical appearance corresponding to it, and we must demonstrate that the appearance does in fact correspond to its concept. This is not, however, to invoke experience in regard to the necessity of the content, and it is even less permissible to invoke what was called intuition, which used to be nothing more than a procedure for representing and fantasizing (and for pipe-dreams) in accordance with analogies, which can be more contingent or more significant, and which only externally stamp determinations and schemata onto objects.

From G. W. F. Hegel, "Naturphilosophie," §246 in *Werke in zwanzig Bänden*, vol. 9 (Part Two of Hegel's *Encyclopedia of the Philosophical Sciences*), eds. Eva Moldenhauer and Karl Markus Michel. Frankfurt am Main: Suhrkamp Verlag, 1971. New English translation by Terry Pinkard.

Zusatz: In theoretical conduct the first thing is (a) that we step back from natural objects, leaving them to be as they are and directing ourselves towards them. Here we start from our sensuous acquaintance with nature. If physics were, however, based solely on perceptions, and perceptions were nothing but the evidence of the senses, then the physical act would consist only in seeing, hearing, smelling, etc., and in this way animals too would be physicists. However, it is a mind (*Geist*) that thinks, sees, hears, etc. Now, when we said that in the theoretical [consideration with regard to nature] we let things go free, this was only in part related to the outer senses, since these are themselves in part theoretical and in part practical; only representing – the intelligence – takes this free stance (*Verhalten*) to things. To be sure, with [representing and intelligence] we can consider things as means; but then cognition is also only a means, not an end in itself (*Selbstzweck*). (b) The second relation of things to us is that things acquire for us the determination of universality, that is, that we transform them into universals. The more thought enters into representation, the more naturalness, singularity, and immediacy vanish from things: by virtue of thought's pushing its way into the wealth of the infinitely multiple configurations of nature, nature's riches are impoverished, its springtimes fade away, and its play of colors turns pale. In the stillness of thought, life's rustle in nature goes silent; her warm abundance, which shapes itself into a thousand delightful surprises, shrivels into dry forms and shapeless universalities resembling a murky northern fog. (c) Both of these determinations are not only opposed to both practical ones, but we also find that the theoretical stance is within itself self-contradictory since it seems immediately to bring about the direct opposite of what it intends. That is to say, we want to cognize the nature that efficaciously exists, not something that does not exist, but instead of leaving nature as she is and taking her as she is in truth, instead of perceiving her, we make her into something quite different. In thinking things, we thereby convert them into some-thing universal; things, however, are singular, and the "lion in general" does not exist. We make them into something subjective, something produced by us, belonging to us, and indeed characteristic of us as human; for natural objects do not think, and they are not representations or thoughts. But according to the second determination, to which previous reference was made, it is precisely this inversion which does take place; in fact, it could seem that what we are starting to do is all at once rendered impossible for us. The theoretical stance begins with the suspension of desire, it is selfless, it lets things alone, it lets them subsist; with this stance [towards nature], we have straightaway established two items, the object and the subject, and their separ-ation – something worldly (*Diesseits*) and something beyond (*Jenseits*). However, our intention is all the more to grasp nature, to conceptually comprehend it, to make it our own so that it is not something alien, something "on the other side" (*Jenseitiges*). Here the difficulty enters onto the scene: How do we subjects cross over to those objects? If we set ourselves to working on how we would leap over this gulf, we are certainly letting ourselves be led astray in thinking of nature by virtue of having made nature, which is an other to us, into something other than what it is. Both theoretical relationships are also directly opposed to each other: We make things into universals, that is, make them our own, and yet as natural objects they are supposed to have a free being for self. This therefore is the point in which the issue has to do with the nature of cognition – this is the interest of philosophy ...

In order to state briefly what is the defect of this idea (*Vorstellung*), it must at once be admitted that there is something lofty in it which at first glance strongly recommends itself. This unity of intelligence and intuition, of the being-within-itself of spirit and its stance to externality, must however not be the beginning but the goal, not be an immediate unity but rather a unity that is produced. A natural unity of thought and intuition is that of the child and the animal, which can at most be called feeling but not mindedness (*Geistigkeit*). Man must have eaten of the tree of knowledge of good and evil and must have gone through the labor and activity of thought in order to be what he is: The being who overcomes this separation of himself from nature. That immediate unity is only abstract truth existing-in-itself, not the efficaciously real truth; for not only must the content be true but the form must also be true. The dissolution of this rift must be said to have the shape of the knowing Idea, and the moments of that dissolution must be sought in consciousness itself. It is not an issue of being given over to abstraction and emptiness, of taking refuge in the nothingness of knowing; rather, consciousness must preserve itself in that we wish through ordinary consciousness itself to refute the suppositions through which the contradiction arose.

The difficulty, that is, with the one-sided assumption of theoretical consciousness that natural objects confront us as persisting and impenetrable objects, is directly refuted by the kind of practical conduct in which the absolutely idealistic faith is rooted, namely, that individual things are nothing in themselves. The defect of desire is, seen from the way it comports itself towards things, not that it is realistic towards them, but that it is all too idealistic. Philosophically genuine idealism consists of nothing else than just in the determination (*Bestimmung*) that the truth about things is that they are as such immediately individual, i.e., are sensuous, only a seeming-to-be, an appearance. One could say of a certain metaphysics that is all the rage today and which maintains that we cannot know things because they are absolutely cut off from us, that not even the animals are so stupid as these metaphysicians, for animals go after things, grasp them, seize them, and consume them. The same determination lies in the second aspect of the theoretical conduct referred to above, namely, that we think natural objects. Intelligence familiarizes itself with things, not of course in their sensuous existence, but by thinking them, positing them with respect to their content within itself; and since it, so to speak, adds form, adds universality to practical ideality, which for itself is only negativity, it gives an affirmative determination to what is negative in individuality. What is universal in things is not something subjective, something up to us; rather it is, in contrast to the transient phenomenon, the Noumenon, what is true, what is objective, what is efficaciously real in things themselves, like the Platonic ideas, which do not exist somewhere far off, but rather exist in individual things as their substantial genera. Not until one does violence to Proteus – that is, not until one does not turn away from the sensuous appearance of nature – is Proteus compelled to speak the truth. The inscription on the veil of Isis, "I am that which was, is, and will be, and no mortal has lifted my veil," melts away before thought. "Nature," [Johann Georg] Hamann therefore rightly says, "is a Hebrew word written only with consonants, and the understanding must add the periods."

Now, if the empirical consideration of nature has this category of universality in common with the philosophy of nature, it sometimes wavers as to whether this universal is subjective or objective; one can often hear it said that these classes and orderings are only made as aids to cognition. This wavering goes still further in the search for distinguishing marks, not in the belief that they are essential objective determinations of things, but that they only serve our convenience to help us to distinguish things. If nothing more than that were at stake, we might, e.g., take the lobe of the ear as the mark of humanity, for no animal has it; but we feel at once that such a determination is not sufficient to cognize what is essential about man. When, however, the universal is determined as law, force, matter, then we cannot allow that it counts only as an external form and a subjective addition; rather, objective reality is attributed to laws, forces are immanent, and matter is the genuine nature of the thing itself. Something similar may also be conceded in regard to genera, for example, that they are not just a compilation of similarities, an abstraction made by us, that they not only have something in common but rather that they are the objects' own inner essence, that the orderings do not only serve to give us an overview but form a hierarchy of nature itself. The distinguishing marks should likewise be what is universal and substantial in the genus. Physics itself looks on these universals as its triumph: one can even say that it unfortunately goes too far in this universalizing...

The philosophy of nature takes the material that physics has prepared from experience for it to the point to which physics has brought it and further reshapes it without having experience be its final test and basis; physics must therefore work into the hands of philosophy in order that the latter may translate the abstract universal transmitted to it into the concept, since philosophy shows how this universal, as a whole that is necessary in itself, emerges from the concept. The philosophical way of exhibiting this is not something capricious, like all at once going about on one's head for a change after having for a long while gone about on one's legs, or all at once seeing our everyday face bedaubed with paint: rather, it is because physics' approach does not satisfy the concept that we have to go further.

What distinguishes the philosophy of nature from physics is, more precisely, the kind of metaphysics used by both of them; for metaphysics is nothing other than the range of the universal determinations of thought, as it were, the diamond net into which we bring all material and thereby first make it intelligible. Every cultured consciousness has its metaphysics, a quasi-instinctual thinking, the absolute power within us of which we become master (*Meister*) only when we make it in turn the object of our cognition. Philosophy in general has, as philosophy, other categories than those that ordinary consciousness has: all formative education (*Bildung*) reduces itself to the distinction among the categories. All revolutions, in the sciences no less than in world history, only come about from spirit's having changed its categories in order to understand and interrogate itself, in order to possess itself, and to come to grasp itself more truly, more deeply, with more inwardness, and more in unity with itself. Now the inadequacy of physics' thought-determinations can be traced back to two points which are closely bound up with each other. (a) What is universal in physics is abstract or only formal; it has its determinations not within itself, that is, it does not pass over into particularity; (b) the determinate content is for that very reason external to the universal and is split up, broken into parts, individualized, isolated, without any necessary connection within itself, and for that reason exists only as finite content. If we have a flower, for example, the understanding considers its particular qualities; chemistry takes it apart and analyzes it. In this way, we separate color, shape of the leaves, citric acid, oil of the ether, carbon, hydrogen, etc.; and now we say that the plant consists of all theses parts ...

Spirit cannot remain at this stage of thinking in the terms framed by reflection carried out in "the understanding" (*Verstandesreflexion*), and there are two ways in which it transcends it. (a) The ingenuous spirit (*der unbefangene Geist*), when it animatedly intuits nature, as in the clever examples we often find Goethe asserting, feels the life and the universal connectedness within nature; it gets an inkling of the universe as an organic whole and a rational totality just as it also feels in individual living things an inward unity within itself; but even if we bring together all those ingredients of the flower, no flower thus emerges. Thus in the philosophy of nature, people have fallen back on intuition and posited it as being higher than reflection; but this is the wrong route to take for one cannot philosophize out of intuition. (b) Intuition must also be [subject to] thought, and the isolated parts must be brought back by thought to simple universality; this unity *as thought* is the concept, which contains the determinate differences, however, as a self-moving unity within itself. The determinations of philosophical universality are not indifferent; it is the universality that sets it own demands and lives up to them (*sich selbst erfüllende Allgemeinheit*), and which at the same time in its diamond-like identity also contains difference.

Genuine infinity is the unity of itself and the finite; and this is now *the* category of philosophy and consequently also of the philosophy of nature. If genera and forces are what is inner in nature, and if over and against this universal, what is external and individual is what is transitory, then it is demanded that there be yet a third stage, the inner of the inner, which, according to what has been said, would be the unity of the universal and the particular ...

In grasping this inward aspect (*dieses Innern*), the one-sidedness of the theoretical and practical stances is superseded and incorporated (*aufgehoben*), and at the same time both determinations are satisfied. The former contains a universality without determinateness, the latter an individuality without a universal; comprehending cognition (*begreifendes Erkennen*) is the middle term in which universality does not remain something on *this* side, in *me*, over and against the individuality of the objects, but rather because it itself takes up a negative stance to things and assimilates them to itself, it equally finds individuality in them, leaves things alone so that they determine themselves freely within themselves. Comprehending cognition is thus the unity of the theoretical and practical stances: The negation of individuality is, as negation of what is negative, the affirmative universality that gives enduring existence (*Bestehen*) to its determinations; for genuine individuality is at the same time within itself a universality ... Since what is inner in nature is none other than the universal, then so are we in our own sphere (*bei uns selbst*) when we have thoughts of this which is what is inner to nature. If truth in its subjective sense is the correspondence of a representation with its object, then truth in its objective sense is the correspondence of the object, the real thing, with itself so that its reality is adequate to its concept. In its essence, the "I" is the concept that is self-identical (*sich selbst Gleiche*), which passes through all things, and which, because it retains mastery over the particular differences, is the universal returning into itself. This concept is at once the genuine Idea, the divine Idea of the universe which alone is what is efficaciously real (*das Wirkliche*). Thus God alone is the truth, in Plato's words, the immortal being whose body and soul are united in a single nature ...

BERGSON

3

BERGSON'S SPIRITUALIST METAPHYSICS AND THE SCIENCES

Jean Gayon

1. Introductory Remarks: Not a "Philosopher of Science"

There can be no doubt that Henri Bergson (1859–1941) was the most influential of all twentieth-century French philosophers. Until about 1960 there was general agreement about this. I probably belong to the last generation of French students who were taught that Bergson was a philosopher as important as Plato, Descartes, or Kant. From my own experience I can confirm that this seemed obvious to all French philosophers, whatever their philosophical orientation. In the past 30 years, however, this situation has changed dramatically. Most philosophers under the age of 50 know little or nothing of Bergson. Until recently, I myself had not read a single line of Bergson since 1966, when I took my first course in philosophy. As a philosopher of science, and, more especially, as a philosopher of biology, I confess that I have never referred to Bergson in any of my writings, talks, or courses. I have never even thought of doing so. My original training in philosophy of science was based on a mixture of neopositivist authors and of three French philosophers who will be invoked at length in the course of the present volume: Bachelard, Canguilhem, and Foucault. It is thus a rather strange experience, for a French philosopher of science, to participate in such a book and write about Bergson. In a sense, this is a good thing. Commenting on Bachelard, Canguilhem, or Foucault would be like evaluating my intellectual parents. Writing about Bergson allows me to be in the ideal anthropological situation of the "foreigner": I am familiar enough with the language and culture of the author to feel empathy, but Bergson's style and worldview lie far away from my philosophical habitat. Thanks to Gary Gutting, I spent a good deal of summer 2002 reading Bergson's writings. I do not claim that what I will say is original from the point of view of the current literature in history of philosophy. In fact, I have deliberately ignored most of this literature, and I have chosen to concentrate on a single question: what kind of relationship did Bergson's philosophy have with science?

As the title of this chapter suggests, I wish to carefully avoid implying that there might be anything like a "philosophy of science" in Bergson's thinking. Bergson was *not* a "philosopher of science" and did not want to be one. To be sure, when he was a student of philosophy his first intention was to devote himself to the philosophy of science, in the conventional sense that this expression began to have in the second half of the nineteenth century (i.e. a critical reflection on the methodology of science). But he quickly decided that he would not engage in that kind of philosophical activity. This change is nicely described in a letter to William James in 1908. That year James gave a series of lectures at Oxford University. Since one of the announced lectures dealt with Bergson, James asked him to provide a *curriculum vitae* and to comment on the key events in his career. Here is an excerpt from Bergson's answer:

> As far as remarkable events are concerned, there have been none in my career, or at least no *objectively* remarkable events. However, subjectively, I cannot stop myself from giving a great deal of importance to the change that took place in my way of thinking during the two years that followed my graduation from the École Normale, between 1881 and 1883. Up until then I remained thoroughly imbued with the mechanistic theories to which I had been attracted at an early age, by reading Herbert Spencer, the philosopher whom I followed more or less without reservation. My intention was to devote myself to what at the time was called "the philosophy of science," and it was with this aim that, after graduating from the École Normale, I had begun to study a number of fundamental scientific notions. It was the analysis of the notion of time, as it intervenes in mechanics and in physics, that overturned all my ideas. To my great surprise, I realized that scientific time has no *duration*, that nothing would change in our scientific knowledge of the world if all of reality took place at once, all in a single instant, and that positive science fundamentally involves eliminating duration.[1]

This letter is remarkable. In a few words, it says that Bergson renounced being a philosopher of science – a critical analyst of scientific knowledge – because science ignored the notion of time, which was the central problem of Bergson's entire philosophy, from the beginning to the end. However, this letter should not be taken to mean that knowledge is divided into two branches: on the one hand, "positive science" (dealing with space and measurable phenomena), and on the other, "philosophy" or "metaphysics" (dealing with time and related notions in Bergson's thinking: duration, mind, qualitative knowledge, intuition, etc.). Bergson's philosophy is undoubtedly one of the most radical pleas in favor of spiritualism in the entire history of philosophy. But it would be a major mistake to believe that Bergson ignored and despised positive science. Quite the opposite. As shown in the Bergson text which follows the present paper, he was so respectful of positive science that he described his own philosophical method with the expression "positive metaphysics."[2] For Bergson, positive science was both the main source of information for the philosopher, and a model. Bergson's major books were all devoted to problems and theories that were directly inspired by major areas of scientific investigation and major modern scientific theories. He spent a tremendous amount of time reading primary sources in psychology, medicine, biology, physics, and sociology, and commenting upon them. But positive science was also a model: by "positive metaphysics" Bergson meant a kind of metaphysics founded upon "facts" and able to correct and rectify itself indefinitely, with the help of "experience." Experience was definitely not for him a good criterion for the characterization of the difference between science and philosophy. This explains why he did not see himself as a "philosopher of science." Bergson was not a philosopher of science because he rejected the very notion of "philosophy of science" as a metadiscourse, either normative, foundational, or interpretive. In his book *Creative Evolution* he was perfectly explicit on this subject:

> At first sight, it may seem prudent to leave the consideration of facts to positive science, to let physics and chemistry busy themselves with brute matter, the biological and psychological sciences with life. The task of the philosopher is then clearly defined. He takes facts and laws from the scientist's hand; and whether he tries to go beyond them in order to reach their deeper causes, or whether he thinks it impossible to go further and even proves it by the analysis of scientific knowledge, in both cases he has for the facts and relations, handed over by science, the sort of respect that is due to a final verdict. To this knowledge he adds a critique of the faculty of knowing, and also, if he thinks proper, a metaphysic; but the matter of knowledge he regards as the affair of science and not of philosophy
>
> But how does he fail to see that the real result of this so-called division of labor is to mix up everything and confuse everything? The metaphysic or the critique that the philosopher has reserved for himself he has to receive, ready-made, from positive science, it being already contained in the descriptions and analyses, the whole care of which he has left to the scientist. Because he did not wish to intervene, at the beginning, in questions of fact, he finds himself reduced, in questions of principle, to formulating purely and simply in more precise terms the unconscious and consequently inconsistent, metaphysic and critique which the very attitude of science to reality marks out ... Here we are not in the judiciary domain, where the description of fact and judgment on the fact are two distinct things ... Form is no longer entirely isolable from matter, and he who has begun by reserving to philosophy questions of principle, and who has thereby tried to put philosophy above the sciences, as a "court of cassation" is above the courts of assize and of appeal, will gradually come to make no more of philosophy than a court of registration, charged at most with wording more precisely the sentences that are brought to it, already pronounced and irrevocable.[3]

This quotation clarifies what Bergson meant by "positive metaphysics." On the one hand, philosophers should be as rigorous as scientists regarding the empirical adequacy of their theories. On the other hand, there is no absolute frontier between science and philosophy: philosophers should feel free to discuss with scientists the content of scientific knowledge. Similarly, it is not enough for philosophy to be a "critique" of scientific knowledge. Bergson's attack on Kant and anything that would resemble "epistemology" (a philosophical reflection upon the foundations and limits of scientific knowledge) is obvious.

These observations clarify the second half of my title: " ... and the Sciences." Although Bergson often used the word "science" as a generic word for a certain kind of knowledge, he could hardly accept the thesis that philosophy or metaphysics was something different in nature from "science" in general. Rather than a general "philosophy of science," philosophers had better examine "the sciences" in detail and confront them or cooperate with them. "Confrontation" and "cooperation" are two key words in Bergson's terminology when he speaks of the relationship between philosophers and scientists.

The first half of my title, "Bergson's Spiritualist Metaphysics," focuses on the most distinctive feature of his thinking. Bergson was indeed a spiritualist. All of his writings are devoted to demonstrating the existence of the mind, the supremacy of the spiritual over "matter," by a careful reflection on various spheres of human knowledge and experience: sensation and perception (psychology), memory (neurology and psychopathology), life (biology), time (physical theory of relativity); but also morals and art. Beyond the variety of objects that Bergson considered, he never stopped repeating and exploring the same basic idea from his first to his last writings – namely: that time is the key problem that had been neglected by all prior philosophers and scientists. By reflecting on time, philosophers can firmly establish the reality and importance of "mind." Most commentators concentrate on the system of concepts that Bergson progressively constructed in order to express this idea. This is a legitimate way of analyzing the thinking of a "great philosopher." What I want to stress here is the relationship of Bergson's spiritualism to his concern for empirical knowledge and for the sciences (plural).

For Bergson, matter and mind are not substances. They are "tendencies" or "forces." These tendencies conflict and collaborate in many areas of human experience and, beyond, of reality. The problem, then, is not whether the mental is reducible or irreducible to the material, or the reverse. The problem is: what is the respective weight of both tendencies in this or that aspect of reality:

> Both "yes" and "no" are sterile in philosophy. What is interesting, instructive, and fertile is "to what extent?" Nothing is gained by stating that two concepts such as mind and matter are external to each other. On the other hand, important discoveries can be made if we start from the point at which two concepts meet, at their common frontier, in order to study the form and nature of their contact. True, the former kind of operation has always attracted philosophers because it is a dialectical process that is effected immediately and on the basis of pure ideas, whereas the latter is a difficult process that can only be carried out progressively, on the basis of facts and experience – experience being precisely the place where concepts touch or interpenetrate.[4]

This quotation is taken from the 1901 paper on "positive metaphysics," which appears below. Later on in the paper, Bergson explains exactly what this program means in relation to his own "spiritualist" philosophy, a term that for once he endorses (though in general he avoids this conventional term). He explains that his work deliberately focused on aspects of reality where the "interpenetration" of "matter" and "mind" is obscure: sensation (psychophysics), aphasia and pathology of memory (neurology), biological evolution. This is typical of Bergson's strategy in the advocacy of spiritualism. For him, the mind/body problem (or more broadly the mind/matter problem) had to be examined in areas where these distinctions were obscure: phenomena of a high degree of material complexity, which can also be interpreted as "lower manifestations of the mind." The spiritualist/materialist debate is uninteresting and sterile if it focuses on the superior psychological faculties, understanding, reason, creative imagination: "j'ai fait descendre l'esprit aussi près que possible de la matière" (I made the mind descend as close as possible to matter).[5]

In his successive books, Bergson went "lower" and "lower" in this downward movement: first sensation (*Les Données immédiates de la conscience*, 1889), then memory (*Matter and Memory*, 1896), then life (*Creative Evolution*, 1907). The 1922 book on Einstein's relativity (*Duration and Simultaneity*, 1922), curiously neglected by most commentators, can be seen as the ultimate attempt to understand what "mind" means on the basis of a reflection upon the meaning of space and time in definite areas of modern empirical science.

The link between Bergson's spiritualism and his interest in empirical science is most explicit in the same 1901 paper (again, reprinted as chapter 4 below):

> I consider that spiritualism must be prepared to descend from the lofty heights to which it has retreated. As long as it remains up there, even if it is in possession of the truth, it will be unable to convince anyone else. I want to replace this age-old game between schools of thought, in which each side develops an abstract concept to its ultimate conclusion in order to then oppose it to the other side's conception, by a broad, progressive philosophy, open to all, in which opinions will test and mutually correct each other through contact with the same experience.[6]

Thus for Bergson, "the 'spiritual' is never left to pure speculation ... For him, reasoning had no importance if it did not lead to facts."[7] Bergson was a "spiritualistic positivist." This is not retrospective interpretation, something that I would formulate because it sounds like a nice paradox. It is the plain expression of the historical fact. Around 1900, "spiritualistic positivism" was the current name of a living tradition among certain French philosophers, such as Jules Lachelier or Émile Boutroux. Like Bergson, who was directly influenced by them, they emphasized a conception of the mind founded on spontaneity, contingency and indeterminism. They also emphasized the importance of time, and, correlatively, they maintained that the notion of free causation was as important as the notion of law, in all areas of the natural sciences.[8] For instance, Boutroux, in his *La Contingence des lois de la nature* (Ph.D., 1874), declared:

> Abandoning the external standpoint from which things appeared as fixed and circum- scribed realities, entering into the deepest part of ourselves and seizing, if possible, our being at its source, we find that liberty is an infinite force. We sense this force each time that we act truthfully. Our acts do not realize it, cannot realize it, and thus we are not ourselves this force. But it exists, because it is the root of our being ... This doctrine of divine freedom explains the contingency that is presented by the hierarchy of forms and general laws of the world ... In turn, lower beings, in their nature and in their progress, remind us, in their own way, of the divine attributes.[9]

These sentences are a good illustration of the kind of "indeterminism" and "positive spiritualism" that flourished in France when Bergson began his philosophical career. Bergson would probably have endorsed all the ideas that are contained in the above quotation (style apart). This helps us understand the cultural background of Bergson's philosophy.

In the subsequent sections of this chapter, I will first examine the meaning of Bergson's "positive metaphysics" in relation to the general state of scientific know- ledge in his time. I will then briefly describe his major "confrontations" with given areas of science. Finally, I will characterize Bergson's general appraisal of "science" – his implicit "philosophy of science"; this can be done by examining his interpretation of notions such as: knowledge, reality, science, causation, law, and fact.

2. "Positive Metaphysics"

We need to go a little further in our characterization of Bergson's provocative phrase, "positive metaphysics." If metaphysics can be "positive," this means that it can be not only a "science" (in the loose sense of justified knowledge), but also an *empirical* science. An empirical science deals with facts, and leaves room for the rectification of its theories. Bergson had no problem with this requirement:

> I see future metaphysics as a science that is in its own way empirical, progressive, obliged like other positive sciences to present the results of its attentive study of reality as merely provisional.[10]

However, there is a problem here. Although metaphysics deals with notions such as "duration," or "unforeseeable novelty," how can this discipline apply the standard methodology of the natural sciences? Can we conceive of a kind of metaphysics that would measure its objects and make predictions? This seems impossible. Bergson himself rejected this perspective. In fact, when he said that metaphysics could claim to be an empirical science, the word "science" did not have the meaning that he usually attributed to it in most of his writings. Innumerable quotations could show that for Bergson "science" was a kind of knowledge defined by the requirement of measuring and predicting the phenomena: "The main object of science is to forecast and measure: now we cannot forecast physical phenomena except on condition that we assume that they do not endure as we do; and, on the other hand, the only thing we are able to measure is space."[11] Science, here, means mathematized knowledge, which enables us to master the material phenomena. In this sense, science and metaphysics have different objects and methods. This is a commonplace in Bergson's philosophy: science is quantitative, metaphysics is qualitative; science relies on spatial schemata, metaphysics does not; science is the work of "intelligence," and is therefore adaptive, pragmatic, and technically oriented, whereas metaphysics relies on "intuition," and aims at a disinterested "understanding" of what is absolute and real. When Bergson develops such ideas, the qualification of metaphysics as "science" is a sheer impossibility.

However Bergson was convinced that the Cartesian or mechanistic ideal of science as mathematized knowledge was no longer the only one. For him, the emergence of biology and the human sciences in the nineteenth century forced scientists and philosophers to recognize that science could not be merely a vast mathematics, that is, a system of symbolic relations between measurable phenomena. The idea of a universal mathematics, a *mathesis universalis*, was definitively finished. Again, the 1901 lecture given to the French Society of Philosophy is highly interesting. On that occasion, he clearly formulated the philosophical import of the emergence and development of the biological and human sciences:

> If the Cartesians (even more than Descartes himself) considered that all in nature that is clear and distinct was due to extension, this is because the discoveries of the astronomers and the physicists of the 16th and 17th centuries, and above all the discoveries of Descartes, had revealed the explanatory power of the idea of extension. Their criterion

of intelligibility was much more empirical than they thought. It corresponded to a thorough deepening of their own experience. But our experience is far greater. It has grown so large that, nearly a century ago, we had to renounce all hope of a universal mathematics. New sciences have been formed on the basis of this renunciation, sciences that observe and experiment without dreaming of producing a mathematical formula. Intelligibility thus gradually extends to new notions, also suggested by experience.[12]

For Bergson, biological evolution and physiology provided empirical evidence in favor of the existence of a certain degree of "indeterminacy," "contingency," and "capacity of choice." Thus, together with an unprecedented development of the psychological and social sciences, biology opens the route to a genuine collaboration between metaphysics and the empirical sciences. Here are the closing words of the 1901 lecture:

> Let us work to grasp experience from as close as we can. Let us accept science with its current complexity, and let us recommence, with this new science as our raw material, an analogous effort to that which the old metaphysicians carried out on a much simpler science. We need to break out of mathematical frameworks, to take account of the biological, psychological, and sociological sciences, and on this broader base construct a metaphysics that can go higher and higher through the continual, progressive, and organized effort of all philosophers, in the same respect for experience.[13]

Bergson's claims about the possibility of a "positive metaphysics" were thus closely related to his evaluation of the state of the sciences in his own time. The biological sciences were absolutely crucial in that respect. For Bergson, they provided thousands of examples of the existence of something in nature that cannot be correctly understood in the mere language of "matter in motion." For this, he had a conventional expression, "*la signification de la vie*" (the meaning of life). By this term, Bergson meant "the insertion of thinking in life" ("*l'insertion de la pensée de la vie*"). If this concept could be empirically documented, then a "positive metaphysics" was possible:

> If this meaning of life can be empirically determined in a way that is increasingly accurate and complete, a *positive* metaphysics – one that is uncontested and which can progress linearly and indefinitely – is possible.[14]

3.   "Confrontation"

I will now briefly characterize and classify the various interactions that Bergson had with given areas of empirical science. With the exception of a few brilliant, but marginal, essays (especially those on laughter and dreams), each of the chief books written by Bergson can be presented as a major "confrontation" between a philosopher and a given area of scientific knowledge. The word "confrontation" is the one he used himself with some solemnity at the beginning of his book on Einstein's theory of relativity.

> We wanted to find out to what extent our concept of duration was *compatible* with Einstein's views on time. Our admiration for this physicist, our conviction that he was

giving us not only a new physics but also certain new ways of thinking, our belief that science and philosophy are unlike disciplines but are meant to complement each other, all this imbued us with the desire and even impressed us with the duty of proceeding to a *confrontation*.[15]

Compatibility, complementarity, confrontation: these words perfectly describe Bergson's attitude towards given areas of science. Note, however, the nuances associated with these words: reciprocal complementarity is desirable, it is an ideal; compatibility is subject to discussion; confrontation is then the normal relationship between the two disciplines. Confrontation implies a collision between two lines of thought. It is not a word that philosophers of science would ordinarily use to describe the kind of relationship that they have with science, or with given areas of scientific knowledge or practice. Philosophers of science would typically say that they evaluate or interpret the scientific structure of a given theory, its rational or nonrational foundations, its significance for the understanding of science and the history of science in general. They would hardly characterize their work as a "confrontation" between their understanding of a given subject (for instance the brain, or the phenomena of sensation and memory, or the concepts of simultaneity) and the scientific treatment of these subjects in given areas of investigation. As far as I know, Hempel, Popper, Bachelard, or Kuhn did not devote entire books to showing that, for instance, the foundations of psychophysics were just wrong, that the theory of cerebral localization and psychophysical parallelism was not proved, that biologists had missed something essential in evolution, and that the current interpretation of Special Relativity in physics was unsound. But this is precisely what Bergson did. He did it not as a specialist in the fields I have mentioned, although he worked hard on each of them, but because he wanted to "confront" scientific evidence and theories with philosophical theories of his own. To have a "confrontation," one needs two sides. Bergson's side is easy to designate: it is a certain conception of time. This conception of time had many consequence for many issues in philosophy, such as perception, knowledge, matter, reality, freedom, God, and morals. It is this conception of time, which Bergson gave the name "duration," which he repeatedly used to "confront" many areas of the scientific knowledge of his time.

Commentators on Bergson generally consider that he wrote four important books:

Essai sur les données immédiates de la conscience [*Time and Free Will: An Essay on the Immediate Data of Consciousness*], 1889. This was his Ph.D. dissertation. First English translation 1912.
Matière et mémoire [*Matter and Memory*], 1896. First English translation 1911.
L'Évolution créatrice [*Creative Evolution*], 1907. First English translation 1911.
Les Deux Sources de la morale et de la religion [*The Two Sources of Morality and Religion*], 1932. First English translation 1935.

This view is certainly correct from the point of view of the genesis of Bergson's philosophical thought. The "essential Bergson" is indeed contained in these four books. It is tempting imagine Bergson's philosophy with respect to scientific knowledge as evolving. Although these books have rather complex structures, and go far beyond specific comment on given areas of scientific knowledge, it is true that each of them is concerned with a major scientific area: psychology (*Time and Free Will* and *Matter and Memory*), neuropathology (*Matter and Memory*), biology (*Creative Evolution*), and social science – especially ethnology (*The Two Sources*). This would even accord with Bergson's prophetic declaration of 1901, when he said that modern metaphysics should take into account not only mathematics and physics, but also "biology, psychology, and sociology" (see quotation above, and n. 13).

However, another presentation is possible, if we take seriously the notion of the "confrontation" of science and philosophy that was so crucial to Bergson. As I have said already, this confrontation was not merely a set of diverse interactions between Bergson's philosophy in general and science in general. If we are looking only for this, almost everything written by Bergson is relevant. The real "confrontation" relates to a single problem, that of the meaning of time. Although all of Bergson's writings have some relation to this problem, not all of them are specifically intended to explore it, and even fewer claim to have made a significant advance on the issue of "duration" (the concept that Bergson introduced into philosophy in order to account for the nature of time). First, the last major book (*The Two Sources of Morality and Religion*) is not a book on duration. It does not present anything new on the subject, nor does it claim to do so. Second, the list of four major books above does not mention *Duration and Simultaneity*, first published in 1922. This book is important here for several reasons. It is the only book which contains the word "duration" in its title. But also, it is a work on the most general physical theory that was available to Bergson, Einstein' Theory or Relativity. Moreover, Einstein is explicitly mentioned in the subtitle of the book. Nowhere else did Bergson ever mention the name of a scientist in the title of a book (or even of a chapter).

Duration and Simultaneity had a strange fate. There was a second edition in 1923, with several important appendices. Four other printings appeared between 1926 and 1931. In spite of this success, Bergson gave up publishing on relativity, and finally forbade the reprinting of his book. He confessed that he was unable to understand the mathematics involved in the Theory of General Relativity. General Relativity was not the subject of *Duration and Simultaneity*, which discussed only Special Relativity. But Bergson came to realize that some of his criticisms (especially the paradox of "asymmetrical aging") could be dramatically affected by taking General Relativity into account. The result of this is that *Duration and Simultaneity* was never reissued before his death. It was also excluded from the centenary volume which gathered his major books and articles in 1959. Similarly, the three volumes of minor texts published in 1957 under the title *Dits et Écrits* also omitted this book, and related articles on relativity. *Duration and Simultaneity* was finally reissued in French in 1968, and is currently in print. In spite of this, commentators generally ignore this book and do not quote from it. The most reasonable explanation for this is that *Duration and Simultaneity* does not offer anything really new from the point of view of Bergson's own elaboration of the notion of duration. He confronts the theory of Special Relativity with his concept of duration as elaborated in previous works. Another possible explanation for the lack of interest shown by the historians of philosophy for the book on Einstein is that it is written in a style quite different from the other books: from the first to the last page of a 285-page book (second extended edition, 1923),

Bergson deals with one single scientific theory, without a single digression. The book is austere, nonmetaphoric, and entirely devoted to a methodological discussion. In other words, were it not for the controversy it introduces with regard to Bergson's own philosophy of time, *Duration and Simultaneity* would look like a traditional book in "philosophy of science." Commentators on Bergson are generally not "philosophers of science."

For our purposes, however, *Duration and Simultaneity* is important. It reveals a kind of hidden research agenda. Consider the first three books in the list above, and add *Duration and Simultaneity*. Each of these books can be considered as a decisive element in a program of exploring the content and empirical plausibility of the concept of duration. The 1889 *Essay on the Immediate Data of Consciousness* (or *Time and Free Will*, English main title) introduces the concept of duration in its most conspicuous form: the psychological problem of "consciousness." *Matter and Memory* (1896), Bergson's masterpiece, broadens the subject: duration is still examined and documented in the sole case of human beings, but the concept now applies to "inferior faculties," which can be unconscious, and are indeed explicitly presented as such by Bergson: inferior manifestations of "mind," with its properties of continuity, unforeseeable novelty, free causation, contingency. In *Creative Evolution* (1807), Bergson moves to biology in general, with special regard to evolution. Here, the concepts of duration and mind find their broadest meaning, life and creation, with discrete but firm allusions to extraterrestrial life and, ultimately, God. At this point, it seems this Bergson has reached the conclusion of his program: mind and matter, although not "substances," exist as "forces" or "tendencies" that conflict at all levels of reality. Science does not escape that conflict: physics explores the manifestations of matter and space, biology and psychology are also concerned with mind and time. *Duration and Simultaneity* (1922), however, shows that Bergson wanted to go a step further. The central thesis of this book is that Einstein's theory of the relativity of space and time is in fact "compatible" with a notion of "universal physical time." The entire book consists in the development of a single argument. Bergson demonstrates, or tries to demonstrate, that the contraction of distance, the dilation of time, and the "dislocation" of simultaneity, although fully valid from an empirical point of view, result from a fictitious theoretical situation where each observer (taken as "real") imagines another "virtual" observer and tries to convert his or her own measurement into the measurements of this "phantasmal" observer. This operation, Bergson says, is absolutely legitimate for the construction of laws of nature that are genuinely invariable, whatever the position of the observer in space and time. But for Bergson, this does not demonstrate that time and motion are relative. The conclusion of the book is that Einstein's relativity does not force us to admit Langevin's paradox of "asymmetrical aging." In 1911, this French physicist had stated that a space traveler would be younger upon his return to earth than his twin who had stayed at home, and who might well be dead by that time. In fact, it is this paradox that led Bergson to write on physical relativity. The paradox of asymmetrical aging was a major problem for a philosopher who asserted that movement in general is real and not relative, and that duration is an internal, immanent, and absolute property of living beings. Bergson's refutation of the paradox of asymmetrical aging deserves a more detailed analysis, for which there isn't space here. What I want to emphasize here is Bergson's objective: his intention was to show that the new physics of Einstein, if it did not illustrate his notion of duration, was at least compatible with it. Einstein's theory was true from the point of view of pragmatic and predictive science, but it was neutral with respect to the philosophical question of the existence of a "universal time."

We can now understand better the program of Bergson's "positive metaphysics." First in psychology, then in neurology, then in evolutionary biology, and finally in physics, he claimed to have convincingly "refuted" a number of theories, or interpretations of theories, that were vitiated because they neglected the role and the meaning of time: the treatment of conscious states as magnitudes (psychophysics), the engramatic interpretation of memory in neurology, the "mechanistic" theories of evolution (especially Darwin). The study of Einstein's relativity was also a necessary part of this philosophical program. Of course, physics, which is the science of matter, could hardly confirm or positively illustrate the philosophy of duration. But it was crucial to Bergson to show that the first theory in the history of physics that really incorporated space and time as *physical* objects was not *incompatible* with his philosophical interpretation of time.

Of course, the positive content of Bergson's metaphysics does not go beyond biology. Bergsonian metaphysics finds its most powerful formulation when dealing with the opposition of the sciences of matter and the sciences of life (biology and psychology), which reveal the extent of mind in nature. As early as 1901 – six years before the publication of *Creative Evolution* – this is already perfectly clear:

> I cannot envisage general evolution and the progress of life throughout the organized world, the mutual coordination and subordination of vital functions in the same organism, the relations that psychology and physiology combined seem to establish between cerebral activity and thought in man, without arriving at the conclusion that life is an immense effort made by thought to obtain something from matter that matter does not want to give up. Matter is inert, it is the seat of necessity, it proceeds mechanically. Thought seems to try and benefit from matter's mechanical aptitude, to use it for *actions*, to thus convert into contingent movements in space and into unforeseeable events in time all that it contains that is of creative energy (or at least that which this energy has which can be enacted and exteriorized).[16]

4.   Bergson's Epistemology and Philosophy of Science

Bergson did not want to be a "philosopher of science." The kind of division of labor implied by this conventional term did not appeal to him. Science was indeed a major partner for philosophy, and probably *the* major partner. But it was not enough for philosophy to be merely a "critique" of scientific knowledge. Philosophy can and should participate in the construction of the content of knowledge. This does not mean that Bergson's philosophy is of no relevance for philosophers of science. Science is so much a central object for Bergson that it is not difficult to identify in his thinking a series of theses, which constitute a sort of implicit "philosophy of science." In this last section, I enumerate some of these characteristic theses. Since my objective is only to sketch out a portrait, I will not go far in the analysis of these theses. It will be enough to formulate them.

Knowledge

Knowledge was one of the most central problems for Bergson. A major and constant objective of his thinking was to show that Kant's critique was wrong. Bergson maintained that our knowledge is able to reach something real and absolute. What was real and absolute is what we perceive. All knowledge is based upon conscious perception, no knowledge can go beyond it. This is not a limitation. Perception is the firm basis and ultimate *horizon* of all knowledge. Bergson's constant philosophical model, in this respect, was Berkeley. In *Duration and Simultaneity* there is a remarkable passage. At a certain point, after a lengthy discussion of the relativistic thesis of the "plurality of times," Bergson states his own view of time: "What, indeed, is a real time, if not a time lived or able to be lived?"[17] Shortly after this sentence, the philosopher explains what he means by "reality": "reality, that is the thing perceived or perceptible."[18] It is no accident that this explanation of the meaning of "reality" comes in the context of a discussion of the notion of time. For Bergson, time as duration is real because it is immediately experienced: we experience the continuous succession of things. In contrast, space is a schema that our spirit constructs for the purpose of effective action on our environment. Space is neither a property of things, nor an a priori condition of our faculty of knowledge. The elucidation of the significance of space and time with respect to "reality" was the most constant and central objective of Bergson's philosophy.

A second aspect of Bergson's epistemology is the inseparability of the theory of knowledge and the theory of life. If the theory of knowledge is a theory of the a priori conditions of knowledge, then it depends narrowly on biological evolution. For instance, space and logic are adaptive devices. Their ultimate significance is that of instrumental schemata that render possible our mastery of the material world. Or, in other words, their ultimate background is not understanding the world, but acting upon the world. Bergson would have felt at home with today's discussions of "evolutionary epistemology" and the "meso-cosmic" constraints that have channeled the evolution of our cognitive faculties in the history of the human species.[19]

Science

If "science" is taken to be the "knowledge of the measurable," it is comparatively easy to situate Bergson in terms of current debates about the cognitive status of scientific theories. Insofar as it deals with measurable magnitudes, scientific knowledge is for Bergson a "symbolic" knowledge, which translates perceptual data into spatial symbols. This kind of knowledge aims at controlling our natural environment. It belongs to a "logic of action." Therefore it cannot pretend to tell us what is "real," but only how we can master our material environment. Science, in this respect, is instrumental, conventional, and pragmatic. This thesis is repeated again and again in all Bergson's works.

Causation and the Laws of Nature

A spectacular illustration of Bergson's conception of knowledge and science can be found in his interpretation of the notions of causation and law. Laws are a perfect example of pragmatically oriented knowledge. Science and technology come from the same source. Both are primarily interested in the constant properties of things around us. These constant properties are what make us able to predict and control the behavior of the world around us. A maximal degree of constancy and universality is thus a guarantee of maximal efficiency. We look for laws because we want to control phenomena.

Bergson was very cautious about not confounding the notions of law and of causation. He defended the idea that the empiricist interpretation of the category of causation was unlikely. The traditional account of causation states that this belief comes from habits generated by the observation of external phenomena. In a remarkable talk of 1900,[20] he argues that this conception is erroneous:

> Physics shows us a growing number of examples of phenomena that are concomitant, or which succeed phenomena with which they are invariably linked. Determined phenomena rarely coexist, nor do they generally succeed determined phenomena in our immediate visual experience, and it is generally thought that causality does not clearly imply either succession or concomitance ... The error of empiricism might be to over-*intellectualize* the general belief in the law of causality, to envisage it in its relation to science, and not in its relation to life.[21]

The true psychological origin of the general belief in causation is in fact our internal experience of voluntary action:

> Rightly or wrongly, we view our intentions and our movements as contingent effects, to a certain extent undetermined, in terms of their cause. It is not the notion of determining causality, but that of free causality, which we find in the simple observation of ourselves.[22]

This internal and practical experience has nothing to do with our perception of the external world. It is a "continuously active habit." This habit is itself "coextensive with and essential to life," just like "the habit of breathing."[23] This metaphor is not innocent. The origin of the belief in causation is not just psychological, it is biological. The notion of causation is indeed rooted in the very structure of the nervous system. The nervous system fundamentally consists of sensorimotor mechanisms, which anticipate the contact of our body with the perceived object. The origin of the notions of "cause" and "effect" is to be found there: the sensory impression is the "cause," the contact with the object is the "effect" (135). Thus causation, in the sense of free causation, has its origins in a "practical belief" common to humans and the higher animals. "Determining causation" [*causalité déterminante*] is something more complex. It presupposes the intervention of abstraction. The belief in determining causation results from the projection of our intuitive feeling of necessity upon external objects. This implies some kind of logical necessity. The belief in external causation is thus a rather complex psychological construction. The important point for Bergson is that it is built upon the more primitive experience of free causation. Action, not passive perception, is the key word.

Ontology ("Furniture of the World")

Bergson did not use the word "ontology" in the sense that we currently give it in philosophy of science (i.e. a reflection upon the entities that constitute the "furniture of the world"). But he definitely had a clear notion of what the natural world is made of. His ontology was an ontology of "forces" or "tendencies," not an ontology of "substances." Hence his sympathy for energeticism, which was quite popular among many contemporary scientists and philosophers. This energeticism was in perfect agreement with Bergson's interpretation of science and causation, an interpretation which emphasized the notions of *adaptation* and *action* rather than that of *disinterested knowledge*.

Physical and Biological Sciences: Two Kinds of "Facts"

I have already stressed the huge role of the biological sciences in the genesis of Bergon's philosophy. I would like to conclude this chapter by pointing out a rather fascinating interpretation that Bergson gave of the duality of the biological and physical sciences. In the discussion that followed the famous 1901 talk in which Bergson advocated the idea of "positive metaphysics," a metaphysics founded on facts, he was criticized for having misused the words "fact," "experience," and "scientific." This objection was formulated by Louis Couturat.[24] Bergson was reproached for having confounded "theory" and "fact." He responded that "facts," although objective in all cases, do not have the same significance in physics and in biology, with respect to their dual relation to nature and to our representations:

> I am quite ready to accept that, at least in the inorganic world, and wherever facts take on a mathematical form, the law determines the fact as much as the fact determines the law. Bodies fell before the time of Galileo, and that is what gave Galileo the idea of discovering the law of falling bodies. But it was the law of falling bodies that made it possible to definitively isolate the phenomenon of falling bodies, and even, more generally, to define the "physical fact" and to raise it to the status of an independent entity. In this sense, is the physical fact largely our creation? But as we rise from the inorganic to the organized, we find ourselves in the presence of facts that are more objectively intended as facts, by nature itself. A living organism is a more or less closed circle, one which is closed by nature.[25]

In other words, in physics facts have more the status of cognitive constructs than they do in biology. The objectivity of physical facts is proportional to our ability to recognize constant and universal relations (or "laws"). In biology, facts are facts because they express something that is genuinely posed or "intended" by nature. I cannot read such declarations without thinking of Georges Canguilhem on the "normativity" of the living (the living pose their own norms).[26] Bergson's reflection on biological facts closely resembles Canguilhem's notion of "normativity of the living [being]." Canguilhem might well have been inspired by Bergson's 1901 intuition. Bergson was indeed suggesting that biological facts, if they have any nomological import, are laws instituted by life itself. Thus biological facts testify to the existence of a certain degree of freedom, a certain degree of spirituality in living organisms as such. I am not sure whether the word "fact" still has any methodological value in this kind of discourse. But I am sure that it was absolutely crucial to what Bergson called "positive metaphysics."

Acknowledgments

The quotations included in this chapter were translated by Matthew Cobb (except for the two excerpts from *Creative Evolution* and *Duration and Simultaneity*). Marjorie Grene and Matthew Cobb kindly revised the style of the entire chapter.

Notes

1 Letter to William James, May 9, 1908, reproduced in Henri Bergson, *Écrits et paroles*, textes rassemblés par R. M. Mossé-Bastide (Paris: Presses Universitaires de France, 1959), II, pp. 294–5. Reissued in Henri Bergson, *Mélanges* (Paris: Presses Universitaires de France, 1972), pp. 765–6.

2 Henri Bergson, "Le parallélisme psycho-physique et la métaphysique positive," *Bulletin de la Société française de Philosophie*, Séance du 2 Mai 1901, pp. 33–4 and 43–57. Reproduced in Henri Bergson, *Écrits et paroles*, I, pp. 139–53. Reissued in *Mélanges*, pp. 503–57.

3 Henri Bergson, *L'Évolution créatrice* [1907] (Paris: Presses Universitaires de France, 80th ed.), pp. 195–6.

4 Bergson, "Le parallélisme psycho-physique et la métaphysique positive," p. 144.

5 Ibid.

6 Ibid.

7 Alexis Philonenko, *Bergson ou la philosophie comme science rigoureuse* (Paris: Les Éditions du Cerf, 1994), p. 393.

8 On this tradition, see for instance Émile Bréhier, *Histoire de la philosophie*, III, *XIXe–XXe siècles* [1st ed. 1932] (Paris: Presses Unviersiaires de France, coll. Quadrige, 1981), p. 891.

9 Émile Boutroux, *La Contingence des lois de la nature* [1874] (Paris: Presses Unviersiaires de France, 1991).

10 Bergson, "Le parallélisme psycho-physique et la métaphysique positive," p. 147.

11 Bergson, *Essai sur les données immédiates de la conscience* [1889] (Paris: Presses Universitaires de France, 1940), p. 173.

12 Bergson, "Le parallélisme psycho-physique et la métaphysique positive," p. 143.

13   Bergson, "Le parallélisme psycho-physique et la métaphysique positive," p. 153.

14   Ibid., p. 139.

15   Bergson, *Durée et simultanéité: A propos de la théorie d'Einstein*, 3rd ed. (Paris: Librairie Félix Alcan, 1923), Préface, p. vii. In English: *Duration and Simultaneity, With Reference to Einstein's Theory*, trans. Leon Jacobson (Indianapolis: Bobbs-Merrill, 1965), p. 5. In this translation, I have substituted "complement each other" for "implement each other." I have also reintroduced the italics of the French edition.

16   Bergson, "Le parallélisme psycho-physique et la métaphysique positive," p. 151.

17   Bergson, *Durée et simultanéité*, p. 107.

18   Ibid., pp. 107–8.

19   For a comprehensive review of the literature on this subject, see G. Vollmer, "What Evolutionary Epistemology is Not," in Werner Callebaut and Rik Pinxten, eds., *Evolutionary Epistemology: A Multiple Paradigm* (Amsterdam: D. Reidel, 1987), pp. 203–21.

20   "Note sur les origines psychologiques de notre croyance à la causalité," Comunication au Congrès international de philosophie, Paris, 1900. Reproduced in *Dits et écrits*, I, pp. 128–37.

21   Ibid., p. 131.

22   Ibid. Bergson draws this idea from Maine de Biran, whom he quotes in this text.

23   Ibid., p. 132.

24   Bergson, "Le parallélisme psycho-physique et la métaphysique positive," intervention de M. Couturat, p. 164.

25   Ibid., réponse de Bergson à Couturat, p. 165.

26   Georges Canguilhem, *Essai sur Quelques Problèmes Concernant le Normal et le Pathologique*, Publications de la Faculté des Lettres de l'Université de Strasbourg, Clermont-Ferrand, Imprimerie "La Montagne," 1943.

4

PSYCHOPHYSICAL PARALLELISM AND POSITIVE METAPHYSICS

Henri Bergson

Monsieur Bergson brought the following points to the attention of the Société de Philosophie:

1. If psychophysical parallelism is neither strict nor complete, if any determined thought does not correspond to an absolutely determined cerebral state, experience will have to map out with increasing accuracy the precise points at which this parallelism begins and ends.

2. If such an empirical study is possible, it will provide an increasingly accurate measure of the separation between thought and the physical conditions in which thought takes place. In other words, it will provide increasingly accurate knowledge about the relation between man the thinking being and man the living being and thus about what we can call *the meaning of life*.

3. If this meaning of life can be empirically determined in a way that is increasingly accurate and complete, a *positive* metaphysics – one that is uncontested and which can progress linearly and indefinitely – is possible. No philosophers, no matter how hostile they might be with regard to purely metaphysical inquiry, that is, an inquiry that is transcendental to life, would deny that our intelligence has the ability to think legitimately and usefully about life itself. If we rise to the level of thought in itself, or of matter in itself, through a construction, which in some respects remains fragile, on the other hand the relation between these terms, and above all the separation between them, are, or can become, facts that are open to observation. A metaphysics that began by molding itself around the contour of such facts would have many of the characteristics of an undisputed science. And it would be open to an indefinite progress, because the increasingly precise determination of the relation of consciousness to its material conditions, by showing us with growing accuracy on what points, and in which directions, and by which necessities our thought is limited, would guide us in the very special effort that we have to make to free ourselves of this limitation.

Henri Bergson, "Psycho-physical Parallelism and Positive Metaphysics," pp. 33–4 and 43–57 from *Bulletin de la Société française de philosophie*, séance du 2 mai 1901. New English translation by Matthew Cobb.

Monsieur Bergson. I will begin by thanking my colleague and friend, Monsieur Belot, for the most interesting critique of my theses that he has just presented. He had sent me a brief sketch, a *schema* of the objections he intended to present. These objections seemed to deal with the general method that I proposed, rather than with the particular applications that I attempted to make or the results to which it led me. I would have preferred that the discussion remained on this terrain. I believe in the efficacy of the method; I would not wish it to be judged on the basis of the incomplete and imperfect results that an isolated thinker has been able to conclude from it. However, because Monsieur Belot appears now to consider that the method is inseparable from its application, I will successively examine the various objections that he raises against both. I hope that my honorable opponent will raise any important points that I miss.

Monsieur Belot is first astonished by the "hypothetical" form in which I set out my conclusions. "We are presented with less of a thesis than a hypothesis" he says. True, each of my three propositions begins with an "If . . . ." The "if" of the final proposition expresses the fact that it is subordinated to the acceptance of the second proposition, and that of the second proposition supposes that the first proposition is accepted. But the first proposition itself begins with an "if": this is undoubtedly what surprises Monsieur Belot. I hasten to point out that I would not have set out the proposition in this form if I had thought that everyone agreed with its starting point, namely that "psycho-physical parallelism is neither rigorous nor complete." I do indeed support this thesis, but although I am convinced myself, I could not speak as though I had convinced everyone else. To my mind, the "if" was thus a courtesy "if" with respect to possible critics. If the "if" was thought to be a timid "if," I would immediately replace it by a "because." I have not the slightest doubt about this: as I will explain later, I am entirely convinced that there is a certain relation between the psychological fact and cerebral activity, a kind of correspondence, but that there is no parallelism whatsoever.

I now come to Monsieur Belot's second preliminary observation. "The problem of the relation of the soul and the body already have pride of place in Cartesian metaphysics," he says. "But whereas the Cartesians were above all concerned to make these relations intelligible, Monsieur Bergson situates himself solely on the terrain of facts." Much could be said about this distinction. I wonder whether the Cartesians, if they were alive today, would still have the same understanding of intelligibility. I believe it is very difficult to say of a notion, on first sight, whether it is intelligible or not. Intelligibility comes to a notion slowly, through its application. The intelligibility of an idea cannot be measured only by its suggestive richness, by the fecundity and the certainty of its application, by the growing number of articulations it enables us to reveal, as it were, in reality, or even by its internal energy. The concept of the differential, which was initially very obscure for the first mathematicians who used it, became, through the use that was made of it, an example *par excellence* of a clear notion, which sheds light on the whole of mathematics. If the Cartesians (even more than Descartes himself) considered that all in nature that is clear and distinct was due to extension, this is because the discoveries of the astronomers and the physicists of the sixteenth and seventeenth centuries, and above all the discoveries of Descartes, had revealed the explanatory power of the idea of extension. Their criterion of intelligibility was much more empirical than they thought. It corresponded to a thorough deepening of their own experience. But our experience is far greater. It has grown so large that, nearly a century ago, we had to renounce all hope of a universal mathematics. New sciences have been formed on the basis of this renunciation, sciences that observe and experiment without dreaming of producing a mathematical formula. Intelligibility thus gradually extends to new notions, also suggested by experience. I do not consider myself to be unfaithful to the method of Descartes when I suggest that this or that Cartesian notion should be revised, just as a Cartesian philosopher, faced with a more supple science, informed by a far greater experience, and ready to accept in natural phenomena an organizational complexity that cannot easily be reduced to a mathematical mechanism, would also doubtless argue that the notion should be revised. A method can be described as a particular attitude of the mind towards its object, a certain adaptation of the form of inquiry to its material. In this case, by conserving the procedures of a method come what may, when the materials on which it operates have radically changed, we would not be remaining faithful to that method. On the contrary, remaining faithful to a method involves constantly remodeling its form on the basis of its subject, in order to maintain the same precision of fit.

Now I arrive at the fundamental question. Monsieur Belot begins by remarking that "the old spiritualism also thought it necessary to insist on a division between the physical and the moral, but it sought this division in terms of the superior faculties, whereas, in the new spiritualism, this division is seen to exist in the lower and unconscious functions." I would first like to point out an error of detail. It is indeed true that I posed the division in terms of the "lower" faculties, but not in terms of the "unconscious" faculties. However, I do not deny the unconscious. It should be said in passing that the idea of the unconscious could serve as a proof of my earlier statement that an idea becomes intelligible through its application. It was widely held 20 years ago (and I have to admit that I myself taught this for many years), that a psychological state is by definition a conscious state, and that the idea of an unconscious psychological state is thus contradictory. Nevertheless I believe that it has become very difficult for whoever has followed the progress of psychology over recent years not to accord a large place to the unconscious in psychological explanations, and even not to recognize that the idea of the unconscious, as we manipulate it, increasingly tends to become a clear idea, as our mind expands and stretches itself, finally embracing this initially-resistant representation. Piecemeal progress no doubt occurs in the sciences, through the growing verification of already accepted principles: but how could important, radical scientific progress occur other than by an effort of intellectual dilation, which leads to some concepts, which had hitherto appeared to verge on contradiction, becoming intelligible? But, I repeat, the unconscious has nothing to do with the present discussion, because I do not claim to measure the separation between the physical and the moral on the basis of *unconscious* psychological facts.

Lower psychological facts do fit the bill, however. And, as Monsieur Belot puts it so well, it is here that we find one of the characteristic traits of what he calls the "new" spiritualism.

I am the first to recognize that the higher mental faculties — understanding, reasoning, creative imagination — are faculties that are unique and essential to human beings. The old spiritualism was right to look in this direction to find the mental characteristics of man. But when, struggling against its materialist opponents, or seeking to determine the relation of the soul and the body, it treated these higher faculties like a barricade and retreated behind them, it was, I think, doubly wrong. It appeared *arbitrary* and it was *unproductive*.

It appeared arbitrary. Its opponents could always point out that the separation it accepted between the psychic and the physical flowed simply from the fact that it dealt with matter in its most rudimentary forms, and with the mind in its most advanced forms. In these circumstances, went the argument, it was all very well to speak about the irreducibility of thought to movement, but if matter was studied at that degree of complexity and mobility at which it imitates certain characteristics of consciousness, and consciousness at that degree of complexity and mobility at which it participates in the inertia of matter, it would be simple to make them coincide. By combining these elementary psychological states, we would be able, through a series of syntheses, to reconstitute the highest manifestations of psychological activity. There is a type of monism, close to materialism, which dualist spiritualism has never been able to refute, precisely because spiritualism was limited to opposing one or the other of the two extreme terms – thought and movement. Dualism considered the extreme ends of the continuum, while monism was situated in the middle: placed on different terrains, how could these two doctrines meet and compare themselves? It seemed to me that there was one way – and only one – of reducing monism: by going onto its own terrain. That is, instead of considering all psychological states, taking the most rudimentary psychological state. This involved demonstrating the existence of a real, observable separation between this state and the physical conditions on which it is based. You can think whatever you like about matter "in itself" or mind "in itself"; you can even accord matter a vague consciousness, an essence that is analogous to mind, in order to reinforce your monist conception of the universe; it nevertheless remains the case that at the point at which true, precise consciousness appears, I can show you something absolutely novel, a certain indeterminacy or contingency, a *capacity to choose*. At this point, should you wish, you could reconstitute the higher activities of the mind, together with the simplest psychological states, and your hypothesis will be disarmed faced with spiritualism, because it will have accepted some of its fundamental premises. In other words, I consider that spiritualism must be prepared to descend from the lofty heights to which it has retreated. As long as it remains up there, even if it is in possession of the truth, it will be unable to convince anyone else. I want to replace this age-old game between schools of thought, in which each side develops an abstract concept to its ultimate conclusion in order to then oppose it to the other side's conception, by a broad, progressive philosophy, open to all, in which opinions will test and mutually correct each other through contact with the same experience.

This explains why I consider that the old spiritualism must have appeared arbitrary. I would add that it was necessarily unproductive, and that the disdain expressed towards it – both yesterday and today – by all men of science flowed above all from this. It was unproductive, precisely because it insisted on considering extreme terms and on purely and simply declaring that mind could not be reduced to matter. Although a declaration of this type might be true (indeed, I would argue it is), it is of no consequence – no more nor less than the opposite statement. Both "yes" and "no" are sterile in philosophy. What is interesting, instructive, and fertile is "to what extent?" Nothing is gained by stating that two concepts such as mind and matter are external to each other. On the other hand, important discoveries can be made if we start from the point at which two concepts meet, at their common frontier, in order to study the form and nature of their contact. True, the former kind of operation has always attracted philosophers because it is a dialectical process that is effected immediately and on the basis of pure ideas, whereas the latter is a difficult process that can only be carried out progressively, on the basis of facts and experience – experience being precisely the place where concepts touch or interpenetrate. It was to carry out this very difficult and very long work that I summoned the philosophers.

I myself tried this approach, to the limited extent that I felt capable of fulfilling it. I initially considered the manifestations of matter not in their simplest forms, in physical facts, but in their most complex form, in physiological facts. And it was not physiological facts in general that I focused upon, but cerebral facts. And not even upon cerebral facts in general, but on a given well-localized and highly-determined fact, which conditions a certain function of speech. I moved upwards, through various complications, up to the point at which the activity of matter brushes against that of mind. Then, through various simplifications, I brought mind down as low as I could, nearly to the level of matter. I left ideas aside in order to focus only on images; of images I retained memories, of memories in general only the memories of words, of the memories of words only the special memories we retain of the sounds of words: this time I was at the frontier, I *almost* touched the cerebral phenomenon in which sound vibration continues. And yet, there was a separation. True, it was no longer that abstract separation between two concepts such as consciousness and movement that can be affirmed a priori – I repeat, nothing can be gained from the reciprocal exclusion of two concepts. It was a concrete and living relation. I saw, at the very moment at which the fact of consciousness has a cerebral concomitant, why and how thought needs to develop through movement in space, all that it contains in terms of possible action, all that it has that can be enacted. I also saw, in the psychological fact that is added to cerebral activity, something that was partly free, partly indeterminate. The part of this fact that could be enacted was strictly determined by its physical conditions, whereas the aspect of the image or the representation of this same fact was much more independent. I considered that this led to the possibility of empirically and progressively determining what I called "the meaning of life," that is, the true sense of the distinction between the soul and the body, as well as the reason why they unite and collaborate. It also seemed to me that this would provide an increasingly better understanding of the very special kind of limitation that life gives to our thought. Had the philosophers not taken as a *relative* knowledge that which is merely a *reduced* knowledge, shrunken and obliged to externalize itself through action before deepening itself through thought? And as the form of this limitation became increasingly apparent, would we not find, with increasing accuracy, in which direction we should make an effort to transcend it? Further, were the obscurities of dualism, the difficulty in establishing such a radical distinction between consciousness and its basis, simply an artificial difficulty and obscurity, the fruit of the limits that the very duality of mind and body imposes on our intelligence? By restricting spiritualism to these extremely narrow boundaries, it seemed to me that we could indefinitely increase its fertility and its force, making it acceptable to those who reject it, bringing to it a theory of knowledge through which it could dissipate its obscurities, and finally to make it the most empirical of doctrines in terms of its method, and the most metaphysical in terms of its results.

Against such metaphysics, or at least against this method, Monsieur Belot raises a series of objections based firstly on the apparent impossibility of rigorously establishing the existence of a definitive distinction between the psychological fact and its physiological substrate. If I am not mistaken, a large part of his argument could be summed up as follows: "Even if you find a separation, nothing proves that future scientific progress will not be able to overcome this separation. You cannot prove the impossibility of parallelism." Of course, I cannot prove the impossibility of parallelism. There is no known or conceivable way of proving the impossibility of a fact. We can prove that a fact is possible, by proving experimentally that it is true, but we cannot, neither by experience nor by reason, prove it is impossible. Nevertheless, I accept that some impossibilities of fact have been well established by science. Since Pasteur, we accept that spontaneous generation is impossible, at least for life as we know it. I realize that this is not a rigorous, absolute, and mathematical certainty. All that Pasteur was able to do was to show to his opponents that, in all the experiments in which they saw evidence for spontaneous generation, living micro-organisms already existed. And we can still ask ourselves if, under other experimental conditions which even Pasteur's opponents had not thought of, we might not see the spontaneous generation of life. However, I repeat, we agree to recognize that Pasteur took his thesis to a degree of probability that scientifically and practically corresponds to certainty. Well, if in dealing with all the questions concerning the relation between the psychic and the physical and in general with all questions of metaphysics, I could arrive at the same or even a comparable degree of certainty as in Pasteur's proposition "There is no spontaneous generation," that would suit me very well.

I fear, however, that it is not this point that separates us, and that, despite yourself, you consider metaphysics to be analogous to mathematics, obliged to deal in the clear simplicity and decisive dogmatism of mathematics. If this is the nature of metaphysics, then we merely need to choose between simple, finished conceptions which we will take to their logical conclusion: this is a closed science, or rather, it is merely an act played out by rival schools that clamber onto the stage where they are applauded one after the other. I see future metaphysics as a science that is in its own way empirical, progressive, obliged like other positive sciences to present the results of its attentive study of reality as merely provisional. I stopped when I found this kind of result. A dozen or so years previously, I had posed myself the following problem: "What do modern physiology and pathology tell an open mind about the age-old question of the relations between the physical and the moral, if we are determined to forget all previous speculations on the question and all statements by experts that are not limited to the pure and simple statement of facts?" And so I started work. I soon realized that I could only arrive at a provisional answer to the question, and indeed only formulate it precisely, if I focused on the question of memory. In dealing with memory, I was led to set out boundaries that became increasingly restricted. After focusing on the memory of words, I saw that even posed this way the problem was too broad, and that it was the memory of the sound of words that posed the question in its most precise and most interesting form. There is a massive literature on aphasia – it took me five years to study it all. And I came to the conclusion that there must be a relation between the psychological fact and its cerebral substrate, a relation that was unlike all the ready-made concepts provided by philosophy. This relation was neither the absolute determination of one by the other, nor the production of one by the other, nor a simple concomitance, nor a strict parallelism, nor, I repeat, any of the relations that can be obtained a priori by using or mixing abstract concepts. It is a *sui generis* relation, which I would formulate (very incompletely) as follows:

In a given psychological state, the enactable part of this state, which is revealed by the attitude or the actions of the body, is represented by the brain: the rest is independent, and has no cerebral equivalent. Thus for a given cerebral state there are indeed corresponding different psychological states, but not any old states. These are psychological states that all have in common the same "motor schema." Many paintings can be placed in a given frame, but not all paintings. This is a higher, abstract, philosophical thought. We cannot conceive this idea without introducing a pictorial representation into it. In turn, we cannot represent this picture without thinking of a drawing that outlines the picture. And we cannot imagine the drawing without also sketching out the movements that would be required to produce the drawing. It is this motor sketch, and this motor sketch alone, that is represented in the brain. If you provide the motor sketch, there is room for the picture. Provide the picture, and there remains even more room for the thought. Thus thought is relatively free and undetermined by the cerebral activity that conditions it, the latter only expressing itself by the motor articulations of the idea, and these articulations may be the same for completely different ideas. And yet this is not complete freedom or complete indeterminism, because a given idea, taken at random, may not represent the desired articulations. To summarize, none of the simple concepts provided by philosophy express the desired relation, but this relation appears to emerge clearly from experience.

But you persist, and you say that this experience is incomplete, and you ask if experience, as it discovers more facts, will not increasingly support the thesis of parallelism. For what reason might one suppose this to be true? Is it scientific to prefer a possible experience, of which we as yet know nothing, to a real experience? Ask yourself now whether your faith in parallelism, your confident expectation in future demonstration, might not simply be a relic of the Leibnizian or Spinozan belief in a universal mechanism. The successors of Descartes, pushing the ideas of their master to their limits, believed in a unique science of nature, in a grand mathematics that could embrace everything. In order not to break this strict chain of cause and effect, they spoke of a parallelism between the psychic and the physical, as though mind and body said exactly the same thing in two different languages. But what would they now think of this conception of nature? Would they be aiming at the same simplicity through science? Would they conceive of intelligibility in the same way? If you remain at an abstract level, if you consider metaphysics to be a linear development of simple ideas, you will no doubt end up supporting the thesis of parallelism, because it expresses immediately, radically, and simply the requirements of the principle of causality itself, expressed in the simplest possible way. But reality is much more complex, and experience much more instructive.

True, I have not been able to enter the brain, to track a mental movement, to measure precisely the separation between this phenomenon and the corresponding psychological state. But even though a truth is by nature empirical, it does not follow that it can immediately be verified empirically. Often we have to skirt round it, approach it from many angles, none of which take us directly to our goal, but the convergence of which effectively outlines the point at which we would arrive. This is how you measure the distance of an inaccessible point, by observing it from all accessible points. There are scientific certainties that hold only due to the accumulation of probabilities. There are *lines of facts*, none of which on its own suffices to determine a truth, but which together determine it by their intersection. In the book that Monsieur Belot has so kindly referred to, I proceeded by the addition of probabilities, by the intersection of "lines of facts." I would have preferred not to speak of this work today. However, I am obliged to say a few words about it, because Monsieur Belot has taken the debate onto this terrain.

The second and third chapters of *Matter and Memory* are indeed devoted to determining the relation that links a psychological state to its cerebral concomitant. But it should not be thought that I focused on proving the thoroughly negative thesis that "there is no parallelism" between them, nor that I based my demonstration on the study of plasticity [*suppléances*] in particular, or on that of localizations in general. Little would be gleaned from a purely negative hypothesis. Furthermore, the question of plasticity is so obscure, the observed facts can be interpreted in so many ways, that I felt it necessary to leave it entirely to one side: I did not pronounce the word, nor speak of the thing. Finally, as to cerebral localization, I did not for an instant think of questioning it, because, on the contrary, I considered it only in those cases it which it has strictly been proven, in the function of speech. For me, the question was posed in a very different way: determining the exact *meaning* of the facts of localization in those cases where localization is certain. Contemporary science has no precise interpretation of these facts taken separately – perhaps they will never have a full interpretation. But it seemed to me that by combining these facts with a large number of others taken from normal or pathological psychology, it would be possible to give a near solution to the problem, a solution that could be increasingly approximated, that is, a scientific solution. By converging lines of facts, by the facts of normal recognition, by the facts of pathological recognition, by mental blindness in particular, and finally and above all by the various forms of sensory aphasia, I was led to the conclusion that the brain contains "motor schemas" of images and ideas, that at every instant it traces out their motor articulations, which, as a result and to a certain extent and in certain ways, condition thought. For example, a theatrical director's stage-book contains the changes of stage position by the actors that occur throughout the scene. But these changes of position only represent a small part of the play, and only determine a small part of the acting. If the brain has the same kind of relation to thought, it follows that there cannot be parallelism or equivalence between cerebral activity and thought.

Thus I do not base the negation of parallelism on negative considerations. Nothing could be concluded from the absence of facts or proofs in favor of parallelism. I do not state that in the absence of facts and proofs we would have the right to affirm, as Monsieur Belot appears to, that the proofs and facts will be discovered as science progresses: one can only adopt this approach if, like many philosophers, one is fully imbued with the Leibnizian or Spinozan idea of a universal mechanism. But we still have the right to reserve our judgment and to wait, irrespective of the fact that philosophy can gain nothing from a purely negative thesis. On the contrary, I tried to formulate a positive thesis, capable of progressive improvement and verification. And I would add that if some thinkers have adopted the hypothesis of parallelism without any discussion, it is not because this hypothesis is the most scientific, but because it is the simplest, and the philosophers of the current century have not taken the trouble to search for another.

It is precisely because my thesis is positive, open to progressive improvement and verification, that it has nothing whatsoever to do with what Monsieur Belot calls "the taking to the limit" of some metaphysical doctrines. The method that I propose does not extract a simple concept from reality (even less so a negative concept such as nonparallelism), and then subject it to dialectical study. Quite the opposite – this method requires a continuous contact with reality. It follows reality in all its sinuosity. It requires that our faculties of observation be so strained that they sometimes surpass themselves (as, for example, when they catch on the rim of the unconscious "pure perception" and "pure memory," both of which are far from being mere construc-tions of the mind, as Monsieur Belot believes). It is composed of corrections, of refinements, of gradual complications. It aspires to form a metaphysics, as certain and as universally recognized in science as in other realms. It must be able to follow as closely as possible the insertion of thought into life, such that the meaning of life will appear clearly and indubitably to everyone.

I am immediately asked to say what is this meaning of life. A formula is required. It is surprising that there is no thesis. But how could I today formulate a definitive conclusion, when the method that I propose requires us to arrive progressively at ideas via the long and hard road of facts? You always want us to proceed like mathematicians, by the a priori development of a simple conception. All I can do is to sum up in a few words the provisional conclusions to which my studies have led me. They are far too vague to teach you much new. And, separated from the reasons and the facts to which they are joined, they will not have the power to attract those who consider life from a different viewpoint.

I will thus state that I cannot envisage general evolution and the progress of life throughout the organized world, the mutual coordination and subordination of vital functions in the same organism, the relations that psychology and physiology combined seem to establish between cerebral activity and thought in man, without arriving at the conclusion that life is an immense effort made by thought to obtain something from matter that matter does not want to give up. Matter is inert, it is the seat of necessity, it proceeds mechanically. Thought seems to try and benefit from matter's mechanical aptitude, to use it for *actions*, to thus convert into contingent movements in space and into unforeseeable events in time all that it contains that is of creative energy (or at least that which this energy has which can be enacted and exteriorized). Slowly and skill-fully, thought piles complication upon complication to make freedom out of necessity, to compose a matter that is so subtle, so mobile, that liberty can balance itself upon this mobility, through a real physical paradox and thanks to a near-exhausting effort. But it is trapped. The whirlwind which placed it there picks it up and sweeps it away. It becomes a prisoner of the very mechanisms it has constructed. Automatism takes over, and, through an inevitable oversight, it forgets the goal it set itself; life, which should have been merely a means to a higher end, becomes entirely consumed in the effort to conserve itself. From the humblest of organisms, to the higher vertebrates who come just before humans, we can see a continually blocked and continually renewed attempt, each time renewed with an art that is increasingly intelligent. Man triumphed – with difficulty and so incompletely that it only requires a moment of relaxation or of inattention for automatism to take over. Nevertheless, Man triumphed thanks to that marvelous instrument that is the human brain. The superiority of this instrument seems to me to be entirely based on the virtually infinite latitude it is given to construct mechanisms that will block other mechanisms. It constructs – not only once, but continually – motor habits to which it then delegates control over lower centers. The faculty of animals to form motor habits is limited. But the human brain accords Man the power to learn an infinite number of "sports." Above all it is an organ of sport and, from this point of view, one could define Man as "the sporting animal." The first of these sports is language, the function that occupies such a huge region of the human brain. Language was the instrument of liberation *par excellence*, despite the later automatism Man inflicts on thought. But, in general, the superiority of our brain resides in the power of liberation it gives us with respect to bodily automatism, by allowing us to create endlessly new habits that will abort the others or not interfere with them. In this sense, we will find nothing in the brain that corresponds to the true operation of thought; and yet it is the human brain that made human thought possible. Without it, the higher faculties of thought could not turn towards matter without being seized by automatism and drowning in the unconscious.

What more can I say? How, on the basis of this as yet vague philosophy of life, could I construct the precise and definitive morality which you seem to wish for? All I can say is that the normal exercise of human activity will increasingly well define itself, through the deepening of life itself. For my part, I still see everywhere a dual direction in the development of this activity. At the same time as thought is inserted into life and concentrates itself on action (which appears to be the object of life), it realizes better its own nature and as a result also its independence from matter. Attachment and release, these are the two poles between which morality oscillates. You ask me to which it should fix itself? I do not see why it should become fixed. If effort is not attached to life, it lacks intensity. If effort does not release itself, at least slightly, and through thought, it lacks direction. You need to know where you are with respect to the first point to have the force to act, and with respect to the second in order to abstract yourself from the prejudices of the moment and to know what is to be done. But one should not go completely to either of these two extremities. I return to the idea that has been the *leitmotif* of my whole response. It is neither interesting nor instructive nor does it conform to the truth, to oppose concepts each of which partly corresponds to reality, because it was necessarily extracted from reality. Philosophy must rather propose their mixture and, if possible, create higher concepts in which previous oppositions can be absorbed.

Let us work to grasp experience from as close as we can. Let us accept science with its current complexity, and let us recommence, with this new science as our raw material, an analogous effort to that which the old metaphysicians carried out on a much simpler science. We need to break out of mathematical frameworks, to take account of the biological, psychological, and sociological sciences, and on this broader base construct a metaphysics that can go higher and higher through the continual, progressive, and organized effort of all philosophers, in the same respect for experience.

CASSIRER

5

ERNST CASSIRER AND THE PHILOSOPHY OF SCIENCE

Michael Friedman

Cassirer began his serious study of philosophy as a student of Hermann Cohen's at the University of Marburg. He worked with Cohen from 1896 to 1899, when he completed his doctorate with a dissertation on Descartes's analysis of mathematical and natural scientific knowledge. This then appeared as the Introduction to Cassirer's first published work, *Leibniz' System in seinem wissenschaftlichen Grundlagen* (1902). Upon returning to Berlin in 1903, Cassirer further developed these themes while working out a monumental interpretation of modern philosophy and science from the Renaissance through Kant, his two-volume *Das Erkenntnisproblem in der Philosophie und Wissenschaft der neuren Zeit* (1906–7). The first volume served as his "habilitation" at the University of Berlin, where he taught as *Privatdozent* from 1906 to 1919.[1]

Cassirer thus began his career as an intellectual historian, one of the very greatest of the twentieth century. Cassirer's *Das Erkenntnisproblem*, in particular, is a magisterial and deeply original contribution to both the history of philosophy and the history of science. It is the first work, in fact, to develop a detailed reading of the scientific revolution as a whole in terms of the "Platonic" idea that the thoroughgoing application of mathematics to nature (the so-called mathematization of nature) is the central and overarching achievement of this revolution. And Cassirer's work is acknowledged as such by the seminal historians Edwin Burtt, E. J. Dijksterhuis, and Alexandre Koyré, who developed this theme later in the century in the course of establishing the discipline of history of science as we know it today. Cassirer, for his part, simultaneously articulates an interpretation of the history of modern philosophy as the development and eventual triumph of what he calls "modern philosophical idealism." This tradition takes its inspiration, according to Cassirer, from idealism in the Platonic sense, from an appreciation of the "ideal" formal structures paradigmatically studied in mathematics, and it is distinctively modern in recognizing the fundamental importance of the systematic application of such structures to empirically given nature in modern mathematical physics – a progressive and synthetic process wherein mathematical models of nature are successively refined and corrected without limit. For Cassirer, it is Galileo, above all, in opposition to both sterile Aristotelian–Scholastic formal logic and sterile Aristotelian–Scholastic empirical induction, who first grasped the essential structure of this synthetic process; and the development of "modern philosophical idealism" in the work of Descartes, Spinoza, Gassendi, Hobbes, Leibniz, and Kant then consists in its increasingly self-conscious philosophical articulation and elaboration.[2]

In both his 1902 and 1906–7 works Cassirer thus interprets the development of modern thought as a whole (embracing both philosophy and the sciences) from the perspective of the basic philosophical principles of Marburg neo-Kantianism, as initially articulated by Cohen (1871). Kant's transcendental method is interpreted as beginning with the "fact of science" – the existence of the sciences in their modern, post-seventeenth-century form – as its ultimate given datum. The task of transcendental philosophy is to take these sciences as they are actually given, and then to seek, by a regressive argument, their ultimate presuppositions or preconditions. Kant had performed this task to perfection for the case of the fundamentally Newtonian mathematical sciences of the seventeenth and eighteenth centuries; and our task, at the end of the nineteenth century, is to generalize and extend Kant's approach so as to embrace the main developments in the mathematical sciences that have occurred since Kant's time. When we do this, the chief alteration that is necessary in the Kantian system is a rejection of Kant's own sharp separation between two independent faculties of the mind: a passive or receptive faculty of sensibility and an active or intellectual faculty of understanding. And in this way, in particular, we avoid the idea that the pure forms of sensibility, space and time, have their own independent structure – the structure, basically, of Euclidean geometry plus Newtonian absolute space and time – given independently of the synthesizing activity of the understanding. Other geometries than Euclid's, and other structures for space and time than Newton's, are therefore perfectly possibly products of the a priori synthesizing activity of thought.

The crucial question, at this point, concerns how we are now to conceive the a priori synthesizing activity of thought – the activity Kant himself had called "productive synthesis." For Cohen and the Marburg School there is no longer an a priori faculty of sensibility, embracing, as in Kant's original system, the basic structures of Euclidean geometry and Newtonian mathematical physics. Indeed, for Cohen and the Marburg School, there is no longer an independent contribution of a posteriori sensibility either – there is no independent "manifold of sensations" that is simply given, entirely independently of the activity of thought, within the pure spatiotemporal form of our sensibility. What there is, instead, is an essentially dynamic or temporal procedure of active generation (*Erzeugung*), as the mind successively characterizes or determines the "real" that is to be the object of mathematical natural science in a continuous serial process. The "real" itself – the true empirical object of mathematical natural science – is in no way independently given as something separate and distinct from this "productive synthesis" of thought; it is to be conceived, rather, as the necessary endpoint or limit towards which the continuous serial process exemplified in modern mathematical natural scientific knowledge is converging. This "genetic conception of knowledge" is the most characteristic contribution of the Marburg School.

For Cohen, the epistemological process in question is modeled, more specifically, on the methods of the infinitesimal calculus.[3] Beginning with the idea of a continuous series or function, our problem is to see how such a series can be a priori generated step-by-step. The modern mathematical concept of a differential (in contemporary terminology, the concept of a tangent vector) shows us how this can be done, for the differential at a point in the domain of a given function literally points us towards its values on the succeeding points. The differential therefore infinitesimally captures the rule of the series as a whole, and thus expresses, at a given moment, the general form of the series valid for all times. It is in this way, in particular, that Cohen interprets the Kantian Anticipations of Perception – the principles governing the categories of reality, negation, and limitation by which Kant himself initiates the determination of the "real" by the understanding. The difference is that Cohen now elevates the Anticipations of Perception so that they, by themselves, contain the sole and entire key to the synthesizing activity of the understanding in general. For this activity is now completely expressed in the step-wise development of a continuous temporal series, representing the methodological progress of the modern mathematical sciences, whose a priori "general form" is now most aptly expressed by the differential of a continuous series or function.

In neither the 1902 nor 1906–7 works does Cassirer diverge in any essential way from the fundamentals of Cohen's point of view. Indeed, Cohen's overriding emphasis on the differential and the methods of the infinitesimal calculus is especially prominent in Cassirer (1902), where Leibniz's great advance on Descartes is explained in terms of the priority of the Leibnizean calculus in relation to Cartesian analytic geometry. It is in Cassirer's next great book, *Substanzbegriff und Funktionsbegriff* (1910), that Cassirer first takes an essential philosophical step beyond Cohen; and it is this book, accordingly, which is perhaps most characteristic of Cassirer's own particular approach to the philosophy of science. For it is here, in particular, that Cassirer, unlike Cohen, engages with the modern developments in the foundations of mathematics and mathematical logic that exerted an overwhelming influence on twentieth-century philosophy of science – as represented, above all, by the philosophy of logical empiricism.

Cassirer (1910) begins by discussing the problem of concept formation, and by criticizing in particular the "abstractionist" theory according to which general concepts are arrived at by ascending inductively from sensory particulars. This theory, for Cassirer, is an artifact of traditional Aristotelian logic, wherein the only logical relations governing concepts are those of superordination and subordination, genus and species – abstractionism then views the formation of such concepts as an inductively driven ascent from the sensory particulars to ever higher species and genera. Moreover, by this commitment to traditional subject–predicate logic, we are also committed, according to Cassirer, to the traditional metaphysical conception of substance as the fixed and ultimate substratum of changeable qualities. A metaphysical "copy" theory of knowledge, according to which the truth of our sensory representations consists in a (forever unverifiable) relation of pictorial similarity between them and the ultimate "things" or substances lying behind our representations, is then the natural and inevitable result.

Cassirer is himself concerned, above all, to replace this "copy" theory of knowledge with what he calls the "critical" theory. Our sensory representations achieve truth and "relation to an object," not by matching or picturing a realm of metaphysical "things" or substances constituting the stable and enduring substrate of the empirical phenomena, but rather in virtue of an embedding of the empirical phenomena themselves into an ideal formal structure of mathematical relations – wherein the stability of mathematically formulated *universal laws* takes the place of an enduring substrate of ultimate substantial "things." Developments in modern formal logic (the mathematical theory of relations) and in the foundations of mathematics contribute to securing this "critical" theory of knowledge in two closely related respects. On the one hand, the modern axiomatic conception of mathematics, as exemplified especially in Dedekind's work on the foundations of arithmetic and Hilbert's work on the foundations of geometry, has shown that mathematics itself has a purely formal and ideal, nonsensory and thus nonintuitive meaning. Pure mathematics describes abstract "systems of order" – what we would now call relational structures – whose concepts can in no way be accommodated within the abstractionist conception. On the other hand, modern scientific epistemology, as exemplified especially in Helmholtz's celebrated *Zeichentheorie*, has shown ever more clearly that scientific theories do not provide "copies [*Abbilder*]" or "pictures [*Bilder*]" of a world of substantial "things" subsisting behind the flux of phenomena. They rather provide mere formal systems of "signs [*Zeichen*]" corresponding via a nonpictorial relation of "coordination [*Zuordnung*]" to the universal law-like relations subsisting within the phenomena themselves.

In explicitly embracing these ideas from the work of Dedekind, Hilbert, and Helmholtz, Cassirer, as I have already suggested, approaches the point of view of early twentieth-century logical empiricism. Indeed, Cassirer takes the modern logic implicit in the work of Dedekind and Hilbert, and explicit in the work of Frege and early Russell, as providing us with our primary tool for moving beyond the abstractionism of Aristotelian syllogistics.[4] The modern "theory of the concept," accordingly, is based on the fundamental notions of function, series, and order (or order structure) – where these notions, from the point of view of pure mathematics and pure logic, are entirely formal and abstract, having no intuitive relation, in particular, to either space or time. Nevertheless, and here is where Cassirer diverges from logical empiricism, this modern theory of the concept only provides us with a genuine and complete alternative to Aristotelian abstractionism when it is embedded within the genetic conception of knowledge. What is primary, once again, is the generative historical process by which modern mathematical natural science successively develops or evolves; and pure mathematics and pure logic only have philosophical significance as elements of or abstractions from this more fundamental process of "productive synthesis" aimed at the application of such pure formal structures in *empirical* knowledge.

It is for this reason, more specifically, that Cassirer decisively rejects the logicist theory of the nature of mathematics – where this theory is understood (as it was by the logical empiricists) as providing a definitive refutation of the original Kantian conception of the synthetic a priori. Mathematics, according to this logicist view, is completely representable within pure formal logic (that is, within modern mathematical logic), and it is therefore analytic not synthetic. Without in any way rejecting the purely formal achievements of modern mathematical logic, Cassirer nonetheless denies that they can possibly show that mathematics is merely analytic in the philosophical sense. For philosophy's distinctive task (the task of epistemology) is developing what Cassirer calls a "logic of objective knowledge":

> Thus a new task begins at that point where logistic ends. What the critical philosophy seeks and what it must require is a *logic of objective knowledge*. Only from the standpoint of this question can the opposition between analytic and synthetic judgments be completely understood and evaluated.... Only when we have understood that the same fundamental syntheses on which logic and mathematics rest also govern the scientific construction of empirical knowledge, that they first make it possible for us to speak of a fixed lawful order among appearances and thus of their objective meaning – only then is the true justification of the principles [of logic and mathematics] achieved.[5]

Since pure formal logic, from a philosophical point of view, is merely an abstraction from the fundamentally synthetic constitution of mathematical natural-scientific knowledge in general, developments in pure formal logic and mathematics, by themselves, can have no independent philosophical significance. They do not and cannot undermine the essentially Kantian insight into the priority and centrality of "productive synthesis."

In now articulating the nature of this "productive synthesis" more precisely, Cassirer no longer follows Cohen in giving overriding importance to the methods of the infinitesimal calculus. Instead, as already suggested, Cassirer employs his more abstract understanding of modern, late-nineteenth-century mathematics to craft a similarly abstract version of the genetic conception of knowledge. What we are concerned with, as before, is the progression of theories produced by modern mathematical natural science in its factual historical development. But we now conceive this progression as a series or sequence of abstract formal structures ("systems of order"), which is itself ordered by the abstract mathematical relation of approximate backwards-directed inclusion – as, for example, the new non-Euclidean geometries contain the older geometry of Euclid as a continuously approximated limiting case. In this way, in particular, we can conceive all the theories in our sequence as continuously converging, as it were, on a final or limit theory, such that all previous theories in the sequence are approximate special or limiting cases of this final theory. This final theory, of course, is only a regulative ideal in the Kantian sense – it is only progressively approximated but never in fact actually realized. Nevertheless, the idea of such a continuous progression towards an ideal limit constitutes the characteristic "general serial form" of our properly empirical mathematical theorizing, and, at the same time, it bestows on this theorizing its characteristic form of objectivity. For, despite all historical variation and contingency, there is, nonetheless, a continuously converging progression of abstract mathematical structures framing, and making possible, all our empirical knowledge. Finally, in accordance with the "critical" theory of knowledge, as opposed to the "copy" theory, convergence, on this view, does not take place towards a mind- or theory-independent "reality" of ultimate substantial "things." Rather, the convergence in question occurs entirely *within* the series of our historically developed models or structures. "Reality," on this view, is simply the purely ideal limit or endpoint towards which the sequence of such structures is mathematically converging – or, to put it another way, it is simply the series itself, taken as a whole.

This same view, considered from a slightly different perspective, also provides Cassirer with a new interpretation of the synthetic a priori. Contrary to the original Kantian conception of the a priori, even the most fundamental principles of

Newtonian mechanics "need not be taken as absolutely unchanging dogmas." Such temporarily "highest" principles of experience – at a given stage of scientific theorizing – may evolve into others, and, in this case, even our most general "functional form" for the laws of nature would undergo a change. Yet such a transition would never entail that "the one fundamental form absolutely disappears while another arises absolutely new in its place." On the contrary:

> The change must leave a determinate stock of principles unaffected; for it is solely for the sake of securing this stock that it is undertaken in the first place, and this shows it its proper goal. Since we never compare the totality of hypotheses in themselves with the naked facts in themselves, but can only oppose *one* hypothetical system of principles to another, more comprehensive and radical [system], we require for this progressive comparison an ultimate constant *measure* in highest principles, which hold for all experience in general. What thought demands is the identity of this logical system of measure throughout all change in that which is measured. In this sense, the critical theory of experience actually aims to construct a *universal invariant theory of experience* and thereby to fulfill a demand towards which the character of the inductive procedure itself ever more clearly presses.[6]

In other words, since induction is essentially a process of generalization, aiming to subsume individual facts under ever more universal laws, it does not rest content with any particular set of laws, but attempts to subsume even the most general laws at a given stage of theorizing (e.g., Newton's laws) under still more general laws, in such a way that the most general laws at an earlier stage are exhibited as approximate special cases of the still more general laws at a later stage. And this implies, for Cassirer, that we must form the idea of an ultimate or limiting set of laws, such that *all* previous stages are approximate special cases of these ultimate laws. It is at this point – and only at this point – that we can actually specify the content of the "universal invariant theory of experience."

It then follows, as Cassirer immediately goes on to point out, that there is no way to determine the specific content of such ultimate principles in advance:

> The goal of critical analysis would be attained if it succeeded in establishing in this way what is ultimately common to all possible forms of scientific experience – in conceptually fixing those elements that are preserved in the progress from theory to theory because they are the conditions of each and every theory. This goal may never be completely attained at any given stage of cognition; nevertheless, it remains as a *demand* and determines a fixed direction in the continual unfolding and development of the system of experience itself.
>
> The strictly limited objective meaning of the "a priori" appears clearly from this point of view. We can only call those ultimate *logical invariant*s a priori that lie at the basis of every determination of a lawlike interconnection of nature in general. A cognition is called a priori, not because it lies in any sense *before* experience, but rather because, and in so far as, it is contained in every valid judgment about facts as a necessary *premise*.[7]

Just as the object of natural-scientific knowledge, on the genetic or "critical" theory, is the never fully realized ideal mathematical structure towards which the entire historical development of science is converging, so the a priori form of scientific knowledge, for Cassirer, can only be determined as that stock of "categorial" principles which, viewed from the perspective of the ideally completed developmental process, are seen (retrospectively, as it were) to hold at every stage. So we do not know, at any given stage, what the particular content of spatial geometry, for example, must be, but we can now venture the well-supported conjecture that some or another spatial-geometrical structure must be present.

This example of spatial-geometrical structure turns out to be a particularly apt and revealing one. For, although Cassirer (1910) appeared prior to the formulation of Einstein's general theory of relativity, Cassirer's next important contribution to scientific epistemology, *Zur Einsteinschen Relativitätstheorie* (1921), is devoted precisely to this revolutionary new theory. The main burden of Cassirer (1921) is to argue that Einstein's theory actually stands as a brilliant confirmation of a purified and generalized version of the "critical" conception of knowledge. For the increasing use of abstract mathematical representations in Einstein's theory entirely supports the idea that "the reality of the physicist stands opposed to the reality of immediate perception as a thoroughly mediated [reality]: as a totality, not of existing things or properties, but rather of abstract symbols of thought that serve as the expression for determinate relations of magnitude and measure, for determinate functional coordinations and dependencies in the phenomena" (1921: 14). And it follows that Einstein's new theory of gravitation can be incorporated within the "critical" conception of knowledge "without difficulty, for this theory is characterized from a general epistemological point of view precisely by the circumstance that in it, more consciously and more clearly than ever before, the advance from the copy theory of knowledge to the functional theory is completed" (1921: 55).

In particular, the fact that Einstein now replaces the geometry of Euclid with a much more general geometry (a geometry of variable curvature depending on the distribution of mass and energy) in no way implies the collapse of a properly-understood "critical" theory of the a priori: "[f]or the 'a priori' of space, which [physics] asserts as the condition of every physical theory, does not include, as has been shown, any assertion about a determinate particular structure of space, but is concerned only with the function of 'spatiality in general,' which is already expressed in the general concept of the line-element ds as such − entirely without regard to its more particular determination" (1921: 101). In other words, according to Einstein's generalized conception of the "line-element" − a conception derived from Riemann's general theory of "n-fold extended manifolds" − we postulate only that space is *infinitesimally* Euclidean (continuously approximating to Euclidean geometry as the regions under consideration grow smaller and smaller); and what Cassirer is now saying is that it is precisely this new conception of the "line-element" that constitutes our current best candidate for an ultimate geometrical invariant.[8]

Cassirer's assimilation of Einstein's general theory of relativity marked a watershed in the development of his thought. It not only gave him the opportunity, as we have just seen, to reinterpret the Kantian theory of the a priori conditions of objective experience in terms of what Cassirer calls a "universal invariant theory of experience," but it also provided him with an occasion to generalize and extend the original Marburg genetic conception of knowledge in such a way that modern mathematical scientific knowledge in general is now seen as just one possible "symbolic form" among other equally valid such forms. Indeed, Cassirer (1921) first announces the project of a general "philosophy of symbolic forms," conceived, in this context, as a philosophical extension of "the general postulate of relativity." Just as, according to the general postulate of relativity, all possible reference frames and coordinate systems are viewed as equally good representations of physical reality, and, as a totality, are together interrelated and embraced by precisely this postulate, similarly the totality of "symbolic forms" – aesthetic, ethical, religious, scientific – are here envisioned by Cassirer as standing in a closely analogous relationship. So it is no wonder, then, that, subsequent to taking up a professorship at Hamburg in 1919, Cassirer devotes the rest of his career to this new philosophy of symbolic forms.

The philosophy of symbolic forms is of the greatest interest and importance in determining Cassirer's relationship to the later development of what we now call the continental tradition – and, more specifically, for understanding his remarkably close relationship with Martin Heidegger.[9] Here, however, I will break off my discussion of the evolution of Cassirer's philosophy, and I will turn, instead, to the question of the relevance of Cassirer's reinterpretation of Kantian epistemology to our contemporary situation in the philosophy of science. It might seem, in particular, that Cassirer's vision of a "universal invariant theory of experience" based on the idea of a continuously converging sequence of mathematical natural-scientific theories has very little relevance to our contemporary philosophical situation. For one of the central points of Thomas Kuhn's theory of scientific revolutions is that all talk of intertheoretical convergence, of the approximate containment of earlier scientific paradigms in later ones, must be rejected. Indeed, one of Kuhn's primary examples is precisely the relationship between Einstein's theory of relativity and Newtonian mechanics, where Kuhn famously denies that Newtonian mechanics *can* be mathematically derived from relativity theory as an approximate special case.[10]

Yet the situation is not as simple as it first appears, and the best way to see this is to take a brief look at the historical background to the development of Kuhn's own historiography. In the Preface to *The Structure of Scientific Revolutions*, first published in 1962, Kuhn portrays how he shifted his career plans from physics to the history of science, and, in explaining his initial intensive work in the subject, he states that he "continued to study the writings of Alexandre Koyré and first encountered those of Emile Meyerson, Hélène Metzger, and Anneliese Maier; [more] clearly than most other recent scholars, this group has shown what it was like to think scientifically in a period when the canons of scientific thought were very different from those current today" (1970: v–vi). Then, in the introductory first chapter, "A Role for History," Kuhn explains the background to his rejection of what he calls the "development-by-accumulation" model, especially as represented by a naive form of empiricism according to which science linearly progresses via the continuous accumulation of more and more observable facts. The background to his rejection of such views, Kuhn explains, lies in what he here calls "the new historiography," as "perhaps best exemplified in the writings of Alexandre Koyré": "[b]y implication, these historical studies suggest the possibility of a new image of science; [t]his essay aims to delineate that image by making explicit some of the new historiography's implications" (1970: 3).

These statements, from our present point of view, are especially intriguing. On the one hand, Kuhn makes it perfectly clear that the most important influence on the development of his historiography is Alexandre Koyré. On the other hand, and this is perhaps not as well understood as it should be, the most important philosophical influence on Koyré's historiography is Emil Meyerson. Moreover, the philosophical perspective shared by both Meyerson and Koyré is diametrically opposed, in most essential respects, to that articulated by Cassirer.[11]

Meyerson, in particular, is vehemently opposed to all attempts to assimilate scientific understanding to the formulation of universal mathematical laws governing phenomena. Indeed, the central thought of his *Identité et réalité*, first published in 1908, is that genuine scientific knowledge and understanding can never be the result of mere lawfulness (*légalité*) but must instead answer to the mind's a priori logical demand for identity (*identité*). And the primary requirement resulting from this demand is precisely that some underlying substance be conserved as absolutely unchanging and self-identical in all sensible alterations of nature. Thus, the triumph of the scientific revolution, for Meyerson, is represented by the rise of mechanistic atomism, wherein elementary corpuscles preserve their sizes, shapes, and masses while merely changing their mutual positions in uniform and homogeneous space via motion; and this same demand for transtemporal substantial identity is also represented, in more recent times, by both Lavoisier's use of the principle of the conservation of matter in his new chemistry and by the discovery of the conservation of energy. Yet, in the even more recent discovery of what we now know as the second law of thermodynamics ("Carnot's principle"), which governs the *temporally irreversible* process of "degradation" or "dissipation" of energy, we encounter nature's complementary and unavoidable resistance to our a priori logical demands. In the end, therefore, Meyerson views the development of modern natural science as progressing via a perpetual dialectical opposition between the mind's a priori demand for substantiality and thus absolute identity through time, on the one side, and nature's "irrational" a posteriori resistance to this demand, on the other.

In the work of Cassirer and Meyerson, then, we find two sharply diverging visions of the history of modern science. For Cassirer, this history is seen as a process of evolving rational purification of our view of nature, as we progress from naively realistic "substantialistic" conceptions, focusing on underlying substances, causes, and mechanisms subsisting behind the observable phenomena, to increasingly abstract purely "functional" conceptions, where we finally abandon the search for underlying ontology in favor of ever more precise mathematical representations of phenomena in terms of exactly formulated universal laws. For Meyerson, by contrast, this same history is seen as a necessarily dialectical progression (in something like the Hegelian sense), wherein reason perpetually seeks to enforce precisely the "substantialistic" impulse, and nature continually offers her resistance in the ultimate irrationality of temporal succession. It is by no means surprising, therefore, that Meyerson, in the course of considering, and rejecting, what he calls "anti-substantialistic conceptions of science," explicitly takes issue with Cassirer's (1906–7) claim that "[m]athematical physics turns aside from the essence of things and their inner substantiality in order to turn towards their numerical order and connection, their functional and mathematical structure."[12] And it is also no wonder that Cassirer (1910), in the course of a discussion of "identity and difference, constancy and change," explicitly takes issue with Meyerson's views by asserting that "[t]he identity towards which thought progressively strives is not the identity of ultimate substantial things but the identity of functional orders and coordinations." Indeed, this passage continues with a typical statement of Cassirer's version of the genetic conception of knowledge:

> These [viz., "functional orders and coordinations"], however, do not exclude the moment of difference and change but succeed in determination only in and with it. Manifoldness as such is not destroyed, but we have only a manifold of another dimension: the mathematical manifold takes the place of the sensible manifold in scientific explanation. What thought demands is not the dissolution of plurality and changeableness as such but rather their mastery by the mathematical *continuity* of serial laws and serial forms.[13]

Thus, in explicit opposition to the view of Meyerson, Cassirer's whole point is that thought does *not* require a "substantialistic" or "ontological" identity over time of permanent "things," but rather a purely mathematical continuity over time of successively articulated mathematical structures.[14]

If I am not mistaken, this deep philosophical opposition between Meyerson and Cassirer receives a very clear echo in Kuhn's theory of scientific revolutions, particularly with regard to the question of continuity over time at the theoretical level. Here Kuhn shows himself, in this respect, to be a follower of the Meyersonian viewpoint, for he consistently gives the question an ontological rather than a mathematical interpretation. Thus, for example, when Kuhn famously considers the relationship between relativistic and Newtonian mechanics, he rejects the notion of a fundamental continuity between the two theories on the grounds that the "physical referents" of their terms are essentially different, and he nowhere considers the contrasting idea, characteristic of Cassirer's work, that continuity of purely mathematical structures is quite sufficient.[15] Moreover, Kuhn consistently gives an ontological rather than a mathematical interpretation to the question of theoretical convergence over time: the question is always whether our theories can be said to converge to an independently existing "truth" about reality, to a theory-independent external world.[16] By contrast, as we have seen, Cassirer definitively rejects this realistic construal of convergence at the very outset: our theories do not (ontologically) converge to a mind-independent realm of substantial things; they (mathematically) converge *within* the historical progression of our theories as they continually approximate, but never actually reach, an ideally complete mathematical representation of the phenomena.

It follows, then, that Kuhn's remarks about continuity and convergence can in no way be taken as a straightforward refutation of Cassirer's position. For Kuhn simply assumes, in harmony with the Meyersonian viewpoint, that there is rational continuity over time in the traditional sense only if there is also substantial identity. Since, as Kuhn argues, the "physical referents" of Newtonian and relativistic mechanics, for example, cannot be taken to be the same, we are therefore faced with the problem of theoretical incommensurability together with all its "irrationalistic" implications. Yet Cassirer, as we have seen, is just as opposed to all forms of naive realism (as well as naive empiricism) as is Kuhn. He instead proposes a generalized Kantian conception, emblematic of what he himself calls "modern philosophical idealism," according to which scientific rationality and objectivity are secured in virtue of the way in which our empirical

~~ f~~med. and thereby made possible, by a continuously evolving
n order to determine whether, and to
ion is tenable, Kuhnian historiography
instead need a more comprehensive
n all of Cassirer's work), in which both
parallel development of the modern
plementary, historical and philosophical
ill find that a satisfactory philosophical
mathematical science requires a mixture
the logical-empiricist tradition and from
)n.[17]

's

development see Friedman (2000), and espe-

hematization is distinctive of Cassirer, and it
om Koyré's reading (see note 14 below).

3   In this ...   n (1883).
4   Here Cassirer places particular emphasis on Russell (1903).
5   See Cassirer (1907: 44–5). Cassirer's main target here is Russell (1903) together with closely related work by Louis Couturat. The issue is rather subtle, however, for Russell (1903) famously maintains that logic itself is *synthetic* a priori; what Cassirer is contesting is the claim that mathematics, precisely because of its reducibility to logic, is independent of all spatiotemporal "productive synthesis."
6   See Cassirer (1910: 355–6): which passage occurs on p. 86 below of the Cassirer selection. The passage is followed by a comparison with the method of invariants in geometry.
7   See Cassirer (1910: 357): which passage occurs on p. 86 below of the Cassirer selection.
8   See Friedman (2000: ch. 7) for a more detailed discussion of Cassirer's conception of the a priori. In particular, I there argue that this conception is best conceived as involving purely *regulative* ideals in the Kantian sense, rather than what Kant himself thinks of as *constitutive* principles. The latter (which include, for Kant himself, the principles of specifically Euclidean geometry) have a definite content that can be specified in advance as necessarily holding for all possible experience, whereas the former only tentatively and indefinitely indicate a direction for guiding future scientific research. For Cassirer, as I read him, we cannot specify the content of spatial geometry in advance, but we can use our current best analysis of this content as a guide for future research in accordance with the picture of progressive convergence articulated in the genetic conception of knowledge.
9   For more on this topic see again Friedman (2000: esp. ch. 8).
10 See Kuhn (1970: ch. 9), as further discussed below (see note 15).
11 For a more detailed discussion of the intellectual background to Kuhn's historiography see Friedman (2003). Koyré (1931) presents his allegiance to Meyersonian philosophy very clearly and explicitly.
12 See Meyerson (1930: 388–9). The quoted passage is from vol. 2 of *Das Erkenntnisproblem*.
13 See Cassirer (1910: 431): which passage occurs on p. 88 below.
14 Note 11 above emphasizes Koyré's allegiance to Meyersonian philosophy, and it is precisely this that underlies Koyré's explicit rejection of Cassirer's more Kantian version of

"Platonism." See, in particular, Koyré (1978: note 123 on p. 223): "E. Cassirer, in his *Erkenntnisproblem*, vol. I, expresses the opinion that Galileo resurrected the Platonist ideal of scientific knowledge; for which it follows, for Galileo (and Kepler), the necessity for mathematizing nature . . . . Unfortunately (at least in our opinion) Cassirer turns Plato into Kant. Thus, for him, Galileo's 'Platonism' is expressed by his giving priority to function and law over being and substance." Whereas Cassirer's "Platonism" is optimistic, based on the possibility of indefinitely progressive mathematical convergence, Koyré's, following Meyerson, is pessimistic, based on an inevitable unbridgeable gap between mathematical theory and physical reality.

15 Kuhn does not and cannot deny, of course, that laws of the same mathematical form as Newton's can be mathematically derived as special cases of relativistic laws. Indeed, the main point of Kuhn (1970: ch. 9, pp. 101–2) is that preservation of mathematical form does not entail preservation of "physical referents": "The variables and parameters that in the Einsteinian [equations] represented spatial position, time, mass, etc., still occur in the [derived equations]; and they there still represent Einsteinian space, time, and mass. But the physical referents of these Einsteinian concepts are by no means the same as those of the Newtonian concepts that bear the same name."

16 See especially Kuhn (1970: 206–7), where Kuhn rejects all talk of convergence over time on the grounds that "[t]here is, I think, no theory-independent way to reconstruct phrases like 'really there'; the notion of a match between the ontology of a theory and its 'real' counterpart in nature now seems to me illusive in principle."

17 For my own first attempts to sketch such an alternative see Friedman (2001).

References

Cassirer, Ernst. 1902. *Leibniz' System in seinen wissenschaftlichen Grundlagen*. Marburg: Elwert.

——. 1906–7. *Das Erkenntnisproblem in der Philosophie und Wissenschaft der neueren Zeit*. 2 vols. Berlin: Bruno Cassirer.

——. 1907. "Kant und die moderne Mathematik." *Kant-Studien* 12: 1–40.

——. 1910. *Substanzbegriff und Funktionsbegriff: Untersuchungen über die Grundfragen der Erkenntniskritik*. Berlin: Bruno Cassirer. Translated as *Substance and Function*. Chicago: Open Court, 1923.

——. 1921. *Zur Einsteinschen Relativitätstheorie. Erkenntnistheoretische Betrachtungen*. Berlin: Bruno Cassirer. Translated as *Einstein's Theory of Relativity*. Chicago: Open Court, 1923.

——. 1923–9. *Philosophie der symbolischen Formen*. 3 vols. Berlin: Bruno Cassirer. Translated as *The Philosophy of Symbolic Forms*. New Haven: Yale University Press, 1955–7.

Cohen, Hermann. 1871. *Kants Theorie der Erfahrung*. Berlin: Dümmler.

——. 1883. *Das Princip der Infinitesmal-Methode und seine Geschichte: ein Kapitel zur Grundlegung der Ekenntniakritik*. Berlin: Dümmler.

Friedman, M. 2000. *A Parting of the Ways: Carnap, Cassirer, and Heidegger*. Chicago: Open Court.

——. 2001. *Dynamics of Reason: The 1999 Kant Lectures at Stanford University*. Stanford: CSLI.

——. 2003. "Kuhn and Logical Empiricism." In T. Nickles, ed., *Thomas Kuhn*. Cambridge: Cambridge University Press.

Koyré, Alexandre. 1931. "Die Philosophie Emile Meyersons." *Deutsch-Französische Rundschau* 4: 197–217.

——. 1978. *Galileo Studies*. Atlantic Highlands, NJ: Humanities Press. Translation of *Etudes Galiléenes*. 3 vols. Paris: Hermann, 1939.

Krois, J. 1987. *Cassirer: Symbolic Forms and History*. New Haven: Yale University Press.

Kuhn, Thomas. 1970. *The Structure of Scientific Revolutions*, 2nd ed. Chicago: University of Chicago Press. Original edition published 1962.

Meyerson, Émil. 1930. *Identity and Reality*. London: Allen & Unwin. Translation from 3rd ed., 1926 (identical to 2nd ed., 1912), of *Identité et réalité*. Paris: Alcan, 1908.

Russell, Bertrand. 1903. *The Principles of Mathematics*. London: Allen & Unwin.

6

From *SUBSTANCE AND FUNCTION*

Ernst Cassirer

The two fundamental moments on which the procedure of induction rests – attaining individual "facts" and connecting these facts into laws – trace back, as has been shown, to one and the same motive of thought. In both cases the task is to elevate constituents from the flux of experience that can serve as *constants* of theoretical construction. Establishing any individual, temporally delimited event already indicates this fundamental tendency – it already requires that we are able to grasp and hold fast certain complexes of conditions that remain the same in the intrinsically variable happening. The scientific explanation of any more developed groups of phenomena by the "isolation" and "superposition" of simpler basic relations then takes this task a step further. We now discover, in the ultimate empirical "laws of nature," *constants of a higher order*, as it were, which elevate themselves above the merely actual existence of individual facts fixed in determinate values of magnitudes. Nevertheless, the general procedure that is here everywhere effective only apparently achieves completion even in this result. The "fundamental laws" of natural science, which first appear to present the final "form" of all empirical happening, now serve in turn, from another point of view, only as the material for further consideration. Even these "second-level constants" are in turn dissolved into variables within the further process of cognition. They are valid only relative to a certain empirical sphere and thus must expect immediately to change their content as soon as this sphere is extended. So we here are confronted with an unceasing progression in which the fixed fundamental form of being and happening that we just believed ourselves to have established now appears to melt away in turn. All scientific thinking is governed and permeated by the demand for unchanging elements, whereas, on the other side, the empirically given continually thwarts this requirement again and again. We grasp permanent being only to lose it once again. What we call science does not appear, from this point of view, as an approximation to some "fixed and abiding" reality, but rather only as a continually renewed illusion, a phantasmagoria, in which a new image momentarily displaces all earlier ones, only immediately to disappear itself and be destroyed by another.

Ernst Cassirer, selections from chapter 5, §III, and chapter 7 of *Substanzbegriff und Funktionsbegriff*. Berlin: Bruno Cassirer, 1910. New English translation by Michael Friedman.

Precisely this comparison, however, indicates a necessary limitation on radical skepticism. Even the images in the individual stream of representations that are here compared to particular phases of science – no matter how many-colored and variously they may follow one another – still always possess a determinate inner form of connection, without which they cannot be conceived as contents of one and the same consciousness. They all stand at least in an ordered *temporal* connection, in a determinate relation of earlier and later; and this suffices to impress on them a common fundamental character throughout all variability of individual formation. No matter how much the individual elements may deviate from one another in their material constituents, they must nevertheless agree in those determinations on which the *serial form* in which they all participate is based. Even in the loosest and laxest succession of terms the preceding term is not destroyed by the appearance of the succeeding; rather, certain fundamental determinations on which the homogeneity and uniformity of the series rests are preserved. In the phases of *science* that follow one another this demand is fulfilled in the fullest and purest fashion. Every change that occurs in the system of scientific concepts simultaneously brings to light the enduring structural elements we must ascribe to this system, for only under the presupposition of just these elements can it be established and described. If we imagine the totality of experience as it exists at any given stage of cognition, this totality is never a mere aggregate of perceptual data but is classified internally in accordance with certain theoretical viewpoints and formed into a unity. It has already been shown that without such viewpoints no single assertion about facts, in particular, no single concrete *determination of magnitude*, would be possible. Thus, if we consider the totality of empirical knowledge at an arbitrary moment, we can present it in the form of a function that reproduces the characteristic relation in virtue of which we can think the individual terms in their mutual dependence.... If it now happens that some completely secure observation does not agree with the determinations that are calculated and expected on the basis of this most general formula, then this formula requires a correction – which, however, does not select indiscriminately any arbitrary element within it but rather stands under a certain *principle of methodological progression*. The reformulation occurs, as it were, "from the inside out": we first reformulate *less general* relations while preserving the more general, and we attempt in this way to restore the continuous agreement between theory and observation. The introduction of intermediate terms subject to new experiments results in the tendency of thought to preserve and "save" the more comprehensive laws, in that the deviant event is now derived as necessary *from these* [laws] *themselves* by the introduction of a new determining individual factor.

The preservation of a general "form" for the totality of experience is therefore clearly evident here; but it also occurs if the necessary revision of "facts" and purely empirical "rules" of their connection encroaches on the *principles* and *axioms* themselves. Even these axioms, such as those Newton places at the head of his mechanics, need not be taken as absolutely unchanging dogmas, but rather hold as temporarily simplest "hypotheses" on which we ground the unity of experience. We do not deviate from the content of these hypotheses so long as some less comprehensive variation, which thus concerns a *derivative* moment, may restore the agreement between theory and experience; but, if this way has proved impassable, then criticism is now referred back to the presuppositions themselves and the demand for their reformulation. Now it is the "functional form" itself that is transformed into another, but this transformation never means that the one fundamental form absolutely disappears while another arises absolutely new in its place. The new form is to contain the *answer* to *questions* that were outlined and formulated within the older one; and this already establishes a logical interconnection between the two and indicates a common forum for judgment to which both are subject. The change must leave a determinate stock of principles unaffected; for it is solely for the sake of securing this stock that it is undertaken in the first place, and this shows it its proper goal. Since we never compare the totality of hypotheses in themselves with the naked facts in themselves, but can only oppose *one* hypothetical system of principles to another, more comprehensive and radical [system], we require for this progressive comparison an ultimate constant *measure* in highest principles, which hold for all experience in general. What thought demands is the identity of this logical system of measure throughout all change in that which is measured. In this sense, the critical theory of experience actually aims to construct a *universal invariant theory of experience* and thereby to fulfill a demand towards which the character of the inductive procedure itself ever more clearly presses. The procedure of "transcendental philosophy" can in this respect be directly compared with geometry: just as the geometer emphasizes and investigates those relations in a certain figure that remain unchanged under certain transformations, so we here seek to ascertain those universal formal elements that are preserved in all change of the particular material empirical content. As such formal elements, which therefore cannot be lacking in any empirical judgment or any system of such judgments, we establish the "categories" of space and time, of magnitude and the functional dependence of magnitudes, etc. Just as we were not required there [in geometry] actually to perform and run through all possible changes of a certain kind in order to establish a conceptual relation as independent of these changes – where it rather sufficed once and for all to view the *direction* of change in order to decide, so the same holds here [in philosophy]. We establish that the *sense* of certain empirical functions is in principle not affected by a change in the material content in which they are clothed: as, e.g., the validity of a spatiotemporal dependence between elements of a happening in general, expressed in *universal causal laws*, remains unaffected by every change in *particular* causal propositions. The goal of critical analysis would be attained if it succeeded in establishing in this way what is ultimately common to all possible forms of scientific experience – in conceptually fixing those elements that are preserved in the progress from theory to theory because they are the conditions of each and every theory. This goal may never be completely attained at any given stage of cognition; nevertheless, it remains as a *demand* and determines a fixed direction in the continual unfolding and development of the system of experience itself.

The strictly limited objective meaning of the "a priori" appears clearly from this point of view. We can only call those ultimate *logical invariants* a priori which lie at the basis of every determination of a lawlike interconnection of nature in general. A cognition is called a priori, not because it lies in any sense *before* experience, but rather because, and insofar as, it is contained in every valid judgment about facts as a necessary *premise*. If we analyze such a judgment we find, along with that which it immediately contains from observational data and which varies from case to case, an abiding enduring store — a system of "arguments," as it were, for which the assertion in question presents a corresponding functional value. *This* fundamental relationship has never been seriously denied by any "empiricism," no matter how resolute . . . .

* * *

. . . The one reality can only be exhibited and defined as the ideal limit of the manifold changing theories; and the postulation of this limit is not optional but unavoidable, in that only through it is the *continuity of experience* produced. No *single* astronomical system may hold for us as the "true" cosmic order, the Copernican no more than the Ptolemaic, but only the totality of these systems as they continuously unfold in accordance with a determinate mode of interconnection. Thus, the instrumental character of scientific concepts and judgments is not here contested: these concepts are valid, not insofar as they picture a given fixed Being, but rather insofar as they contain an outline for possible postulates of unity which must be progressively confirmed in their application to empirical material. However, the very *instrument* that conduces to unity, and thus to the truth of what is thought, must be fixed and secure within itself. Were it not to possess in itself a certain stability, then no secure and enduring use of it would be possible — it would crumble to pieces at the first attempt and dissolve into nothing. We do not require the objectivity of absolute things, but rather the objective determinateness of the *path of experience itself.*

The real *content* of what is thought — that to which cognition penetrates — therefore corresponds to precisely the active *form* of thinking in general. The same task is here posed in the realms of both rational and empirical cognition. In the process of cognition itself there arises and solidifies the thought of a fundamental store of ideal relations, which as such remains identical to itself and unaffected by the contingent changing circumstances of psychological apprehension. The assertion of some such constancy is essential to every act of thought as such; the difference between the various levels of cognition depends only on the manner in which the *proof* of this assertion is produced. So long as we remain within the realm of purely logical and mathematical propositions we possess a totality of truths, joined together in a fixed fashion, which rest unchangingly within themselves. Each proposition is always what it is initially — it can be enlarged by others added to it but not reformulated in its own content. Purely empirical truth, however, appears in principle to be estranged from this determinacy: it is tomorrow different from what it was yesterday, and thus signifies only a fleeting halt in the change of representations which we grasp only immediately to give it up in turn. Nevertheless, the two motifs finally join together, despite all opposition, into a single type of knowledge. We can dissolve the absolutely enduring moment from the passing moment and oppose the two only in an abstraction, for the proper *concrete* task of cognition consists in making the enduring fruitful for the passing. The store of eternal truths becomes a means for gaining a footing in the very realm of change. The changing is considered as if it were enduring, in that we attempt to understand it as the result of universal theoretical laws. Although the difference between the two factors can never completely disappear, the entire movement of cognition consists in a continual equalization from one to the other. The changeableness of the empirical material is in no way manifested as only an obstacle, but equally as a positive advancement of knowledge. The opposition between mathematical theory and the totality of presently known observations would be insuperable if we were dealing with rigid, unalterable givens on both sides. The possibility of overcoming this conflict only opens up insofar as we become conscious of the conditioned nature of our empirical cognition and thereby of the formability, as it were, of the empirical material with which cognition operates. The harmony of what is given and what is demanded results from our investigating the given anew in light of our theoretical demands and thereby extending and deepening its concept. The enduring existence of the ideal forms no longer has a purely static meaning but equally and above all a dynamical one: it is not so much in *being* as in logical *use*. The ideal interconnections of which logic and mathematics speak are the abiding lines of direction orienting experience itself in its scientific formation. This *function* they continuously fulfill is their enduring and inalienable content, which maintains and confirms itself as identical in all changes of the contingent empirical material.

Identity and difference, constancy and change, therefore also appear from this perspective as interconnected logical moments. To decree an absolute essential opposition between them would be to destroy not only the concept of Being but also that of Thought . . . . If this correlative *double form of the concept* is misunderstood, then an insuperable gulf between cognition and phenomenal reality must immediately open up. We are then confronted again with the fundamental conception of Eleatic metaphysics, which has in fact experienced an interesting and significant revival in modern epistemological investigations. In order to understand reality with our mathematical-physical concepts, we must – it is now concluded – *annihilate* it first in its own nature, in its manifoldness and changeableness. Thought tolerates no inner inhomogeneity or alterability in the elements from which it constructs its form of Being. The manifold physical qualities of things therefore dissolve in the single concept of the ether, which is itself nothing but the hypostatization of empty featureless *space*; the living intuition of the temporal passage of events rigidifies for it into the persistence of ultimate absolute constants. To explain nature thus means to destroy it as nature, as a manifold and changing totality: the eternally homogeneous and immovable "sphere of Parmenides" constitutes the ultimate goal which all natural science unconsciously approaches. It is only thanks to the circumstance that reality resists the strivings of thought and finally sets thought insuperable limits that it maintains itself vis-à-vis the logical levelling of its content – so that in the perfection of *knowledge* Being itself does not disappear.[1] As paradoxical as this consequence may appear, it is precisely and rigorously derived from the assumed explanation of the intellect and its characteristic function. But this very explanation requires a limitation. The identity towards which thought progressively strives is not the identity of ultimate substantial things but the identity of functional orders and coordinations. These, however, do not exclude the moment of difference and change but succeed in determination only in and with it. Manifoldness as such is not destroyed, but we have only a manifold of another dimension: the mathematical manifold takes the place of the sensible manifold in scientific explanation. What thought demands is not the dissolution of plurality and changeableness as such but rather their mastery by the mathematical *continuity* of serial laws and serial forms. For establishing this continuity thought needs the perspective of difference just as much as the perspective of identity: this perspective, too, is not forced upon it simply from outside but is grounded in the character and the task of scientific "reason" itself. The analysis of given, sensible individual qualities into a plurality of elementary *motions*, whereby the reality of the "impression" becomes the reality of a "vibration," shows that the path of research does not merely consist in traveling from plurality to unity, from motion to rest. Rather, the opposite direction – the overcoming of the apparent constancy and simplicity of perceptual things – is no less justified and necessary. Only through this overcoming can we attain the new sense of identity and duration that lies at the basis of scientific laws. The complete concept of Thought thereby reestablishes the harmony of Being: the inexhaustibility of the task of science is no indication of its unsolvability in principle, but rather contains within itself the condition and stimulus for its ever more complete solution.

Note

1   See E. Meyerson, *Identité et réalité* (Paris: Alcan, 1908), esp. pp. 229ff.

HUSSERL

7

SCIENCE AS A TRIUMPH OF THE HUMAN SPIRIT AND SCIENCE IN CRISIS: HUSSERL AND THE FORTUNES OF REASON

Richard Tieszen

The reason for the failure of rational culture, as we said, lies not in the essence of rationalism itself but solely in its being rendered superficial, in its entanglement in "naturalism" and "objectivism."

–Husserl, "The Vienna Lecture," 1935

Husserl's later philosophy contains an extensive critique of the modern sciences. This critique of what Husserl calls the "positive," "naïve," or "objective" sciences has been very influential in Continental philosophy and, in particular, in the retreat from holding science and technology up as models for philosophy. Philosophers like Heidegger, Merleau-Ponty, Ricoeur, Habermas, and Derrida, along with many others on the Continent, have been influenced by this part of Husserl's thought. There is also in Husserl's work, however, a very grand view of the value and possibilities of science, provided that science is understood appropriately and in a broad sense as a theory of the many forms of reason and evidence. It cannot be adequately understood as a narrow, technical, naturalistic, and "one-sided" specialization of one sort or another. Indeed, in its broadest sense, science would coincide with the rigorous exercise of philosophical reason. In Husserl's writings we find (1) reflections on the nature of science as a whole, (2) studies of particular sciences, especially geometry, arithmetic, logic, and natural science, but also (3) a far-reaching analysis of how modern scientific culture has fallen away from its higher calling. I will discuss each of these aspects of Husserl's philosophical thinking. (For some earlier studies of these topics see e.g. Gurwitsch 1974; Heelan 1989; Ströker 1979, 1987, 1988.)

The critique of the modern sciences in Husserl's later work did not spring forth *ex nihilo*. There are seeds of it in his earliest work. The *Philosophie der Arithmetik* (*PA*), for example, is already premised on the view that only the philosopher can provide the kind of deeper reflection on arithmetic that the mathematical technician either cannot or will not provide. In this first book we are presented with an analysis of how this science is founded on the everyday experience of groups of sensory objects, how abstraction from this basis and a kind of formalization is required to get the science of arithmetic off the ground. Arithmetic is built up from "higher" cognitive activities that are founded on ordinary perceptual acts. The idea of analyzing "origins" of concepts, which Husserl inherited from some of his teachers (especially Brentano), is already at work here, and Husserl will continue to insist throughout his career on the value of this kind of analysis. Ideas of this type will figure into his later charge that the modern sciences have forgotten their origins in the everyday practices of the lifeworld.

By the time of the *Logical Investigations* (*LI*) Husserl tells us that philosophers have the right, indeed the duty, to critically examine the foundations of the sciences. Philosophy and, in particular, phenomenology, is needed to supplement the sciences. After 1907 or so phenomenology is portrayed in ever more detail as the transcendental and a priori "science" that investigates the essence of the sciences (as well as other domains of human experience). Scientists themselves are technicians who build up theories and methods without insight into the essence of these theories and methods or into the conditions for their possibility. Scientists are concerned more with practical results and mastery than essential insight. For just this reason, the sciences are in need of continual epistemological reflection and critique of a sort that only the philosopher can provide. Scientists are oriented toward their objects but not toward the scientific thinking itself in which the objects are given. It is the phenomenologist who will study the essential features of this thinking. Husserl pictures the work of the philosopher and the scientist as mutually complementary (Husserl 1973, "Prolegomena to Pure Logic," §71). The philosopher does what the scientist cannot and will not do if she is to practice her science, and the scientist does what the philosopher cannot do *qua* philosopher.

Phenomenological philosophy, it is thus argued, has the right to investigate the sciences and subject them to critical scrutiny. With the "transcendental" and "eidetic" turn of phenomenology we see, in particular, more and more development in Husserl's critique of naturalism, empiricism, naive "objectivism," positivism, and related positions. Psychologism and a kind of evolutionary biologism about logic, as forms of empiricism, are already subjected to criticism in the "Prolegomena to Pure Logic" of 1900. All of these views entail a kind of relativism about science and logic that Husserl rejects. Toward the end of his career he rejects historicism for similar reasons.

While many of the ideas that figure into Husserl's later critique of the sciences can be found in his earlier writings, the language of the "crisis" and the "danger" of the sciences does not emerge in his publications until the 1930s. If Husserl had his worries about various aspects of the sciences earlier on, he evidently did not yet see the sciences as being at a point of crisis. From the early thirties on, however, the idea of a crisis moves from the margins to the center of his thinking. There has been, Husserl declares, a "superficialization" of reason in the modern sciences. The problems with the modern sciences are being unjustly imputed to reason itself. Reason is under attack. The real problem, however, is that the modern sciences themselves can be practiced and are being practiced, in a manner of speaking, without reason. In the earlier writings the "superficialization" of reason in the positive sciences had not yet impressed Husserl so deeply. The abstractions, idealizations, formalizations, quantifications, technization, mechanization, specialization, and other activities required by the modern sciences had not yet been plumbed so deeply with respect to their potentially negative consequences for human existence. In developing his work in this direction Husserl wanted to speak to the more general crises of human existence in European culture that were all around him at the time. He was no doubt responding to some extent to other philosophers with whose work he was engaged, like that of his own student Martin Heidegger. Unlike Heidegger, however, he saw himself as trying to preserve the value and possibilities of human reason in dark times. Irrationalism and feelings of meaninglessness were breaking out everywhere, and science, supposedly the very embodiment of reason, seemed powerless to stop it. Husserl thought that there was a fundamental difference, however, between the existing positive sciences and "universal, responsible science."

I will open my treatment of Husserl's views on science in this chapter with a consideration of his pre-*Crisis* writings on the subject. In these writings we find general reflections on science, but also specific analyses of arithmetic, geometry, logic, and natural science. These analyses are of lasting interest in their own right, quite independently of his later worries about the dangers to civilization that arise when the positive sciences are not seen in their proper perspective. I will then make a transition to the specific arguments on and analyses of the crisis of the modern sciences.

1.   Arithmetic, Geometry, Logic, and the Science of All Possible Sciences

The grand view of the sciences mentioned a moment ago appears for the first time in Husserl's publications in the *LI* of 1900. By this time Husserl had already worked in mathematics, completing a doctoral thesis on the calculus of variations. He had published his *PA* and had continued to work on material related to his philosophy of arithmetic. He had also been working on the foundations of geometry. In the *LI*, however, the scope broadens considerably. The focus is now on logic, but on logic in the very broad sense of a theory of reason, a *Wissenschaftslehre*. Leibniz, Bolzano, and Lotze in particular are cited as the philosophers who have seen most deeply into logic as the theory of science or as the "science of all possible sciences." Husserl was to continue to develop these ideas on logic and science in all of his later publications, from the *LI* to the *Ideas Pertaining to Pure Phenomenology and a Phenomenological Philosophy* (in three "books" or parts) up through the *Formal and Transcendental Logic* and other writings. In what follows I will concentrate on the later, more mature views on the science of all possible sciences. Arithmetic, geometry, real analysis, physics, and all of the other sciences are subsumed under this conception. Indeed, what Husserl has in mind, at least as an ideal, is a unified and systematic conception of the sciences of the sort found in many of his rationalist predecessors.

Much of Husserl's work on science, like that of other rationalists, is focused on the exact sciences. Pure logic occupies a central role because it studies the most fundamental ideas that underlie all of the sciences. Logic, as the study of reason in a very broad sense, is a condition for the possibility of any science. All testing, invention, and discovery rest on regularities of form, and it is the science of logic that focuses on form. All of the sciences will require logic but they will then add to the basic forms and structures of logic, filling them out in various ways. As we will see, Husserl elaborates on these ideas quite extensively.

Husserl tells us that he is interested in the old and venerable idea of pure logic as *mathesis universalis*. One already finds this idea in philosophers like Descartes and Leibniz. *Mathesis universalis* includes the idea of a mathematics of judgments. Mathematicians had always been focused on their own objects in different domains of mathematical thought, but they had not focused on judgments themselves. Logicians, on the other hand, had focused on judgments and their logical properties and relations to one another. *Mathesis universalis* should, among other things, unite the two. After all, the objects of mathematics (and logic) are referred to and given through judgments. Indeed, all sciences are composed of sets of judgments and it is by virtue of these judgments that they refer to their own objects and states of affairs.

Now it is "pure" logic that we are to think of as the science of all possible sciences. Husserl quotes Kant's words on logic with approval: we do not augment but rather subvert the sciences if we allow their boundaries to run together (Husserl 1973, "Prolegomena to Pure Logic," §2). Logic is its own autonomous, a priori subject. In particular, it ought to be kept distinct from psychology, anthropology, biology, and other empirical sciences. The view that logic was concerned with mental processes and entities, as these would be studied in empirical psychology, was especially prevalent at the time. Frege railed against this psychologism about logic and by 1900 Husserl also subjected it to extensive criticism. Logic is a formal, deductive, and a priori discipline and as such it is distinct from all of the empirical sciences. Husserl's critique of psychologism in particular broadens to include any effort to found logic on an empirical science. To found logic on an empirical science would be to involve it in a relativism that is in fact foreign to it.

Pure logic is not about "real" mental processes or mental entities. Rather, it is concerned with "ideal" meanings. Husserl's critique of psychologistic and other empiricist views of logic is underwritten by an ontology that recognizes both real and ideal objects. Real objects, in the first instance, are objects to which temporal predicates apply, in the sense that the objects come into being and pass away in time. They have a temporal extension. Some real objects also have spatial extension. Spatial predicates apply to them. Thus, an ordinary physical object is one to which both spatial and temporal predicates apply. By way of contrast, a thought process has a temporal but not a spatial extension. Ideal objects and truths have neither temporal nor spatial extension. In his later writings Husserl shifts from saying that they are atemporal or nontemporal to saying that they are "omnitemporal." One does not somehow locate them outside of time altogether but holds that they exist at all possible times. As one would expect, ideal objects are also acausal. Thus, we seem to be presented with a rather platonic view of ideal objects, but this "platonism" becomes rather nuanced after Husserl takes his turn into transcendental idealism around 1908. In these later writings Husserl speaks of how the sense of every existent is constituted in the subjectivity of consciousness. This means that the sense of ideal objects and truths as transcendent, nonmental (and, hence, in a sense, as mind-independent), partially given, omnitemporal and acausal is itself constituted in the subjectivity of consciousness (see e.g. Husserl 1969, §94).

Pure logic is concerned with ideal meanings and it is judgments that express these ideal meanings. Judgments refer to objects or states of affairs by way of their meanings. The meanings expressed by judgments are ideal and the objects to which we are referred by judgments may themselves be either real or ideal. The objects of pure logic and pure mathematics are ideal. Judgments have a form and a "matter" (or content). Two judgments with different "matters," for example, may have the same form: "This house is red" and "This table is blue" both have the form "This S is P" (this is only a partial formalization). Among other things, pure logic will therefore need to track features of judgments and other types of expressions, the ideal meanings expressed by judgments, the objects referred to, and the form and matter in each case. Husserl's view of all of this, inaugurated in the *LI* and developed up through the *Formal and Transcendental Logic* (*FTL*), is mapped out in his stratification of "objective formal logic" into three levels. I will give a brief overview of the conditions for the possibility of science that are included in this three-fold stratification.

A fundamental condition for the possibility of any science is that it operate with judgments that are not only formally well-formed but that express unified or coherent meanings. Thus, at the bottom or first level of any possible science we have a priori or "universal" grammar. This is not to be a psychologistic or anthropological science. Husserl thinks there will be a priori rules of grammar for both the form and matter of expressions. The basic idea at this level is to lay out the rules and methods for determining whether or not a string of signs (words) is meaningful. This will require purely formal grammar, in order to distinguish formally well-formed strings of signs from strings that are not well-formed. Here Husserl seems to have in mind what we now think of in laying out the grammar of purely formal languages. Using one of his own examples, we could say that "This S is P" meets the relevant formal conditions whereas a string like "This is or" does not. Husserl distinguishes simple from complex meanings and he wants to know the rules for forming judgments as meaningful wholes from meaningful parts. Already in the Fourth of his *Logical Investigations* he had applied the theory of parts and wholes from his Third Investigation to this question and concluded that there must be a priori laws of grammar. What is needed, in addition to a purely formal grammar, is a theory of semantic categories to determine which substitutions of matter in the forms will give us something meaningful and which will give us mere nonsense (*Unsinn*). Consider, for example, the following two substitutions for "This S is P": "This tree is green" and "This careless is green." The former is a judgment. It expresses a unified meaning. The latter, however, is simply nonsense. Each part of it is meaningful but the whole formed from the parts is not. An expression from the wrong category has been substituted in place of "S." Roughly speaking, a categorial grammar would allow nominal material to be replaced by nominal material, adjectival material by adjectival material, relational material by relational material, and so on. Adjectival material could not be freely replaced by nominal material, and so on. Husserl allows that false, foolish, or silly judgments may result from the substitutions permitted but they would still be meaningful judgments. We might, for example, obtain a judgment like "This blue raven is green." This is not meaningless (nonsense) but it is false. Indeed, it is an a priori material absurdity or inconsistency, on the assumption that nothing that is blue all over can be green. On the other hand, "This careless is green" is nonsense.

Level two of "objective formal logic" is the "logic of noncontradiction" or "consistency logic." The strings of words admitted at level one express unified meanings. The next question we can ask about such (sets of) judgments is whether or not they are consistent. Here we also need to track formal and material versions of this question. A form like "There is an S and there is no S" is formally contradictory. It is a purely formal a priori absurdity. There may also be judgments that are not formally contradictory but that are instead materially inconsistent, like "This blue raven is green." The latter expression is meaningful but Husserl says it is countersensical (*Widersinn*). It is a material a priori absurdity. Thus, there are both formal and material a priori absurdities. A judgment like "This raven is orange," on the other hand, is meaningful and is not a material a priori absurdity. It is a synthetic *a posteriori* judgment, the kind of judgment whose truth or falsity requires sense experience.

Husserl in fact distinguishes synthetic a priori from synthetic a posteriori judgments along these lines (see Husserl 1973, Investigation III, and Husserl 1982, §10 and §16). Both require material concepts but the former depend on purely a priori relations among such concepts. For example, if an object x is blue all over then it cannot be green. If an object x is red then x is spatially extended, but the converse does not hold. If an object x is red then it is colored, but again the converse does not hold. Sense experience could not tell against such truths. What Husserl calls "regional ontologies" (see below) are made up of synthetic a priori truths for different domains of cognition. Analytic a priori truths, on the other hand, are purely formal. They depend on purely formal a priori laws. Husserl thus speaks of purely formal logic as "apophantic analytics." ("Apophansis" is the Greek term for judgment or proposition.) The ontological correlate of purely formal logic is formal ontology. In this connection, Husserl also distinguishes formalization from generalization or, if you like, formal from material abstraction. Abstracting the form from a material proposition is different from the kind of abstraction involved in moving from species to genus. The relationship of the form of a proposition to its material instances does not require concepts with content in the same way that these are required in setting out genus/species relationships. For example, a concept lower in a species/genus hierarchy (e.g., being red) implies all of the concepts (e.g., being extended) above it (but not the converse). It is for reasons of this type that there can be synthetic a priori judgments and regional ontologies.

At the second level of logic we should aim to distinguish judgments that are countersensical from those that are not. If a judgment is countersensical it is meaningful but it is not possible that there are objects corresponding to it. There could be no state of affairs that corresponded to it. On the other hand, if it is consistent then it is possible that there are corresponding objects or states of affairs.

Husserl develops at this level the idea of purely formal "apophantic" logic(s). Apophantic logic would just be the logic of forms of judgments, broadly conceived. It would presumably include forms of any kinds of meaningful judgments. One might speak of different logics here but Husserl seems to have in mind some unified conception of these in one overarching logic. Consistent apophantic logic(s) would stand as a condition for the possibility of any science since we cannot have sciences that contain formally contradictory judgments.

Apophantic logic is conceived of in terms of axiomatic formal systems. Husserl sometimes refers to this level of consistency logic as "consequence logic." We view logic as an axiomatic formal system in which one derives consequences of the axioms on the basis of formal rules of inference. The idea that judgments express meanings by virtue of which they refer to objects or states of affairs is mirrored at this second level of logic, in the following sense: if a set of judgments (including a formal system) is consistent then it is possible that there are objects or states of affairs corresponding to the set of judgments. Husserl calls the "ontological correlate" of a consistent formal axiomatic system a manifold. Since it is the correlate of the mere forms of judgments it constitutes a purely formal ontology. In apophantic logic we are focused on the forms of judgments and the features of judgments as expressions of meanings, but in formal ontology we are focused on the possible objects or states of affairs referred to by such judgments, solely with respect to their form. In formal ontology we are concerned with the most general and formal notions of object, state of affairs, property, relation, whole, part, number, set, and so on. In the particular sciences these notions would be "materialized" or specified in different ways. Geologists, for example, would speak about particular kinds of objects (e.g., rocks, mountains) and particular kinds of parts, properties, and relations of these objects. Physicists would specify their objects and properties differently from geologists or biologists but would still be using the notions of object, part, property, and relation. In the different sciences we would have different regional ontologies. (For studies of the natural sciences in particular see Hardy and Embree 1993; Heelan 1983; Kockelmans 1993 and Kockelmans & Kisiel 1970.)

We can ask not only whether formal systems are consistent but also whether they are complete. Husserl introduced the notion of a "definite" formal system early on in this thinking and it seems to be the notion of a formal system that is both consistent and complete. He frequently mentions how he and Hilbert arrived at such a notion independently of one another (see e.g. Husserl 1982, §72). Husserl also introduces the notion of a definite manifold as the ontological correlate of a definite formal system. It is the definite, purely formal "world" that corresponds to a definite axiomatic formal system.

At the third level we have what Husserl calls "truth logic" (Wahrheitslogic). At this level we are concerned with the truth or falsity of judgments. If judgments are consistent then it is possible that there are objects corresponding to the judgments and it is possible that the judgments are true. For judgments to be considered true, however, it appears that we need more than mere consistency. Truth or falsity is determined by intuition, or by what Husserl calls "meaning-fulfillment." Intuition takes place in sequences of acts carried out through time and it provides evidence that there are objects corresponding to judgments or, on the other hand, it can show us how the intentions expressed by our judgments are frustrated. There are different degrees and types of evidence: clear and distinct, adequate, apodictic. If we have evidence for a state of affairs then we can return again and again to precisely the same state of affairs. Moreover, the idea of intersubjective confirmation is built into Husserl's conception of evidence.

Thus, Husserl pictures level 2 as being concerned with the senses or meanings of judgments apart from the "truth" or "falsity" of the judgments (see Husserl 1969, ch. 5). It is concerned with mere meaning-intentions and their forms. But when we are concerned with the truth or falsity of judgments, with meaning-fulfillment, we are at the level of truth-logic. There is, evidently, a notion of "truth" (and "exist-ence") operative at level 2 but, relative to level 3, it is merely a notion of possible truth (and possible being). At one point, Husserl says it is a notion of truth that simply means "derivable from the axioms," given whatever set of axioms we start with. At level 3, however, we need to include structures of evidence and meaning-fulfillment.

It thus appears that at level 3 we need not only judgments and their consistency but also some kind of fulfillment procedure for making the object or state of affairs referred to by the judgment present. We need a judgment plus an intuitive ("verifica-tion" or "construction") process for finding the object or state of affairs. Consistency alone does not guarantee this. Husserl seems to hold that consistency does not by itself imply existence but only possible existence. This notion of possible existence may suffice, in particular, in purely formal mathematics. It does not, however, imply that there is an intuition of objects or states of affairs referred to by mathematical judgments (see below). Something more is needed. In previous work I have discussed how we might understand this idea of judgment + fulfillment procedure in the case of natural numbers and finite sets (see Tieszen 1989).

If we now consider the relationships between the levels in this broad conception of logic as the science of all possible sciences, we see that each level, starting from the lowest, is a condition for the possibility of the next level. In this account of the possibility of science the higher levels presuppose the lower levels. We could not have a true scientific theory unless we have a consistent scientific theory, and consist-ency itself presupposes that we have meaningful expressions, but meaningfulness does not itself guarantee consistency and consistency does not itself guarantee (intuition of) truth.

The relationship, in fact, is somewhat more nuanced. It appears that level 2, with its emphasis on consistency, can serve where intuition fails. As long as a set of judgments is consistent we can feel secure in operating with the judgments and they can have an important role to play in science, even if we can have no intuitions corresponding to the judgments. Husserl emphasizes this in connection with theories of various kinds of numbers (e.g., negative, real, complex) and with n-dimensional Euclidean and non-Euclidean geometries where $n > 3$. He in fact cites Riemann, Grassmann, and others as central influences on his own conception of manifold theory. Thus, consider Euclidean geometry of two or three dimensions. This geom-etry existed for many years as a "material" eidetic (a priori) science before it was formalized. The objects of Euclidean geometry were taken to be exact and ideal objects. They are idealizations of objects given to us in everyday sensory perception. The objects of everyday sensory intuition are inexact. Husserl says that the essences of such sensory objects are imprecise or "morphological" (see Husserl 1982, §74). An idealization of the shapes given to us in everyday perception takes place through the conception of greater and greater perfectings of the shapes until we form the concep-tions of ideal circles, triangles, etc. Euclidean geometry can thus be viewed as an idealization of structures found in everyday sensory perception. Once Euclidean geometry is *formalized* we obtain for the first time a Euclidean manifold of two or three dimensions. The foundation is then in place for generalizing and constructing n-dimensional Euclidean or even non-Euclidean manifolds. These latter constructions arise out of free mathematical imagination by playing with and making variations on the Euclidean manifold. In this manner Husserl wants to account for the origin and constitution of different "spaces" and "geometries," starting from the space of everyday perceptual intuition. He wants to allow complete freedom to devise new such constructions and to regard the resulting mathematics as perfectly legitimate, subject only to maintaining consistency.

Now as soon as we go beyond three-dimensional geometry it does indeed seem to be the case that our spatial intuitions begin to fail. But what if we could show that the resulting geometries were indeed consistent? Then there would be nothing illogical or incoherent about our conceptions (intentions) in these cases. The existence of relative consistency proofs of different geometries by way of various kinds of models seems to lie behind Husserl's conception of manifolds and formal theories. If we generalize this perspective to any part of mathematics or logic, or to science as a whole, then we seem to have some conception of what is supposed to be covered by level 2 of the theory of science.

Once we obtain new manifolds and formal systems through imagination we can reinterpret them any way we like. We are free to reinstantiate them with "matter" in a variety of ways. Although Husserl does not discuss the topic, one can say that this is just what happened in the case of the application of non-Euclidean manifolds in relativistic physics, where we see a surprising application of what had previously appeared to be merely conceptual or symbolic mathematics (see Becker 1923 and Weyl 1918).

Of course we do not always know that a given set of judgments is consistent. In some cases it may be a very long time before we know, if ever. We also know, based on the work of Gödel, that there are certain limitations on providing consistency and completeness proofs for mathematical theories. It seems that Husserl already held that level 3 considerations can help us in science even if we do not get an answer to our question at level 2 whether a set of judgments is consistent. That is, we have an intuitive basis for some parts of mathematics that can provide some security or reliability. We can sometimes appeal to (possible) intuition of mathematical objects for our evidence. Husserl's studies of arithmetic and Euclidean geometry show that he thought these sciences had a foundation in intuition. In effect, we start with shapes given to us in everyday perception or with our awareness of groups of everyday perceptual objects and trace out the cognitive acts of abstraction, idealization, reflection, and formalization that would lead to the Euclidean manifold or to the natural-number manifold.

In his earliest studies of arithmetic Husserl held that most of our arithmetic knowledge was purely symbolic. Only a small part of it was actually intuitive. In fact, we cannot make number determinations much beyond five in single acts of intuition. Everything else amounts to purely symbolic thinking that does duty for the objects and laws that we cannot make present to ourselves in intuition. At this point in his career Husserl placed great emphasis on the nature and use of formal symbol systems in mathematics. These early studies abound in investigations of purely algorithmic methods, formal systems, and manifold theory (see e.g. Husserl 1979, 1983). By the time of the *LI* Husserl's conception of intuition broadened and was explicated in terms of the distinction between mere meaning-intention and meaning-fulfillment. He also now distinguished the "static" from the "dynamic" fulfillment of a meaning-intention. On this view there could be at least partial fulfillment of intentions to larger numbers and one can begin to speak about what is in principle fulfillable and what is not. Natural numbers themselves were now clearly taken to be ideal objects. In any case, the distinction between the intuitive and the merely symbolic was always present in Husserl's philosophy of science. Some parts of mathematics, logic, and even natural science would evidently always have to be considered merely symbolic or conceptual. (We might keep the "symbolic" and the "conceptual" separate here, reserving the former term for the merely algorithmic "games-meaning" of signs and the latter term for the full meaning-intention of signs.)

Within this architectonic, Husserl, as we said, distinguishes formal ontology from various regional ontologies. Regional ontologies are a priori ontologies of particular regions of being. When forms are filled in with "material" or content we are considering specific kinds of objects, properties, relations, parts, wholes, sets, and so on. All of the different specific scientific theories would have different regional ontologies. Transcendental phenomenology itself is a regional science. It is the science of consciousness with all of its various structures and characteristics. Each science has its own objects, with their own properties, etc. Husserl is generally an antireductionist about the specific sciences. Part of the reason for this derives from his view that judgments in the various sciences express their own meaning-intentions and that it is by virtue of these meaning-intentions that we are directed toward objects. Reductionist schemes may very well fail to respect differences in meaning-intention, along with the different implications and purposes reflected in these. For certain scientific purposes they might, if feasible, be useful. Eliminative reductionism in particular might diminish the many aspects or perspectives under which the world could be viewed and investigated. Eliminativism could hinder scientific work and perhaps even be dangerous, blinding us to important phenomena.

One very broad distinction among the sciences, as has been suggested, is that some are eidetic or a priori and some are empirical. Formal and regional ontologies are a priori, dealing respectively with the formal (analytic) and material (synthetic) a priori. Husserl also says that among the eidetic sciences some are exact and some are inexact. Mathematics and logic are exact. They trade in exact essences and exact objects. Among the inexact eidetic sciences Husserl counts transcendental phenomenology. It is, Husserl says, a "descriptive" science that deals with inexact or "morphological" essences, like the essence "consciousness." Unlike empirical psychology, however, it is not a causal-explanatory science. It does not seek causal generalizations. Underneath this rationalistic superstructure of the eidetic sciences we have the empirical sciences, the sciences that depend essentially on the evidence of sense experience and induction. Husserl's rationalism comes across quite clearly in many of his writings. The empirical sciences of course contain various imprecise, vague, or inexact components. They deal in probabilities and contingencies, not necessities. While mathematics and logic set the standard for what is clear, distinct, and precise, the empirical sciences deal with indistinct or vague typifications of or generalizations from sense experience. The empirical sciences depend on the inexact essences associated with sensory objects. Even though the empirical sciences may be vague in this way and trade on various contingencies they will presuppose various kinds of essences and essential truths. It is this latter "material" a priori domain that will form in each case the subject matter of a regional ontology (which then has lying behind it the purely formal level). To work at developing a regional ontology is not to engage in the work of the empirical sciences. Rather, it is to explore and map out the essences and essential laws pertaining to a given domain. This, in effect, gives us the a priori conditions for the possibility of the empirical science of the domain. The phenomenologist, for example, is to do this for the domain of human consciousness. As such, she is not simply doing empirical psychology. Husserl returns again and again in his work to the topic of how phenomenology is distinct from empirical psychology.

2. Phenomenological Philosophy and the Foundation of the Sciences

The science of all possible sciences focuses on the expression of meanings in judgments, and the reference to objects or states-of-affairs by way of these meanings. At the different levels and in different domains of inquiry we are to distinguish judgments from nonjudgments, consistent judgments from inconsistent judgments, and fulfilled from frustrated or unfulfilled judgments. The scientist may deal with her judgments and objects but she *abstracts* from the role of the human subject in the sciences. She does not consider the scientific thinking itself in which her objects are given. We have the following picture:

scientific judgment J, expresses a meaning

↓   by which it refers to

[object or state of affairs]

This is really only part of a larger whole, however, that includes human subjects as those who are doing the judging, thinking, remembering, and so on. The whole picture is more like this:

Subject S thinks that J, where J an expresses a meaning

↓ by which the subject refers to

[object or state of affairs]

The positive or "objective" sciences deal with only part of a larger whole, and a dependent part at that. Husserl calls the broad conception of logic or of science in which we do not omit the role of the human subject "transcendental logic." As in Kant's conception, it is a "logic" or "science" in which we do not abstract from the possible experience of human beings. The expression "transcendental logic" is in fact often used as another expression for transcendental phenomenology itself. Transcendental phenomenology is to be the science of the subjective and intersubjective side of experience, of consciousness and its object-directedness in any domain of conscious experience. It is to be the science of the essential features and structures of consciousness that provides the philosophical foundation of the sciences. If we are really interested in the conditions for the possibility of science, we cannot forget that, at bottom, it is the human subject who makes science possible. It is human subjects who bring about or constitute the sciences over time and who hand the sciences down from generation to generation.

On Husserl's analysis there are different types and levels of consciousness. Science is built up from the lifeworld experience of human subjects on the basis of acts of abstraction, idealization, reflection, formalization, and so on. The most basic, founding experiences are the everyday lifeworld activities, practices, and perceptions of people. The lifeworld, Husserl says, is "the intuitive surrounding world, pregiven as existing for all in common" (Husserl 1970a, §§33–4). All of our activities, including our loftiest sciences, presuppose the everyday practical and situational truths of the lifeworld. Our praxis and our prescientific knowledge in the lifeworld play a constant role in all of our activities. The lifeworld was there for us before science, and even now human beings do not always have scientific interests. We have the intuited, everyday world that is prior to theory, and then the various theories that are built up from this basis. Science thus presupposes the lifeworld as its starting point and cannot therefore replace this world or substitute something else for it. Human subjects are the meaning-givers and interpreters who produce the sciences from this basis and who can choose to be responsible about them or not.

Various fields of "objective" science must therefore be correlated with the subjective if we are to do justice to the sciences. To see things whole we must deal with both objectivity and subjectivity. Otherwise science is naive and one-sided. We would have only the positive, objective sciences in which the human subject is not remembered in the scientist's work. The human subject we are talking about here, moreover, should not be the reduced, desiccated subject that is presented to us by the *objective* sciences. It should not be the subject as interpreted through these sciences, which are themselves already founded on more basic forms of human experience. As we said, the positive "objective" sciences are not foundational but are themselves founded on our lifeworld experience. In this manner, Husserl turns the table on those who think of the sciences and technology as providing the fundamental ways of knowing, understanding, and being, or who would value the positive sciences and technology above all else. It is this turn of thought, this critique of scientism, that has resonated with so many Continental philosophers for so long, from existentialists up through various postmodern philosophers.

Husserl was arguing long before philosophers like Thomas Nagel that the positive sciences cannot in principle understand or do justice to the human subject (see e.g. "The Vienna Lecture" in Husserl 1970a). They are "objective" sciences that make of the human subject a kind of scientific object, an objectified subject. Husserl's view thus has many implications for psychology and the social or human sciences, as can be seen in the secondary literature on the subject (see e.g. Natanson 1973).

In order to obtain the objective sciences one *abstracts* away from many features of experience. To abstract and to idealize is automatically to simplify, leaving behind some of the complexity and richness of concrete experience. Of course in the sciences such simplifications and idealizations help us to get a grip on things and make the work manageable. The simplification serves a purpose and might sometimes be put to good use but it should not be forgotten that it does not give us the complete picture. In the objective sciences one abstracts, for example, from the qualitative features of experience to focus on quantitative features. Calculation and calculational techniques come to the fore. Primary qualities of phenomena are highlighted and secondary qualities are marginalized. As part of this abstraction one tends to focus on formal or structural features of experience, as distinct from contentual features. This shift to form or structure also involves a shift away from the full meaning of experience. Empirical, third-person observation is prized above all else while eidetic, consciousness-oriented observation is ignored or dogmatically held to be impossible. Add to this the specialized character of each of the positive sciences and we can see that, at best, they can treat of only part of the human subject, the outer shell as it were. The subject with his or her "lived body" is completely overlooked (see e.g. Husserl 1989 on the "lived body"). Indeed, much of what is so important about human subjectivity and intersubjectivity will in principle be missed by the methods of the positive sciences. It is in the very nature of the methods of the objective sciences that they cannot get at human subjectivity with all of its first-person qualitative, contentual, and meaning-giving features.

For Husserl it does not follow that there cannot be a "science" of human subjectivity. Transcendental, eidetic phenomenology is supposed to be (or to become) just such a science. It seeks essential, not accidental or contingent, features of human subjectivity. It does not itself seek empirical facts about human subjectivity but is concerned with necessary conditions for the possibility of such facts. It seems undeniable, for example, that human consciousness exhibits intentionality. How could we deny that a belief is always a belief about something, that it is always object-directed, unless we were already in the grip of an objective (founded) science that tells us there can be no such thing as intentionality (or even as belief)? One of the dangers of the positive sciences is just that they can blind us to phenomena that would otherwise be obvious. They can create prejudices of various types. One unconsciously accepts various background assumptions that are passed along in the positive sciences without subjecting them to critical scrutiny. It is phenomenology that should uncover such assumptions and critically analyze them.

If we can say that human consciousness exhibits intentionality then we can also say that the human conscious subject is directed toward certain objects or states of affairs. We can distinguish different modes of consciousness (e.g., believing, knowing, remembering, desiring) and investigate and compare these modes with respect to their essential features. In further exploring features of consciousness we can see that acts of consciousness involve a kind of meaning-giving, that they are always perspectival, that they will have inner and outer horizons and certain Gestalt qualities, that there is always a time structure to consciousness (including a retention–protention structure and a secondary memory structure), and that the human subject is an embodied subject with particular forms of bodily intentionality. We see that perceptual observation is underdetermined by sensation (or "hyletic data"), that there is a kind of hylomorphism in perception, and so on. Phenomenology is to investigate all of this material in detail. In connection with the positive sciences in particular it will investigate the types of consciousness in which science has its origins. In such a genetic, constitutional phenomenology one investigates the various founding and founded acts of consciousness that make science possible.

At the most fundamental level of experience transcendental phenomenology is to investigate the lifeworld and its structures (Husserl 1970a, §36). We can say that the world is, prescientifically, already a spatiotemporal world in which there are bodies with particular shapes and material qualities (e.g., color, warmth, hardness). There is no question here of ideal mathematical points, or "pure" straight lines, of the exact-ness belonging to geometry, and the like. It is a world in which certain shapes would have stood out in connection with practical needs. It would have been a world in which various relations between objects would have been noticed, and in which there would have been some awareness of causality. Primitive forms of measurement would have emerged, and so on. In works like *The Crisis* Husserl therefore begins to plumb the formal and most general features of the lifeworld. This project was already underway in earlier manuscripts in which Husserl investigated the constitution of material nature and the constitution of animal nature (see Husserl 1989).

The investigation of the meaning and the manner of being of the lifeworld can itself be "scientific" in a broad sense. This will be a notion of "science" that does not systematically exclude our subjectivity as human beings situated in history. There is a difference, Husserl says, between "objective" science and science in general (Husserl 1970a, §34). Philosophy itself, as rational and critical investigation, may have as its goal the ideal of becoming scientific (Husserl 1965). This ideal is very much alive even now, except that it has been transmuted into the naturalism that dominates our age. Phenomenological philosophy, however, should aim to be a science that encompasses both objectivity and subjectivity and their relation to one another. Given this view of phenomenological philosophy as science, along with its relation to the other sciences, we see that, in the end, it is transcendental phenomenology itself that is to be the science of all possible sciences.

3.   The Crisis of the Modern Sciences

The "crisis" of the modern sciences has resulted from the attempt of these sciences to be "merely factual" and to place increasing value on technization, mathematization, formalization, and specialization at the expense of other aspects of experience. There has been, Husserl argues, a positivistic reduction of science to mere factual science. The crisis of science, in this sense, is its loss of meaning for life. What is the meaning of science for human existence? The modern sciences seem to exclude precisely these kinds of questions. The physical sciences abstract from everything subjective and have nothing to say in response to such questions. The "human sciences," on the other hand, are busy trying to exclude all valuative positions and are attempting to be merely factual. They also fail to speak to such questions. It is for just these reasons that there is hostility in some quarters toward the modern sciences. Such human questions, however, were not always banned from the realm of science, especially if we think of the earlier aspirations of philosophical reason to become science. Positiv-ism, however, "decapitates philosophy" (Husserl 1970a, §3). Various forms of skepti-cism about reason have set in. Skepticism insists on the validity of the factually experienced world and finds in the world nothing of reason or its ideas. Reason itself becomes more and more enigmatic under these conditions. In attempting to combat such trends one need not resort to naive and even absurd forms of rationalism. Rather, one needs to find the genuine sense of rationalism.

The modern sciences are made possible by mathematics and, in particular, by formal mathematics of the type that we see in algebra, analytic geometry, and so on. In *The*

Crisis and related writings Husserl focuses on Galileo's mathematization of nature and the role of "pure geometry" in making modern natural science possible. With Galileo, nature becomes a mathematical manifold. What is involved in this mathematization of nature? Husserl discusses this at some length. It depends on the rise of pure geometry with its idealizations and its exactness. Galileo inherits pure geometry and with it he begins to mathematize nature. In interpreting nature through pure geometry we begin to view it in a different manner. It is no longer the "nature" of prescientific, lifeworld experience. Various idealizations of nature are involved in seeing nature in terms of pure geometry. In the process we leave out or abstract from some of the aspects of the lifeworld experience of nature. The "plenum" of original intuitive experience, however, is not fully mathematizable. Husserl thus says that the mathematization of nature is an achievement that is "decisive for life" (Husserl 1970a, §9f.).

With mathematics one has at hand the various formulae that are used in scientific method. It is understandable, Husserl says, that some people were misled into taking these formulae and their "formula-meaning" for the true being of nature itself. This "formula-meaning," however, constitutes a kind of superficialization of meaning that unavoidably accompanies the technical development and practice of method. With the arithmetization of geometry there is already a kind of emptying of its meaning. The spatiotemporal idealities of geometric thinking are transformed into numerical configurations or algebraic structures. In algebraic calculation one lets the geometric signification recede into the background. Indeed, one drops it altogether. One calculates, remembering only at the end of the calculation that the numerals signify multitudes. This process or method of transformation eventually leads to completely universal "formalization" and to the kind of formalism we see applied in so many places in the modern sciences.

Various misunderstandings arise from the lack of clarity about the meaning of mathematization (Husserl 1970a, §9i). For example, one holds to the merely subjective character of specific sense qualities. All concrete phenomena of sensibly intuited nature come to be viewed as "merely subjective." If the intuited world of our life is merely subjective then all the truths of pre- and extra-scientific life that have nothing to do with its factual being are deprived of value. Nature, in its "true" being, is taken to be mathematical. The obscurity is strengthened and transformed all the more with the development and constant application of pure formal mathematics. "Space" and the purely formally defined "Euclidean manifold" are confused. The true axiom is confused with the "inauthentic" axiom (of manifold theory). In the theory of manifolds, however, the term "axiom" does not signify judgments or propositions but forms of propositions, where these forms are to be combined without contradiction.

In the same vein, Husserl speaks of emptying the meaning of mathematical natural science through "technization." Through calculating techniques we can become involved in the mere art of achieving results the genuine sense and truth of which can be attained only by concrete intuitive thinking actually directed at the subject matter itself (Husserl 1970a, §9g). Only the modes of thinking and the type of clarity that are indispensable for technique as such are in action in calculation. One operates with symbols according to the rules of the game (as in the "games meaning" Husserl had discussed in *LI*, Investigation I, §20). Here the original thinking that genuinely gives meaning to this technical process and gives truth to the correct results is excluded. In this manner it is also excluded in the formal theory of manifolds itself. The process whereby material mathematics is put into such formal logical form is perfectly legitimate. Indeed, it is necessary. Technization is also necessary, even though it sometimes completely loses itself in merely technical thinking. All of this must, however, be a method that is practiced in a fully conscious way. Care must be taken to avoid dangerous shifts of meaning by keeping in mind always the original bestowal of meaning upon the method, through which it has the sense of achieving knowledge about the world. Even more, it must be freed of the character of an unquestioned tradition whose meaning has been obscured in certain ways. This technization in which one operates with symbolic concepts often admits of mechanization. All of this leads to a transformation of our experience and thought. Natural science undergoes a far-reaching transformation and there is a covering over of its meaning.

As an example of some of Husserl's ideas here, suppose I give you the formula:

$$P(A_1/B) = \frac{P(A_1) \times P(B/A_1)}{(P(A_1) \times P(B/A_1)) + (P(A_2) \times P(B/A_2))}$$

and ask you to compute the values of $P(A_1/B)$ in case after case where the values of the terms on the right-hand side of the equation are supplied and they always fall between 1 and 0. It is obviously possible to carry out this kind of input–computation–output procedure without knowing anything about what the formula is or means, what the numbers represent, what the origins of the formula are, what the purpose of this task is, and so on. Of course this might be only a small subroutine of a much more extensive routine, and there is a very definite sense in which one needn't know what any of it is about. Indeed, the entire process could be computerized. There is no need to reflect on or to understand the meaning of the formula. One might be carrying out computations about the space shuttle, or nuclear weapons, or economic aspects of homelessness. One could be quite blind about all of this and still carry out all of the calculations. Indeed, the goal may now simply be to solve such computational problems quite independently of what the computations are about. All of one's energies may go into this end and it is possible to become quite submerged in this kind of technical work, as happens in the case of many engineering problems. There is of course conscious directedness in all of this, but it is quite different from being directed toward the objects to which the formula is being applied. The objects are now the signs that are being manipulated in the calculation. In this manner there can be a complete displacement of concern (see Tieszen 1997).

The skills and abilities associated with calculation and other forms of technical work will obviously be valued all the more in this kind of environment. This in turn fosters specialization in and professionalization of domains of scientific work. Technical knowledge and pragmatic success take center stage, while other forms of knowledge and understanding will tend to be marginalized. A pragmatic instrumentalism is the natural outcome of this shift in values and goals. This might be a pragmatic instrumentalism in which efficiency and the control and domination of nature are first principles, an instrumentalism that tends to view nature and everything in it as resources or "input" for scientific/technological processes.

There is clearly potential here for a kind of alienation from reality, a distancing from the basic foundations of knowledge and understanding. This is an alienation made even worse by forgetting about these origins. If we think of Husserl's slogan "Back to the things themselves," then it is an alienation from "the things themselves." One might say that the possibility of such alienation attends the possibility of inauthentic ways of thinking and being, as opposed to more authentic ways. All of this arises with the abstractions and idealizations that make science possible. It arises from taking something as an independent whole that is really only a dependent part of a larger whole. It is to recede from a more holistic perspective.

In the modern natural sciences there is, Husserl thus says, a surreptitious substitution of the mathematically structured world of idealities for the only real world, the one that is actually given through perception: the everyday lifeworld. This substitution already occurred as early as Galileo, and it was subsequently passed down through the generations (Husserl 1970a, §9h). What has happened is that the life-world, which is the foundation of the meaning of natural science, has been forgotten. A type of naiveté has developed. Galileo is at once both a discovering and a conceal-ing genius.

While the techniques and methods of the sciences are handed down through the generations, their true meanings, as we have said, are not necessarily handed down with them. It is the business of the philosopher and phenomenologist to inquire back into the original meanings through an eidetic analysis of the sedimentation involved. There is a "historical meaning" associated with the formations of the sciences but, as mentioned earlier, Husserl is not interested primarily in empirical history. He is interested in finding the a priori unity that runs through all of the different phases of the historical becoming and the teleology of philosophy and the sciences (Husserl 1970a, §15). Husserl says that it is a ruling dogma that there is a separation in principle between epistemological elucidation and historical explanation, or between epistemological and genetic origin. Epistemology cannot, however, be separated in this way from genetic analysis. To know something is to be aware of its historicity, if only implicitly. Every effort at explication and clarification is nothing other than a kind of historical disclosure. The whole of the cultural present implies the whole of the cultural past in an undetermined but structurally determined generality. It implies a continuity of pasts that imply one another. This whole continuity is a unity of traditionalization up to the present. Here Husserl speaks about unity across difference on a global scale, not just in the case of the unities through difference that arise for us in our own personal cognitive lives. Of course in the latter case too there is a temporality and a "history." Anything historical has an inner structure of meaning. There is an immense structural a priori to history.

Historicism is the view that there could be no such historical a priori, no supertem-poral validity. It claims that every people has its own world. Every people has its own logic. Husserl responds by pointing out some of the background assumptions that are necessary for factual historical investigation to occur at all (see Husserl 1970a, appen-dix VI). These are things that we must know or assume before we can even get started with any factual historical investigation. In spite of all the indeterminacy in the horizon of "history," it is through this concept or intention that we make our histor-ical investigations. This is a presupposition of all determinability. But what kind of method can we use to make apparent to ourselves the universal and a priori features? We need to use the method of free variation in imagination in which we run through the conceivable possibilities for the lifeworld. In this way we remove all bonds to the factually valid historical world. We determine what is necessary and invariant through all of the contingencies and variations.

In the case of Euclidean geometry, for example, Husserl says that only if the necessary and most general content (invariant through all conceivable variation) of the spatiotemporal sphere of shapes is taken into account can an ideal construction arise that can be understood for all future time and thus be capable of being handed down and reproduced with an identical intersubjective meaning. Were the thinking scientist to introduce something time-bound into her thinking, something bound to what is merely factual about her present, her construction would likewise have a merely time-bound validity or meaning. This meaning would be understandable only to those who shared the same merely factual presuppositions of understanding. Geometry as we know it would therefore not be possible.

Husserl thus argues against a relativism that would deny the ideal, omnitemporal character of sciences like geometry, arithmetic, and logic, but he also argues against any absolutism that would cut off the truths of such sciences from their relation to human subjects. The objective and the subjective can only be properly understood in relation to one another, each conditioned by the other. One can say the same thing about the real and the ideal.

The things taken for granted in the positive sciences should, according to phenomenological philosophy, be viewed as "prejudices." As ideas and methods become sedimented over the years they become, quite literally, part of a prejudged mass of conditions and assumptions. In this manner various obscurities arise out of a sedimentation of tradition. One should subject such prejudgments again and again to critical, rational judgment. The genetic investigation of phenomenology is thus meant to allow us to become aware of such prejudices and to enable us to free ourselves of various presuppositions. It is therefore supposed to be the deepest kind of self-reflection aimed at self-understanding. Husserl sets up such presuppositionlessness as an infinite, regulative idea that we may never reach but which we should nonetheless never abandon. Husserlian hermeneutics would thus be rather different in some ways from that, for example, of Heidegger or Gadamer.

4.   Conclusion

These kinds of reflections lead to a host of issues that have distinguished continental thinking about science from analytic philosophy of science. Much of analytic philosophy of science has tended to focus on questions that are more internal to the scientific enterprise. There have been periods during which some of the important questions raised by Husserl's critique of the sciences would simply never have occurred to those working in analytic philosophy of science. Consider those circumstances, for example, in which the historical dimensions of science have been overlooked or devalued or in which the hypotheses and explanatory schemes of the positive sciences have been viewed as nonhermeneutical or value-free. Husserl's work, by way of contrast, opens onto many issues involving the broader social, political, historical, and ethical dimensions of science and technology.

Husserl can be read as providing a corrective to scientism, to the view that it is only the positive sciences that supply knowledge, understanding, and truth. He is not arguing that the positive sciences do not have their place or do not have any value. It is rather that we need to see them in their proper perspective. We need the correct balance. If scientism is extreme in one direction, then a view that completely rejects a place for the positive sciences is too extreme in the other direction. Scientism can be a kind of irrationalism, and it can in its own way be blind, but there can also be an antiscientific kind of irrationalism and blindness. It seems to me that neither form of irrationalism would be acceptable to Husserl. It is Husserl's emphasis on reason that distinguishes him both from many analytic philosophers and from many continental philosophers.

The positive, objective sciences both reveal and conceal. Phenomenological philosophy, aiming toward science in a broader sense, tells us to retain what is revealed by the positive sciences (subject to critical scrutiny, responsibility, and broader values) and to reveal what is concealed by the positive sciences. This would be rational enlightenment at its best, for we would then be casting the light of reason as widely as possible.

Acknowledgments

I benefited from the comments of many of the participants in the *Science and Continental Philosophy* conference organized by Gary Gutting and held at the University of Notre Dame in September 2002. I especially thank Ernan McMullin, Gary Gutting, Karl Ameriks, Michael Friedman, Han-Jörg Rheinberger, Terry Pinkard, and Simon Critchley.

References

In addition to the items cited below, there are many other important works on Husserl's views on the sciences. Lack of space, however, prohibits citation of additional references here.

Becker, O. 1923. "Beiträge zur phänomenologischen Begründung der Geometrie und ihrer physikalischen Anwendungen," *Jahrbuch für Philosophie und phänomenologische Forschung* 6: 385–560.

Gurwitsch, A. 1974. *Phenomenology and the Theory of Science*. Evanston, IL: Northwestern University Press.

Hardy, L. and Embree, L., eds. 1993. *Phenomenology of Natural Science*. Dordrecht: Kluwer.

Heelan, P. 1983. *Space Perception and the Philosophy of Science*. Berkeley: University of California Press.

——. 1989. "Husserl's Philosophy of Science," in J. N. Mohanty and W. McKenna, eds., *Husserl's Phenomenology: A Textbook*. Lanham, MD: Center for Advanced Research in Phenomenology/University Press of America.

Husserl, E. 1965. "Philosophy as Rigorous Science," trans. Q. Lauer in *Phenomenology and the Crisis of Philosophy*. New York: Harper, 71–147. Originally published in 1911.

——. 1969. *Formal and Transcendental Logic*, trans. D. Cairns. The Hague: Nijhoff. Originally published in 1929.

——. 1970a. *The Crisis of the European Sciences and Transcendental Phenomenology*, trans. D. Carr. Evanston, IL: Northwestern University Press. Originally published in 1936.

——. 1970b. *Philosophie der Arithmetik*. Husserliana XII. The Hague: Nijhoff. Contains materials from 1890–1901.

——. 1973. *Logical Investigations*, vols. I, II. Translation of the 2nd ed. by J. N. Findlay. London: Routledge and Kegan Paul. First edition published in 1900–1; 2nd rev. ed. in 1913/21.

——. 1979. *Aufsätze und Rezensionen (1890-1910)*. Husserliana XXII. The Hague: Nijhoff.

——. 1980. *Ideas Pertaining to a Pure Phenomenology and to a Phenomenological Philosophy*, Third Book, trans. T. Klein and W. Pohl. The Hague: Nijhoff.

——. 1982. *Ideas Pertaining to a Pure Phenomenology and to a Phenomenological Philosophy*, First Book, trans. F. Kersten. The Hague: Nijhoff. Originally published in 1913.

——. 1983. *Studien zur Arithmetik und Geometrie*. Husserliana XXI. The Hague: Nijhoff. Materials from 1886–1901.

——. 1989. *Ideas Pertaining to a Pure Phenomenology and to a Phenomenological Philosophy*, Second Book, trans. R. Rojcewicz and A. Schuwer. The Hague: Nijhoff.

Kockelmans, J., 1993. *Ideas for a Hermeneutical Phenomenology of the Natural Sciences*. Dordrecht: Kluwer.

—— and Kisiel, T., eds. 1970. *Phenomenology and the Natural Sciences*. Evanston, IL: Northwestern University Press.

Natanson, M., ed. 1973. *Phenomenology and the Social Sciences*, 2 vols. Evanston, IL: Northwestern University Press.

Ströker, E. 1979. *Lebenswelt und Wissenschaft in der Philosophie Edmund Husserls*. Frankfurt am Main: Vittorio Klostermann.

——. 1987. *The Husserlian Foundations of Science*. Lanham, MD: Center for Advanced Research in Phenomenology/University Press of America.

——. 1988. "Husserl and the Philosophy of Science," *Journal of the British Society for Phenomenology* 19: 221–34.

Tieszen, R. 1989. *Mathematical Intuition: Phenomenology and Mathematical Knowledge*. Dordrecht: Kluwer.

——. 1997. "Science within Reason: Is There a Crisis of the Modern Sciences?," in M. Otte and M. Panza, eds., *Analysis and Synthesis in Mathematics*. Dordrecht: Kluwer, 243–59.

Weyl, H. 1985. *Space, Time, Matter*. New York: Dover. Originally publ. 1918.

8

From *INTRODUCTION TO THE LOGICAL INVESTIGATIONS* and from *THE CRISIS OF EUROPEAN SCIENCES AND TRANSCENDENTAL PHENOMENOLOGY*

Edmund Husserl

From *The Introduction to the Logical Investigations*

"Pure logic," in its most comprehensive[1] extension characterizes itself by an essential distinction as *"mathesis universalis."* It develops through a step-by-step extension of that particular concept of formal logic which remains as a residue of pure ideal doctrines dealing with "propositions" and validity after the removal from traditional logic of all the psychological misinterpretations and the normative-practical goal positings [*Zielgebungen*]. In its thoroughly proper extension it includes all of the pure "analytical" doctrines of mathematics (arithmetic, number theory, algebra, etc.) and the entire area of formal theories, or rather, speaking in correlative terms, the theory of manifolds [*Mannigfaltigkeitslehre*] in the broadest sense. The newest development of mathematics brings with it that ever new groups of formal-ontological laws are constantly being formulated and mathematically treated which earlier had remained unnoticed. *"Mathesis universalis"* as an idea includes the sum total of this formal *a priori*. It is, in the sense of the "Prolegomena," directed toward the entirety of the "categories of meaning" [*Bedeutungs-kategorien*] and toward the formal categories for objects correlated to them or, alternatively, the *a priori* laws based upon them. It thus includes the entire *a priori* of what is in the most fundamental sense the "analytic" or "formal" sphere – a sense which receives a strict specification and clarification in the third and sixth investigations. I would like to mention, incidentally, that the propriety of this delimitation can be made evident first in the naive-natural perspective and that it must retain its value in any case also for those who reject my "idealism" on psychologistic grounds. Thus no psychologistic empiricism *à la* Mill can change the fact that pure mathematics is a strictly self-contained system of doctrines which is to be cultivated using methods that are essentially different from those of natural science. He has

(i) Edmund Husserl, pp. 28–31 from *Introduction to the Logical Investigations*, ed. Eugen Fink, trans. Philip J. Bossert and Curtis H. Peters. The Hague: Martinus Nijhoff, 1975. Reproduced by kind permission of Kluwer Academic Publishers; (ii) Edmund Husserl, pp. 48–57 (sections 9h and 9i) from *The Crisis of European Sciences and Transcendental Phenomenology*, ed. David Carr. Evanston, IL: Northwestern University Press, 1970. Reproduced by kind permission of Northwestern University Press.

to recognize the present distinction even if he wants to reinterpret it later on. And, conversely, even an adherent of psychologism [*Psychologist*][2] could, if there were as yet no geometry, understand and approve the demonstration of the necessity of such a discipline in its fundamental peculiarity as opposed to natural science – he would just interpret it subsequently in his own style.

The *mathesis universalis* in its so-to-say naive as well as in its technical forms, as it has been grounded in the natural-objective orientation and can then be further cultivated, has at first no common cause with epistemology and phenomenology – just as little as ordinary arithmetic (a subdivision of *mathesis universalis*) has. However, if it also assumes the problem of the phenomenological "elucidation" in the sense of the "Prolegomena" and of the second volume, if as a consequence of this it learns from the sources of phenomenology what the solution is to the great riddles which here as everywhere arise from the correlation between being and consciousness, if in the process it learns as well the ultimate formulation of the meaning of concepts and[3] propositions (a formulation which only phenomenology is capable of providing): then it will have transformed itself from the naive into the truly philosophical pure logic, and it is in this sense that the pure logic is spoken of in philosophical contexts as a philosophical discipline. Examined precisely (and in harmony with the most recent accounts in my *Ideas*), it is not a mere coupling of phenomenology of knowledge with natural-objective *mathesis* but is rather an application of the former to the latter.[4] In a similar manner the physical theory of nature, for example, changes from mere natural science in the traditional sense of "positive" science into philosophy of nature by introducing the epistemological problematic belonging to it and its step-by-step solution through phenomenology. This describes, however, nothing other than a physics which has been both philosophically deepened and enriched by all of the problems concerning the correlation of physical being and cognitive subjectivity, a physics in which the experiencing subjectivity that performs the methodological achievement of objective knowledge does not remain scientifically anonymous and in which the methodical as well as the pertinent basic concepts and basic propositions are developed from the very beginning in ultimate methodological originality. Philosophical physics does not begin, as does the naive–positive physics, with vague concepts and then proceed in naively practiced methodical technique. It is from the very beginning a science that comprehends itself radically and that justifies right up to the end its constitution of sense and being [*Sinn- und Seinskonstitution*]. So the task is everywhere the same: to transform the merely positive sciences into "philosophical" ones or, where new sciences have to be established, to establish them from the very outset as "philosophical." Above all, philosophy means not irrelevant, speculative mysticism but rather nothing other than the ultimate radicalization of *rigorous* science. To be sure, positive science itself is trying to realize this ideal, but in its abstract one-sidedness, which is blind to the correlation in cognition, it is incapable of satisfying this ideal.[5]

At the same time, one can now understand my paralleling the idea of a "pure" – or better, "formal" – logic with the regional[6] ontologies parallel to it. Just as formal logic refers to the formal idea of object, to the "something in general." and can be characterized as formal ontology, so the regional ontologies refer to the highest material classes of objects whatsoever which (in the *Ideas*) are designated as "regions." If one chooses to call even these *a priori*, regional disciplines "logics," then, for example, Kant's pure science of nature, expanded to a universal ontology of nature in general, would have to be characterized as a logic of nature. Geometry, as a self-contained ontological discipline, would fit into it as the logic of pure (idealized) spatiality. Then, there corresponds again to every such "naive" logic to be constructed in the "natural-objective" orientation a "philosophical" logic that is an epistemologically and phenomenologically clarified one or one that is phenomenologically grounded from the very beginning. Whereas for pure logic in the sense of the present work (an "analytics" understood in the broadest and radical sense) only certain of the most general cognitive-formations [*Erkenntnisgestaltungen*] enter the picture for purposes of phenomenological elucidation, by contrast, for the material ontologies, in addition to the general cognitive-formations, also the corresponding cognitive-formations specifically related to the subject matter are to be drawn into the elucidating consideration of essences. Thus, for the ontology or logic of a possible (merely physical) nature in general, the basic forms of the subjective modes of knowledge used in constituting nature are introduced, and for the ontology of the soul or the mind, the forms related to its constitution are included.

From *The Crisis of European Sciences and Transcendental Phenomenology*

h)   *The Life-World as the Forgotten Meaning-Fundament of Natural Science*

But now we must note something of the highest importance that occurred even as early as Galileo: the surreptitious substitution of the mathematically substructed world of idealities for the only real world, the one that is actually given through perception, that is ever experienced and experienceable—our everyday life-world. This substitution was promptly passed on to his successors, the physicists of all the succeeding centuries.

Galileo was himself an heir in respect to pure geometry. The inherited geometry, the inherited manner of "intuitive" conceptualizing, proving, constructing, was no longer original geometry: in this sort of "intuitiveness" it was already empty of meaning. Even ancient geometry was, in its way, τέχνη, removed from the sources of truly immediate intuition and originally intuitive thinking, sources from which the so-called geometrical intuition, i.e., that which operates with idealities, has at first derived its meaning. The geometry of idealities was preceded by the practical art of surveying, which knew nothing of idealities. Yet such a pregeometrical achievement was a meaning-fundament for geometry, a fundament for the great invention of idealization; the latter encompassed the invention of the ideal world of geometry, or rather the methodology of the objectifying determination of idealities through the constructions which create "mathematical existence." It was a fateful omission that Galileo did not inquire back into the original meaning-giving achievement which, as idealization practiced on the original ground of all theoretical and practical life – the immediately intuited world (and here especially the empirically intuited world of bodies) – resulted in the geometrical ideal constructions. He did not reflect closely on all this: on how the free, imaginative variation of this world and its shapes results only in possible empirically intuitable shapes and not in exact shapes; on what sort of motivation and what new achievement was required for genuinely geometric idealization. For in the case of inherited geometrical method, these functions were no longer being *vitally* practiced; much less were they reflectively brought to theoretical consciousness as methods which realize the meaning of exactness from the inside. Thus it could appear that geometry, with its own immediately evident a priori "intuition" and the thinking which operates with it, produces a self-sufficient, absolute truth which, as such – "obviously" – could be applied without further ado. That this obviousness was an illusion – as we have pointed out above in general terms, thinking for ourselves in the course of our exposition of Galileo's thoughts – that even the meaning of the application of geometry has complicated sources: this remained hidden for Galileo and the ensuing period. Immediately with Galileo, then, begins the surreptitious substitution of idealized nature for prescientifically intuited nature.

Thus all the occasional (even "philosophical") reflections which go from technical [scientific] work back to its true meaning always stop at idealized nature; they do not carry out the reflection radically, going back to the ultimate purpose which the new science, together with the geometry which is inseparable from it, growing out of prescientific life and its surrounding world, was from the beginning supposed to serve: a purpose which necessarily lay *in* this prescientific life and was related to its life-world. Man (including the natural scientist), living in this world, could put all his practical and theoretical questions only to *it* – could refer in his theories only to it, in its open, endless horizons of things unknown. All knowledge of laws could be knowledge only of predictions, grasped as lawful, about occurrences of actual or possible experiential phenomena, predictions which are indicated when experience is broadened through observations and experiments penetrating systematically into unknown horizons, and which are verified in the manner of inductions. To be sure, everyday induction grew into induction according to scientific method, but that changes nothing of the essential meaning of the pregiven world as the horizon of all meaningful induction. It is this world that we find to be the world of all known and unknown realities. To it, the world of actually experiencing intuition, belongs the form of space-time together with all the bodily [*körperlich*] shapes incorporated in it; it is in this world that we ourselves live, in accord with our bodily [*leiblich*],[7] personal way of being. But here we find nothing of geometrical idealities, no geometrical space or mathematical time with all their shapes.

This is an important remark, even though it is so trivial. Yet this triviality has been buried precisely by exact science, indeed since the days of ancient geometry, through that substitution of a methodically idealized achievement for what is given immediately as actuality presupposed in all idealization, given by a [type of] verification which is, in its own way, unsurpassable. This actually intuited, actually experienced and experienceable world, in which practically our whole life takes place, remains unchanged as what it is, in its own essential structure and its own concrete causal style, whatever we may do with or without techniques. Thus it is also not changed by the fact that we invent a particular technique, the geometrical and Galilean technique which is called physics. What do we actually accomplish through this technique? Nothing but *prediction* extended to infinity. All life rests upon prediction or, as we can say, upon induction. In the most primitive way, even the ontic certainty[8] of any straightforward experience is inductive. Things "seen" are always more than what we "really and actually" see of them. Seeing, perceiving, is essentially having-something-itself [*Selbsthaben*] and at the same time having-something-in-advance [*Vor-haben*], meaning-something-in-advance [*Vor-meinen*]. All praxis, with its projects [*Vorhaben*], involves inductions; it is just that ordinary inductive knowledge (predictions), even if expressly formulated and "verified," is "artless" compared to the artful "methodical" inductions which can be carried to infinity through the method of Galilean physics with its great productivity.

In geometrical and natural-scientific mathematization, in the open infinity of possible experiences, we measure the life-world – the world constantly given to us as actual in our concrete world-life – for a well-fitting *garb of ideas*, that of the so-called objectively scientific truths. That is, through a method which (as we hope) can be really carried out in every particular and constantly verified, we first construct numerical indices for the actual and possible sensible plena of the concretely intuited shapes of the life-world, and in this way we obtain possibilities of predicting concrete occurrences in the intuitively given life-world, occurrences which are not yet or no longer actually given. And this kind of prediction infinitely surpasses the accomplishment of everyday prediction.

Mathematics and mathematical science, as a garb of ideas, or the garb of symbols of the symbolic mathematical theories, encompasses everything which, for scientists and the educated generally, *represents* the life-world, *dresses it up* as "objectively actual and true" nature. It is through the garb of ideas that we take for *true being* what is actually a *method* – a method which is designed for the purpose of progressively improving, *in infinitum*, through "scientific" predictions, those rough predictions which are the only ones originally possible within the sphere of what is actually experienced and experienceable in the life-world. It is because of the disguise of ideas that the true meaning of the method, the formulae, the "theories," remained unintelligible and, in the naïve formation of the method, was *never* understood.

Thus no one was ever made conscious of the radical problem of *how* this sort of naïveté actually became possible and is still possible as a living historical fact; how a method which is actually directed toward a goal, the systematic solution of an endless scientific task, and which continually achieves undoubted results, could ever grow up and be able to function usefully through the centuries when no one possessed a real understanding of the actual meaning and the internal necessity of such accomplishments. What was lacking, and what is still lacking, is the actual self-evidence through which he who knows and accomplishes can give himself an account, not only of what he does that is new and what he works with, but also of the implications of meaning which are closed off through sedimentation or traditionalization, i.e., of the constant presuppositions of his [own] constructions, concepts, propositions, theories. Are science and its method not like a machine, reliable in accomplishing obviously very useful things, a machine everyone can learn to operate correctly without in the least understanding the inner possibility and necessity of this sort of accomplishment? But was geometry, was science, capable of being designed in advance, like a machine, without[9] an understanding which was, in a similar sense, complete – scientific? Does[10] this not lead to a *regressus in infinitum?*

Finally, does this problem not link up with the problem of the instincts in the usual sense? Is it not the problem of *hidden reason*, which knows itself as reason only when it has become manifest?

Galileo, the discoverer – or, in order to do justice to his precursors, the consummating discoverer – of physics, or physical nature, is at once a discovering and a concealing genius [*entdeckender und verdeckender Genius*]. He discovers mathematical nature, the methodical idea, he blazes the trail for the infinite number of physical discoveries and discoverers. By contrast to the universal causality of the intuitively given world (as its invariant form), he discovers what has since been called simply the law of causality, the "a priori form" of the "true" (idealized and mathematized) world, the "law of exact lawfulness" according to which every occurrence in "nature" – idealized nature – must come under exact laws. All this is discovery-concealment, and to the present day we accept it as straightforward truth. In principle nothing is changed by the supposedly philosophically revolutionary critique of the "classical law of causality" made by recent atomic physics. For in spite of all that is new, what is essential in principle, it seems to me, remains: namely, nature, which is in itself mathematical; it is given in formulae, and it can be interpreted only in terms of the formulae.

I am of course quite serious in placing and continuing to place Galileo at the top of the list of the greatest discoverers of modern times. Naturally I also admire quite seriously the great discoverers of classical and postclassical physics and their intellectual accomplishment, which, far from being merely mechanical, was in fact astounding in the highest sense. This accomplishment is not at all disparaged by the above elucidation of it as τέχνη or by the critique in terms of principle, which shows that the true meaning of these theories – the meaning which is genuine in terms of their origins – remained and had to remain hidden from the physicists, including the great and the greatest. It is not a question of a meaning which has been slipped in through metaphysical mystification or speculation; it is, rather, with the most compelling self-evidence, the true, the only real meaning of these theories, as opposed to the meaning of being a *method*, which has its own comprehensibility in operating with the formulae and their practical application, technique.

How what we have said up to now is still one-sided, and what horizons of problems, leading into new dimensions, have not been dealt with adequately – horizons which can be opened up only through a reflection on this life-world and man as its subject – can be shown only when we are much further advanced in the elucidation of the historical development according to its innermost moving forces.

i)   *Portentous Misunderstandings Resulting from Lack of Clarity about the Meaning of Mathematization*

With Galileo's mathematizing reinterpretation of nature, false consequences established themselves even beyond the realm of nature which were so intimately connected with this reinterpretation that they could dominate all further developments of views about the world up to the present day. I mean Galileo's famous doctrine of the merely subjective character of the specific sense-qualities,[11] which soon afterward was consistently formulated by Hobbes as the doctrine of the subjectivity of all concrete phenomena of sensibly intuitive nature and world in general. The phenomena are only in the subjects; they are there only as causal results of events taking place in true nature, which events exist only with mathematical properties. If the intuited world of our life is merely subjective, then all the truths of pre- and extrascientific life which have to do with its factual being are deprived of value. They have meaning only insofar as they, while themselves false, vaguely indicate an in-itself which lies behind this world of possible experience and is transcendent in respect to it. [. . .]

k)   *Fundamental Significance of the Problem of the Origin of Mathematical Natural Science*[12]

Like all the obscurities exhibited earlier, [the preceding] follow from the transformation of a formation of meaning which was originally vital, or rather of the originally vital consciousness of the task which gives rise to the methods, each with its special sense. The developed method, the progressive fulfillment of the task, is, as method, an art (τέχνη) which is handed down; but its true meaning is not necessarily handed down with it.

And it is precisely for this reason that a theoretical task and achievement like that of a natural science (or any science of the world) – which can master the infinity of its subject matter only through infinities of method[13] and can master the latter infinities only by means of a technical thought and activity which are empty of meaning – can only be and remain meaningful in a true and original sense *if* the scientist has developed in himself the ability to *inquire back* into the *original meaning* of all his meaning- structures and methods, i.e., into the *historical meaning of their primal establishment*, and especially into the meaning of all the *inherited meanings* taken over unnoticed in this primal establishment, as well as those taken over later on.

But the mathematician, the natural scientist, at best a highly brilliant technician of the method – to which he owes the discoveries which are his only aim – is normally not at all able to carry out such reflections. In his actual sphere of inquiry and discovery he does not know at all that everything these reflections must clarify is even in *need* of clarification, and this for the sake of that interest which is decisive for a philosophy or a science, i.e., the interest in true knowledge of the *world itself, nature itself*. And this is precisely what has been lost through a science which is given as a tradition and which has become a τέχνη, insofar as this interest played a determining role at all in its primal establishment. Every attempt to lead the scientist to such reflections, if it comes from a nonmathematical, nonscientific circle of scholars, is rejected as "metaphysical." The professional who has dedicated his life to these sciences must, after all – it seems so obvious to him – know best what he is attempting and accomplishing in his work. The philosophical needs ("philosophicomathematical," "philosophicoscientific" needs), aroused even in these scholars by historical motives to be elucidated later, are satisfied by themselves in a way that is sufficient for them – but of course in such a way that the whole dimension which must be inquired into is not seen at all and thus not at all dealt with.

Translator's Notes

1   Reading "*umfassendsten*" for "*unfassendsten*."
2   The English term "psychologist" is rendered as "*Psychologe*" in German; the German term "*Psychologist*" designates one who is an adherent of the logical or epistemological position

of psychologism, while "*Psychologe*" designates simply one trained in the science of psychology.

3   Reading "*und*" for "*une*."

4   The German text in *Tijdschrift* reads: "*eine Umwendung der ersteren in die letztere*" ("a transformation of the former into the latter") which is clearly a transcription or printing error that says the opposite of what Husserl means. Husserl's 1913 shorthand manuscript – upon which our translation of this sentence is based – reads: "*eine Anwendung der ersteren an die letzteren*." Thus, natural-objective *mathesis* is transformed into philosophical *mathesis* through the application of phenomenology.

5   These last sentences express the basic theme which Husserl developed in his 1911 *Logos* article "Philosophy as Rigorous Science."

6   Reading "*regionalen*" for "*rationalen*." "*Rational* ontologies" does not make sense in this context and is obviously a transcription or printing error. Two lines later, the German reads "*regionalen*" with respect to the same idea.

7   *Körper* means a body in the geometric or physical sense; *Leib* refers to the body of a person or animal. Where possible, I have translated *Leib* as "living body" (*Leib* is related to *Leben*); *Körper* is translated as "body" or sometimes "physical body." In cases where adjectival or adverbial forms are used, as here, it is sometimes necessary to insert the German words or refer to them in a footnote.

8   *Seinsgewissheit*, i.e., certainty of being.

9   Reading *ohne ein* for *aus einem*.

10   Reading *führt* for *führte*.

11   This doctrine is perhaps best expressed in Galileo's *Il Saggiatore* (*The Assayer*). See *Discoveries and Opinions of Galileo*, trans. Stillman Drake (Garden City: Doubleday Anchor Original, 1957), pp. 274 ff.

12   There is no section "j." In German one does not distinguish between "i" and "j" in an enumeration of this sort.

13   I.e., the infinite pursuit of its method.

HEIDEGGER

9

HEIDEGGER ON SCIENCE AND NATURALISM

Joseph Rouse

Heidegger's reputation in the Anglophone philosophical world suggests that he was unsympathetic to the natural sciences, and generally unconcerned with the philosophy of science. I shall argue to the contrary, not merely that Heidegger made significant contributions to philosophical understanding of the sciences, but that philosophy of science was at the center of his project and its development throughout his career. To capture this centrality, I will examine Heidegger's conception of science and its relation to philosophy at two key points in his philosophical work.

Science and Philosophy in *Being and Time*

Relations between philosophy and the sciences had taken on new urgency for many early twentieth-century philosophers. The dramatic successes of physics and chemistry, and the recent emergence of rigorous scientific disciplines in the human and life sciences, gave plausibility to philosophical naturalism. Naturalists believed that the concerns and subject-matter traditionally accorded to philosophy could be better served by empirical research in the relevant sciences. Heidegger's Freiburg predecessor Edmund Husserl exemplified a widespread philosophical response to the perceived threat of naturalism. Husserl (1981) argued against naturalists that the empirical sciences could not account for their own normativity. In particular, no empirical science could establish the meaning and validity of scientific claims themselves. A radically different kind of philosophical science was therefore needed, one that did not merely describe how the world *is*, but showed how it *must* or *ought* to be understood.

In his early work, Heidegger endorsed Husserl's call for philosophy as a "rigorous science" more fundamental than any empirical science could be. He nevertheless radically reconceived the nature of this philosophical science and its consequent relation to other disciplines. For Husserl and most of his contemporaries, the sciences sought knowledge, and philosophy was to assess and secure the sense and justification of scientific claims to knowledge. Heidegger conceived philosophy instead as "fundamental ontology." The sciences are ways of dealing with entities (beings), whereas philosophy's task was to inquire about being.

Heidegger's distinction between being and entities has been widely misunderstood. We can first approach it as a critical response to the dominant conception of philosophy as epistemology (theory of knowledge). Epistemologists treat knowledge as a relation between two entities, a knower and an object known. The task is then to understand how knower and known ought to be related to obtain genuine knowledge. Heidegger thought that there are unexamined, erroneous presuppositions underlying any such conception of knowers as a special kind of entity (mind, consciousness, language-speaker, or rational agent), and of knowledge as a relation between entities, insisting that "we have no right to resort to dogmatic constructions and to apply just any idea of being and actuality to this entity [that we ourselves are], no matter how 'self-evident' that idea may be" (1953: 16). In posing the question of being (of what it means to be, or of the intelligibility of entities as entities), Heidegger sought to circumvent these unexamined assumptions about knowledge or consciousness, and engage in a more radical philosophical questioning. The "being" of an entity is its intelligibility as the entity it is. Heidegger's point is that we should avoid assuming that the intelligibility of entities is itself an entity (such as a meaning, an appearance, a concept, or a thought).

Heidegger's questioning about being made an initial move in the direction of naturalism. His philosophical contemporaries had sought philosophical grounding for the sciences in some realm of necessity independent of the contingencies of the world in which we happen to find ourselves. No facts about this world could tell us what we ought to do or think. Hence, philosophers looked to the extraworldly realms of pure logic or transcendental consciousness as the locus of philosophical inquiry. Logic was not an empirical science of how people actually reason, but a study of the purely formal structures or norms that actual thinking only imperfectly realizes. Husserlian transcendental consciousness was likewise not a contingent psychophysical object available to empirical psychology, but a realm of pure meanings that become accessible only when concern with worldly existence is temporarily suspended. Heidegger rejected any such suspension of "natural" involvement in the world. He proposed instead to begin with our "average, everyday" activities, and "to work out the idea of a 'natural conception of the world'" as the starting point for philosophical interpretation (1953: 43, 52).

A natural conception of the world needed working out, because philosophical preconceptions about knowledge and the mind had supposedly blocked understanding even of our most familiar, unreflective activities. Husserl, for example, assimilated everyday life and scientific knowledge as both exemplifying a "natural attitude" that was antithetical to philosophy. Nor was Husserl alone in thus taking all ordinary human activities to involve cognition or knowledge. Adapting the term "Dasein" for our way of being, Heidegger proposed instead to start philosophical reflection with Dasein's most ordinary, familiar dealings with our surroundings.

> [Dasein] never finds itself otherwise than in the things themselves, and in fact in those things that daily surround it. It *finds itself* primarily and constantly *in things* because, tending them, distressed by them, it always in some way or other rests in things. Each one of us is what he pursues and cares for. (1982: 159)

Most philosophers talk about mental acts or propositional attitudes (perceiving, judging, desiring) as our basic way of relating to our surroundings. Heidegger talked more encompassingly of Dasein's various dealings with or comportments toward entities, and he challenged the presumption that such comportments always at least implicitly involve mental or linguistic representation. In our everyday comportment, we understand the entities we encounter, but Heidegger construed understanding as a kind of practical competence rather than cognition or mental representation (1953: 143). Cognition and knowledge were derivative from ("founded upon") such every-day practical understanding.

Heidegger's central claim in *Being and Time* was that any such understanding of entities presupposes an understanding of being. John Haugeland (1998) offers a useful analogy to unpack this seemingly obscure claim. One cannot encounter a rook as such without some grasp of the game of chess. In Heidegger's terms, the "discovery" of chess entities (pieces, positions, moves, or situations) presupposes a prior "disclosure" of chess as the context in which they could make sense. The "being" of rooks or knights is their place within the game; that is what confers their intelligibility as the entities they are. The game itself only makes sense in turn, however, as a possible way for us to comport ourselves within the world. In all such comportments toward particular entities, then, what we most fundamentally understand is the world as a significant config-uration of possible ways for Dasein to be, and our own being-toward those possibilities:

> What understanding, as an *existentiale* [an essential structure of our way of being], is competent over is not a "what," but being as existing... . Dasein is not something occurrent which possesses its competence as an add-on; it is primarily being-possible. (1953: 143)[1]

The difficult point to grasp here is Heidegger's claim that the "world" (the situation or context) whose disclosure allows us to discover various kinds of entities is not itself an entity or a collection of entities. If we ask *what* there is, there are only the various and sundry entities we can discover, and nothing more. But we can only discover them at all because we understand being, and thereby belong to an historically specific situation that is a meaningful configuration of possible ways for us to be (what Heidegger called "world").[2]

We can now ask how Heidegger (in *Being and Time*) conceived of science and its relation to philosophy. There are three central themes in Heidegger's early philosophy of science: Heidegger's proposal for an "existential conception of science," his con-ception of how and why fundamental ontology is fundamental for the sciences, and his account of the ontological significance of science in general as the discovery of the occurrent (*Vorhanden*). There are naturalistic overtones to all of these themes, even though Heidegger remained adamantly opposed to naturalism as usually understood.

An "existential" conception of science was called for because "sciences, as human comportments, have [Dasein's] way of being" (1953: 11). Dasein's way of being is future-oriented; it "presses forward into [its] possibilities," and does so out of concern for its own being. Dasein's most basic relation to itself is not self-consciousness, but care: Dasein is "the entity whose own being is at issue for it" (1953: 42), such that everything it does is understood in response to that issue.[3] Heidegger contrasted such an existential conception of science to "the 'logical' conception which understands science with regard to its results and defines it as an 'inferentially interconnected web [*Begründungszusammenhang*] of true, that is, valid propositions'" (1953: 357). Heidegger thus gave priority to scientific research as something people do, rather than to scientific knowledge as something acquired and assessed retrospectively. Science is not primarily the accumulation of established knowledge, but is always directed ahead of itself toward possibilities it cannot yet fully grasp or articulate.

The contrast in conceptions of philosophy's contribution to science was therefore sharp between Heidegger's project of fundamental ontology and traditional logical and epistemological approaches. The latter, he thought, "lag behind, investigating the standing a science happens to have" so far (1953: 10). Such an approach belies the futural orientation of scientific research, and thus utterly misunderstands what matters in science. Heidegger thought that

> the authentic [*eigentlich*] "movement" of the sciences takes place in the more or less radical and self-transparent revision of their basic concepts. The level of a science is determined by the extent to which it is *capable* of a crisis in its basic concepts. (1953: 9)

To the extent that a philosophy of science defined its normative task according to the already accepted orientation of a particular scientific discipline, it sought to secure what science itself should seek to surpass. Heidegger thought philosophy could instead contribute "a productive logic, in the sense that it leaps ahead, so to speak, into a particular region of being, discloses it for the first time in the constitution of its being, and makes the structures it arrives at available to the positive sciences as guidelines for their inquiry" (1953: 10).

Heidegger thought philosophy could do this because the sciences, like any other human activities, must work from a prior understanding of the being of the entities they deal with. Such understanding amounts to a practical grasp (*not* an already articulated description) of what entities are involved, how to approach them in revealing ways, and what would amount to success in dealing with them. The discovery and articulation of what there is in a particular scientific domain draws upon and further develops this prior disclosure of the being of those entities. What philosophy could do was to interrogate a particular science's prior understanding of being within its domain. Such interrogation of being inquires into the a priori conditions of the possibility both of sciences that investigate specific types of entities and of the ontologies that ground them (1953: 11), but it would not itself be an a priori inquiry.[4] Heidegger noted with approval (1953: 9–10) the extent to which many of the contemporary scientific disciplines (specifically mathematics, physics, biology, the historical sciences, and theology) were themselves engaged in renewed reflections upon their conceptual foundations, and he saw such developments as appropriately philosophical turns within those disciplines. Philosophical ontology should be continuous with such philosophical reflection within the sciences. His explicit models for philosophical ontology were the contributions of Plato, Aristotle, and Kant. In the latter case, he thought,

> the positive outcome of Kant's *Critique of Pure Reason* lies in its contribution to working out what belongs to any nature whatsoever, not in a "theory" of knowledge. (1953: 10–11)

Kant's work along these lines was not "prior" to Newton, but a philosophical (onto-logical) engagement with the project of Newtonian physics.[5]

Philosophy would have something distinctive to contribute to ontological reflection within any particular science, for reasons. Most important, the "regional" disclosure of being within any particular scientific domain was itself supposedly dependent upon a prior understanding of being in general. Just as to understand rooks one needs to understand chess, and to understand chess, one has to grasp it as a possible mode of Dasein as being-in-the-world, so Heidegger thought that the disclosure of the being of entities within any particular scientific domain would presuppose an understanding of being in general. Until this more basic understanding of being had been clarified, any regional ontology, "no matter how rich and tightly linked a system of categories it has at its disposal, remains blind and perverted from its ownmost aim" (1953: 11).

The second reason Heidegger thought that philosophy *per se* was indispensable to science was to combat the "normalizing" tendency inherent in scientific research itself.[6] Science always presupposes an understanding of being, but the scientific project of discovering what and how entities are within its domain obscures the understanding of being that makes research possible. Its determined focus upon the *entities* it investigates takes for granted the understanding of being that provides that focus. Thomas Kuhn's (1970) subsequent account of normal science eloquently expresses that tendency in scientific work which Heidegger thought made it inevitably dependent upon philosophical questioning (regardless of whether those who raise such questions are institutionally identified as scientists or philosophers). For Kuhn as for Heidegger, "normal" science avoids controversy over fundamentals in order to develop with greater detail and precision its unquestioned conceptual and practical grasp of a domain of entities. Left to their own devices, both thought, the sciences suppress any such fundamental questioning. When such questioning becomes unavoidable due to the breakdown of positive research into a particular domain, scientists do not then take up ontological inquiry for its own sake, but seek only to reconstitute their ability to attend carefully to entities without having to inquire into their being. Where Kuhn and Heidegger diverged was that Kuhn endorsed this closing off of ontological inquiry, whereas Heidegger did not.

Heidegger did not, however, see scientific normalization as a merely contingent and possibly objectionable psychological tendency or social pressure. It was instead an essential ontological dimension of science. Here we encounter the final theme in Heidegger's early philosophy of science, the connection between science and "occurrentness" (*Vorhandenheit*) as a mode of being. Although Heidegger insisted even in *Being and Time* that being was not itself an entity, there could still be a science of being (fundamental ontology) because there were fundamental distinctions articulable within the understanding of being itself. These fundamental distinctions were not those that differentiate the *domains of entities* to be studied by positive sciences (nature, mathematics, language, history, and the like), but instead mark different *ways of being*, of intelligibility as entities. The most basic distinction is between the being of Dasein (being-in-the-world) and "innerworldly" ways of being. Heidegger was not always terminologically careful in distinguishing the entities whose way of being is Dasein, from Dasein itself, but the distinction is crucial. Heidegger was not interested in doing empirical anthropology, but only in understanding being.

Heidegger first distinguished Dasein's way of being from the "occurrentness" of things (such as a mind, soul, ego, body, or person). Heidegger then argued, however, that the entities we deal with in our ordinary, everyday lives are not things in this sense either. Equipment cannot be understood as individual entities defined by their properties. Something can only be a hammer, in his familiar example, through its "relation" to nails, boards, carpentry, and ultimately those human activities for which hammering and fastening are integral. The interrelations among equipment are more ontologically basic than the individuation of pieces of equipment:

> Strictly speaking, there is no such thing as *an* equipment... . [Equipmental] "things" never show themselves initially for themselves, so as to fill out a room as a sum of real things. What we encounter as closest to us, although unthematically, is the room. (1953: 68)

Moreover, equipment works best when we don't have to think about it at all, and can focus on the task at hand (what is ahead of us). The being of equipment is not the occurrentness of a self-standing entity with properties, but the availability of such normally tacit functionality.

One kind of equipment does have to call attention to itself, however. Signs only function when we take notice of them. Signs themselves still have the being of equipment, signifying only by belonging to a larger practical context. Assertions, however, are signs that can allow things to show up in a different way. Assertions point out entities and make them communicable. Heidegger thought that assertion is in this respect derivative from everyday practical involvement: talk about things pre-supposes a more basic practical understanding.

The ontological significance of science for early Heidegger was bound up with linguistic assertion as a derivative mode of interpretation. To this extent, Heidegger's early philosophy of science remained quite traditional. Science *describes* entities, and in doing so strips them of their ordinary human significance. Some assertions point out entities within a local, practical situation. In science, however, we can discover entities shorn of their practical involvements, as merely occurrent. We then talk about a hammer not in its appropriateness and availability for a task at hand, but as an object with mass and spatiotemporal location. It thereby acquires a new mode of intelligibility. Its local, contextual involvements are replaced by a theoretical contextualization:

> What is decisive for the development [of mathematical physics] ... lies in *the mathematical projection of nature itself*. This projection discovers in advance something constantly occurrent (matter), and opens the horizon to look for guidance to its quantitatively determinable constitutive aspects (motion, force, location, and time). (1953: 362)

This ontological understanding of theoretical interpretation served two roles. The disclosure and theoretical articulation of entities as occurrent was shown to be a genuine, truthful accomplishment of empirical science. This accomplishment, how-ever, was shown to be doubly dependent upon philosophical ontology.

In the most obvious dependence, science and cognition more generally were shown to be derivative modes of understanding. Assertions about occurrent entities are intelligible only because of Dasein's prior practical immersion in a world of significantly interrelated equipment. Fundamental ontology was then needed to clarify the relation between the assertions of theoretical science and the more basic understanding of being they presuppose. But scientific assertion was also supposedly derivative in a more troubling way. Assertions can correctly "point out" entities as occurrent. But assertions also thereby indispensably allow what-is-said (*das Geredete*) to be passed on in "idle talk" (*Gerede*) that obscures understanding. Assertions are "ambiguous" in the sense that the same assertion can be made with or without understanding, and most important, with or without responsibility to what is being talked about. In making understanding communicable, assertion also makes possible a mere semblance of understanding.

The affinities between Heidegger's early philosophy of science and naturalism are significant, but rarely noticed. I have already mentioned Heidegger's insistence upon the worldly character of human understanding, rejecting any recourse to an immaterial realm of pure logic or transcendental consciousness as the locus of thought's accountability to norms. Heidegger's account of assertion goes further in this direction, however. Most of Heidegger's philosophical contemporaries, impressed by the need to understand error and talk about non-existent things, posited meanings as intermediaries between thought and things. We can talk and think about what does not exist, or falsely about what does exist, because our grasp of meanings is more basic than our acquaintance with things. Heidegger rejected such appeals to semantic intermediaries. Assertions "point out" entities themselves, not meanings:

> The assertion ["the picture on the wall is hanging crookedly"] . . . in its ownmost meaning is related to the real picture on the wall. What one has in mind is the real picture, and nothing else. (1953: 217)

Like many naturalists today, Heidegger accounted for linguistic articulation by situating talk first and foremost within a larger pattern of interaction with things, rather than within a linguistic or theoretical structure. Error is a holistic relation to the entities that actually show up within our interactions, not a direct grasp of meanings that fail to represent anything correctly. Heidegger differed from today's advocates of a causal theory of reference in taking our more basic dealings with our surroundings to be practical-normative rather than causal. They make common cause, however, in construing language as interaction with the world around us rather than as an abstract, formal structure of meanings which connects to the world only indirectly.

For Heidegger, however, the claim that assertion is a comportment toward entities themselves gives heightened and ironic significance to the possibility of repeating what is asserted. By making what-is-said communicable, assertions can become distant from the entities that they point out and are accountable to. The proximate grounds for one's assertions then become not the entities themselves, but other assertions. There are two distinct ways in which such "idle talk" interposes other assertions rather than the entities talked about as what is primarily understood in an assertion. Most obviously, one's assertions can be grounded in testimony: I can make an assertion not from my own understanding of how things stand, but as merely passing on what others say, with the anonymous authority of what "one" says about such matters. But assertions can also be grounded inferentially upon other assertions, such that their authority is mediated by complex networks of other claims. These two forms of the interdependence of assertions are closely intertwined, for developing and sustaining complex networks of belief requires sharing and passing on what others say.

The indispensability of such inferential networks for scientific understanding highlights Heidegger's insistence that his account of idle talk is not altogether disparaging. His point was not to reject articulated theoretical understanding, but to recognize that precisely in developing more extensively articulated theoretical networks, the sciences also run the risk of becoming more invested in their own vocabularies and theories than in the things to be understood. Contrary to the sciences' familiar fallibilist image, Heidegger worried that the development of science continually *closes off* the possibility that entities themselves might resist our familiar ways of encountering and talking about them. For Heidegger, science needed philosophy in order to remain "in the truth." The greatest danger in science was not error (which is more readily correctable by further inquiry), but the emptiness of assertions that are closed off from genuine accountability to entities.[7] Thus, Heidegger insisted that truth as correct assertion was grounded in a more fundamental sense of truth as "unhiddenness": correctness alone would not yield genuine understanding unless the entities themselves were continually wrested away from burial in mere talk. We can now better see the connection between Heidegger's account of science as the discovery of entities as occurrent, and his insistence upon the need to ground science in fundamental ontology. In its focus upon the cognitive discovery of the occurrent, science inevitably pulls us away from its own "highest" possibility, which was a readiness for and openness to crisis in its basic concepts, brought on out of fidelity to the entities in question. Only in "philosophically" turning away from involvement with and idle talk about entities, toward the understanding of being within which those entities are disclosed, could science remain open to truthful disclosure of things themselves.

Heidegger thus implicitly distinguished naturalism in philosophy from scientism. He joined naturalists in arguing, against his neo-Kantian, phenomenological, and logical positivist contemporaries, that philosophy must begin from and remain within the horizon of our "natural" involvement with our surroundings in all its material and historical concreteness.[8] No transcendence of this world was permissible. Yet in his early work, Heidegger argued against most naturalists that the empirical sciences were derivative from (rather than constitutive of) the requisite "natural conception of the world." In its turn toward theoretical articulation, science was more akin than opposed to traditional philosophical approaches in obscuring or blocking natural understanding.

Yet despite Heidegger's early commitment to philosophical renewal as a counter to scientism, this supposedly philosophical task of sustaining the sciences' truthful openness to "the things themselves" was inclined toward naturalism in yet another way. Opponents of naturalism typically assign a distinctive subject-matter to philosophy, such as epistemology, logic, semantics, or transcendental consciousness, whereas naturalists tend to emphasize the continuity between philosophy and the sciences. Heidegger's conception of fundamental ontology inclined more in the latter direction. In thinking about the being of the entities discovered in the sciences, we do not think about something *else*. Being is not itself an entity, but only the disclosure of entities as intelligible. Philosophical reflection would not take us away from the subject-matter of the sciences, but would instead aim to bring one back afresh to "the things themselves" in their essential disclosedness. For Heidegger, Aristotle's sustained reflections upon biology or Kant's upon mechanics were not a failure yet to distinguish philosophy clearly from science, but instead recognized philosophy's highest calling.

There were nevertheless tensions within Heidegger's early philosophy of science which pointed toward fundamental difficulties within his project as a whole. Heidegger's project of fundamental ontology was an ahistorical, transcendental-philosophical inquiry into human existence as essentially historical and worldly. In their turn to formal structures of pure logic or transcendental consciousness, Heidegger thought his philosophical opponents had irrevocably severed their connection to the worldly phenomena they aspired to account for. Heidegger adamantly opposed any comparable formalization of his own ontological categories. The in-order-to-for-the-sake-of relations that articulate the being of what is available (*Zuhanden*) can, he admitted,

> be grasped formally in the sense of a system of relations. But ... in such formalizations the phenomena get leveled off so much that their real phenomenal content may be lost ... The "in-order-to," the "for-the-sake-of," the "with-which" of involvement ... are instead relationships in which concernful circumspection as such already dwells. (1953: 88)

It was not so clear, however, *why* the essential structures of fundamental ontology did not also evanesce into ahistorical, immaterial formal relations.[9] As a telling example, what is the relation between the concrete, "circumspective"[10] scientific practices through which existing, historical scientific-Dasein articulates nature theoretically, and the abstract ontological category of science as the theoretical discovery of the occurrent? More generally, how were the differences between ways of being (Dasein, availability, or occurrentness) relevant to the ontological determination of scientific domains such as nature or history? How and why, for example, should the human sciences' investigations of human beings as entities be determined by an understanding of Dasein as our way of being? Likewise, what is the relation between us as cases of Dasein and us as biological or physical entities? Finally, Heidegger's account of science incorporated an ontologically decisive but concretely elusive "changeover" from "the understanding of being that guides concernful dealings with entities" to "looking at those available entities in a 'new' way as occurrent" (1953: 361). This changeover involves both a shift from contextual communication (hammers that are "too heavy" or "misplaced") to thematic assertions about mass or location in spacetime as occurrent properties, and from practical understanding to "the mathematical projection of nature." Yet Heidegger merely asserted such a changeover without an adequate phenomenological description of how it occurred. This changeover from Dasein's practical familiarity with linguistic signs as "equipment for indicating" to explicit, decontextualized assertion was likewise both central and obscure in Heidegger's early philosophy of language.

Art, Technoscience, and Modernity

Within a decade after *Being and Time*, Heidegger significantly revised his philosophical project. His changing views of science and naturalism were central to that reconception, evident in the two lead essays in *Holzwege*: "Origin of the Art-Work" and "Age of the World Picture." These essays marked the abandonment of fundamental ontology as an ahistorical account of essentially different ways of being. Heidegger instead addressed how "metaphysics grounds an age ... through a specific interpretation of entities and through a specific conception of truth" (1950: 69). The centrality of science to Heidegger's rethinking is suggested by his list of "essential phenomena of modernity": it is no accident that science and technology head the list, nor that this crucial overview of his reconceived project occurs in an essay on science.

I shall nevertheless begin with an essay ostensibly devoted to the third "equally essential modern phenomenon, [in which] the art work becomes the object of mere subjective experience, ... an expression of human life" (1950: 69). The significance of the artwork essay for understanding science is highlighted by its implicit juxtaposition with Plato's *Republic*. The *Republic* identified the "light" within which entities are disclosed as the Form of the *Good*, which was to be the ultimate object of knowledge. Heidegger agreed that all intelligibility is normative, but rejected both Plato's identification of this "lighting" with an entity (a Form), and its disclosure through rational intuition rather than discursive articulation.[11] He also objected to the ahistorical, formalist, and disembodied or immaterial character of Platonic intelligibility. These points of contrast came together in Plato's insistence that poetry and other arts, at two removes from the Forms, cover up the intelligibility of entities (their being or "illumination") by turning the soul in the wrong direction, toward what is sensuous, material, and merely signifying.[12] Heidegger proposed that the poetic, discursive, and sensuous working of art *is* what holds open the possibility of meaningful differences between "birth and death, disaster and blessing, victory and disgrace, endurance and decline" (1950: 42). Art does so not through scientific cognition of what is timeless and immaterial. Art "works" instead by holding open an historically specific, discursively articulated, material ("earthy") world, through its normative grip upon the comportments of those through whom it "works."

The artwork essay also contained an implicit critique of *Being and Time* on several counts that are important for my purposes. First, Heidegger had repeatedly tried to get behind dealings with entities to show the understanding of being that guides them, but his early phenomenology of equipment, equipmental malfunction, signs, and anxiety only indicates such understanding under conditions of its *breakdown*. The "working" of artworks supposedly allows the clearing/lighting of entities to show itself when it works, and not just when it breaks down.[13] Scientific discovery of entities as occurrent, by contrast, had supposedly emerged in response to such practical failures.

The second challenge to *Being and Time* emerged in Heidegger's emphasis upon "the conflict between world and earth" rather than "the worldliness of the world" as that wherein we understand and deal with entities. He thereby implicitly acknowledged that *Being and Time* failed to show *why* the contextual interrelatedness of worldly significance resists mathematical formalization. "Earth" expresses both materiality and resistance to intelligibility. Concerning materiality, Heidegger asked,

> What is the essence ... of that in the work that one usually calls the work materials? Because it is determined by usefulness and serviceability, equipment takes into its service that of which it consists: the matter. Stone is used and used up in making equipment, e.g., an ax. It disappears in usefulness. The material is all the better and more suitable the less it resists perishing in the equipmental being of the equipment. By contrast, the temple-work, in setting up a world, does not let its material disappear, but rather lets it come forth for the very first time in the opening of the work's world. (1950: 34–5)

Art's accentuation of materiality also highlights the limits of human understanding, in ways that implicate scientific as well as practical intelligibility. If light (both literally and metaphorically) is that which allows entities to appear, then "earth" is the opacity that simultaneously limits and enables such appearances (for if there were only transparency, nothing could be illuminated).

> If we attempt [a penetration into a stone's heaviness] by smashing the rock, it still does not display in its pieces anything inner that has been opened up. The stone has immediately withdrawn again into the same dull pressure and bulk of its pieces. If we try to grasp this in another way, by placing the stone on a balance, we merely bring its heaviness into the form of a calculated weight. This perhaps very precise determination of the stone remains a number, but the weight's burden has escaped us... . Earth thus allows every penetration into it to shatter itself. (1950: 35–6)

If "world" denotes the significant interrelations within which entities are understandable (whether through the practical interrelatedness of in-order-to-for-the-sake-of relations, or the inferential relations among assertions), then "earth" marks the unsurpassable limits of such holistic intelligibility. As Samuel Todes once put a similar point, "*what* the facts are is made luminous by theory, but *that* these are the facts is plunged by theory into a darkness just as extraordinary as the light shed on their nature" (2001: 270). Artworks brought forth what was simultaneously the enabling condition for and the limit to any articulated understanding.

Another relevant challenge to Heidegger's early work came in his reconception of language. In *Being and Time*, language showed up first in its practical, contextual use as "equipment for indicating," and then in assertion as a derivative form of interpretation that presupposes prior practical competence. The artwork essay claimed instead that

> Language itself is poetry in the essential sense. But since language is the happening in which entities first disclose themselves for humanity as entities, therefore poesy – poetry in the narrower sense – is the most original form of poetry in the essential sense... . Building and portraying, by contrast, always happen already and only in the opening afforded by saying and naming. (1950: 61)

Clearly, language became central to Heidegger's account of the clearing of being within which entities can be understood, but language was also reconceived. The key point for our purposes is the shift from the transparency of assertion to the unsurpassability of language as poetic. Assertions are comportments toward a task at hand or a described entity. The specific words used matter only in pointing out the right issue or feature in the right way; any other words that would function similarly would do. In this respect, availability was the predominant way of being of linguistic signs in *Being and Time*. Poetry is not paraphrasable, however; like the material of other artworks, poetic language does not disappear. Much more needs to be said about this point than I can develop here, but clearly Heidegger came to believe that the historical specificity and sonorous/graphic materiality of language was integral to human understanding.

Finally, both *Holzwege* essays implicitly showed Heidegger's abandonment of fundamental ontology as a successor to Husserl's conception of philosophy as rigorous science. The attempt to articulate essential differences among ways of being, and thus to do ontology as a philosophical "science," was supplanted by an historicized understanding of the intelligibility of entities. Attentiveness to the difference between entities and their intelligibility was no longer expressed by a *philosophical* classification of essentially different ways of being. Not only did Dasein's way of being thereby lose its ontological centrality, but occurrentness and availability also lost their standing as fundamental ontological categories. That in turn meant that science could no longer have the ontological significance of discovering entities as occurrent.

This abandonment of fundamental ontology significantly transformed Heidegger's phenomenology of science, developed most extensively in "Age of the World-Picture." Having lost its general ontological significance, science was reconceived as an essential phenomenon of modernity. From the perspective laid out in that essay, the earlier account of science as the discovery of the occurrent looks like idle talk, a groundless passing on of familiar sayings about science as cognition or as justified assertion. In place of this residue of traditional theories of knowledge, Heidegger characterized modern science instead as research, a relentless activity that encompasses its practitioners "within the essential form of the technologist in the essential sense" (1950: 78). Modern science was not the suspension of practical concern with entities, but its intensification.

Heidegger retained from *Being and Time* the idea that what was decisive for modern science was the "mathematical projection of nature," but he radically shifted his conception of what such a projection accomplished. Throughout, Heidegger had used the term 'mathematical' to indicate not the use of mathematics *per se*, but more generally the determination of entities by a prior understanding of their being.

> *Ta mathēmata* means for the Greeks that which man knows in advance in observing entities and dealing with things: the corporeality of bodies, the vegetable character of plants, the animality of animals, the humanness of man. (1950: 71–2)

In *Being and Time*, the "mathematical" character of physics was expressed in the "projection that discovers in advance something constantly occurrent (matter), and opens the horizon to look for guidance to its quantitatively determinable constitutive aspects (motion, force, location, and time)" (1953: 362). This way of allowing entities to show themselves disentangled them from their practical involvements so as to

thematize them as objects. On his revised view, the mathematical projection of physical entities did not disentangle them from practical involvement, but instead intensi-
fied and more stringently governed dealings with them. The mathematical projection
of nature determined the researcher's way of approaching and dealing with a domain
of entities:

> Every forging-ahead [*Vorgehen*] already requires a circumscribed domain in which it
> moves. And it is precisely the opening up of such a domain that is the fundamental
> process [*Grundvorgang*] in research. This is accomplished, insofar as within a region of
> entities, e.g., nature, a determinate configuration of natural processes [*Naturvorgänge*] has
> been projected. This projection sketches out beforehand the way that a cognizant
> forging-ahead must bind itself to the domain opened up. This binding commitment is
> the rigor of research... . This projection of nature is secured, insofar as physical
> research binds itself to it in each step of its questioning. (1950: 71, 72–3)

Heidegger presented such a rigorously self-binding moving ahead within a projected
domain of entities as the first essential characteristic of science that has been transformed into research.

A second distinctive feature of research is that its movement must be guided by a
distinctive way of proceeding. As it advances further into a projected domain, research must be open to variation and novelty among the phenomena discovered, and
yet must also sustain the generality and objectivity of its overall conception. This dual
demand accounts for the centrality of natural laws in modern scientific explanation:

> Only within the purview of the incessant-otherness of change does the rich particularity
> of facts show itself. But the facts must become objective. The forging-ahead [of science]
> must therefore represent the changeable in its changing, holding it steady while never-
> theless letting the motion be a motion. The stasis of facts in their continuing variation is
> regularity (*Regel*). The constancy of change in the necessity of its course is law. Facts first
> become clear as the facts they are within the purview of regularity and law. Empirical
> research into nature is intrinsically the putting forward and confirming of regularities and
> laws. (1950: 73–4)

This process of unifying manifold phenomena under more general laws simultaneously extends and legitimates the projection of nature that governs ongoing research:

> Explanation, as a clarification on the basis of what is clear, is always ambiguous. It
> accounts for an unknown by means of a known, and at the same time confirms that
> known by means of that unknown. (1950: 74)

The facts receive their definitive determination through their subsumption under law,
whose authority is secured by its success in accounting for a multitude of facts.

Heidegger thought that the turn to experimental science was a consequence of this
novel way of proceeding rather than its basis. Only because nature had been reconceived as the unification of diverse events under law could the creation of new
phenomena in the laboratory be thought to yield fundamental insights rather than just
a proliferation of curiosities.

Experiment begins with the laying down of a law as its basis. To set up an experiment means to represent a condition under which a definite configuration of motions is trackable in the necessity of its course, i.e., of being controlled in advance by calculation. (1950: 74)

This shift has less to do with the projection of nature as a distinctively law-governed domain, however, than with a more general imperative of research itself. In Heidegger's view, all modern research methods, from experimentation to historical source criticism, depend upon a comparable play between a more general explanatory scheme and the particular objects or events subsumed within it. Research inevitably forms specialized disciplines, each seeking to pursue its characteristic explanatory scheme as far as it can go.

Heidegger took this relentless extension of its explanatory frameworks to be the third fundamental characteristic of modern science, as *enterprise* (*Betrieb*).[14] What determines the direction of scientific research is not the significance of the results to be sought, but the need to secure and expand the enterprise of science itself:

The way of proceeding [*Verfahren*] through which individual object-domains are conquered does not simply amass results. Rather, with the help of its results, it adapts itself for a new forging-ahead ... . This having to adapt itself to its own results as the ways and means of an onward-marching way of proceeding is the essence of research's character as enterprising. (1950: 77)

In *Being and Time*, Heidegger had worried that the sciences' efforts to develop inferentially interconnected networks of assertions tended to obscure their accountability to the entities they thereby discovered. In the "World-Picture" essay, an analogue to that tendency has become the defining modus operandi of scientific research. Supplying the incessant demands of the research enterprise for new problems to work on, and new material, conceptual, and institutional resources to apply to those problems, takes precedence over the disclosure and discovery of entities:

What is taking place in this extending and consolidating of the institutional character of the sciences? Nothing less than securing the precedence of their way of proceeding [*Verfahren*] over the entities (nature and history) that are being objectified in research at that time. (1950: 78)

The enterprising character of modern science also transforms its participants. Researchers are not scholars. Their characteristic virtues are not erudition but incisiveness, not reflection but constant activity, not insight but effectiveness in getting the job done.

What is the "job" of modern science, however? In *Being and Time*, science was understood to be for the sake of discovering entities as occurrent. The crucial task for a philosophy of science was then to guide scientific interpretations of entities according to regional ontologies consonant with the insights to be gained from fundamental ontology. The modern orientation of science as research presented in "Age of the World-Picture" undermines any such philosophical governance, however.[15] It seeks to maximize the flexibility of the research enterprise itself, unconstrained by any prior accountability to a domain of entities:

The predilection imposed by the actual system of science is not for a contrived and rigidly interrelated unification of the content of object-domains, but for the greatest possible free but regulated flexibility in initiating and switching the leading task of research at any given time. The more exclusively science isolates itself for the complete conduct [Betreibung] and mastery of its work process, and the more unapologetically its enterprises [Betriebe] are transferred to research institutes and professional schools, the more irresistibly do the sciences consummate their modern essence. (1950: 79)

What makes a research task important is not the intrinsic significance of what it will thereby discover about the world, but the extent to which such discovery might open new vistas for further research. Here Heidegger emphasized a kinship between modern science and technology, but not simply because of the application of knowledge or the use of technology within research. Rather, he saw them conjoined in relentlessly overriding any external accountability that might constrain the expansion of their own capacities for calculation and control. There is and can be no further "for-the-sake-of-which" for modern scientific research; it orders and calculates in order to expand the domain of research, by making entities more fully and extensively calculable. Here, ironically, was a different parallel to Plato: Heidegger characterized the research enterprise as like the tyrant's soul, driven by its own insatiable aspiration to power.

What does this conception of the telos of modern science then imply for the task of a philosophy of science? In abandoning fundamental ontology and the ahistorical conception of science as discovering entities as occurrent, Heidegger did not give up the aspiration to place science at the center of a large *philosophical* story about truth and being. The convergence of science and technology was conceived as an essential phenomenon of modernity, and thereby as a focus for *metaphysical* reflection. Technoscience allowed entities to show themselves as calculable and orderable, and thereby revealed what is at stake in modernity as the impending loss of any meaningful differences. The working of art that could enable a particular human community to sustain and hold it responsible to something genuinely at stake in its activities was being closed off. The source of this tendency was supposedly not just a sociological drive toward professional autonomy for scientific institutions, but a metaphysical transformation of the intelligibility of entities.

Heidegger characterized the modern world metaphysically as the "age of the world-picture":

World-picture, when understood essentially, does not mean a picture of the world but the world conceived as picture. What is in its entirety [entities] is now taken to be first and only insofar as it is set in place by human representation and production. (1950: 82)

Note well that Heidegger was not here endorsing an idealist or constructivist thesis about how entities come to exist. He was instead claiming that the being of entities (their intelligibility as entities) is now determined by the demands of human thought and action. That is not merely a distinctive understanding of other entities as objects, but also of humans as subjects. The link between the two can be seen in the conception of accountability to entities in terms of "objectivity," or correct representation. The ideal of objectivity is to allow the object to show itself as it is, unchanged by our

ways of conceiving or dealing with it. But what is thereby determined is not something about the object, but about how we ought to deal with it. Taking the right
stance toward it or employing the right methods is decisive for whether it shows itself
rightly. Human representation and praxis thereby become the arbiter of what is real.

This conception would seem to place human beings in an exalted position. Our
norms and goals govern what is in its entirety. But Heidegger thought that sense of
mastery was illusory. The relentlessly conjoined objectification of entities and subjec-
tification of our accountability to them inevitably turned such accountability itself
into a further object (a "value") for a subject. Values then need clarification and
objective assessment in turn, but their objectification as values to be chosen undermines any authority they could have over the choice.

> Value appears to express that one is positioned toward it so as to pursue what is most
> valuable, and yet that very value is the impotent and threadbare disguise of the objectiv-
> ity of entities having become flat and backgroundless. No one dies for mere values.
> (1950: 94)

The loss of any accountability beyond ourselves, and hence of the possibility that
what we do could make a significant difference, was what supposedly conjoined
science and technology with the subjectivization of art and the holy as essential
phenomena of modernity.

This historicized conception of philosophy as metaphysics sustained Heidegger's
consistently negative assessment of the sciences' capacity to understand their own
significance and normativity. Science as such could not uncover its "essence," the
metaphysics of the world as picture which made the transformation of science into a
research enterprise seem appropriate and inevitable. Only philosophical reflection
could hold open the possibility of an alternative understanding. This claim depended
upon a contentious distinction between science and philosophy, however. In lectures
contemporaneous with the *Holzwege* essays, Heidegger acknowledged that Galileo
and Newton, or Heisenberg and Bohr, were doing philosophy rather than "mere"
science.[16] In one of his last essays (1969), Heidegger moved further in the direction of
naturalism, abandoning any effort to retain *philosophical* authority over the sciences.
He concluded that philosophy, the tradition of inquiry spanning from Plato to
Nietzsche, reached its legitimate completion by developing into "independent sciences which interdependently communicate" (1969: 64). Lest we fail to get the
point, Heidegger referred explicitly to Husserl's rejection of psychologism and historicism in "Philosophy as Rigorous Science." In effect, late Heidegger endorsed Quine's
naturalistic rejection of "first philosophy." Yet even here, Heidegger retained a consistent orientation. This shift occurred not because Heidegger acknowledged the
possibility of understanding the normative significance of the sciences from within,
but because he denied that philosophical metaphysics provided a sufficiently independent basis for such understanding.

Yet a more sanguine naturalism might still be an unthought possibility within
Heidegger's work. Heidegger's account of science as research has some plausibility in
light of the "Second Scientific Revolution," through which the Baconian sciences of
chemistry, heat, electricity, magnetism, and, later, biology and geology were brought

within the purview of disciplined, experimental science. Capacities for experimental manipulation and theoretical modeling have indeed expanded enormously over the past two centuries, and have reinforced one another. Yet Heidegger's characterization of this expansion as governed only by an unconstrained opportunism overlooks the extent to which only a few phenomena within these domains matter scientifically. Which phenomena are at issue in a given field (or even a particular research program) has frequently shifted over time, with accompanying shifts in what is at stake in such work. Suppose such shifts are driven not merely by a concern to keep the research enterprise going, but by an implicit concern for what it really matters to understand scientifically and why. One might then envision a more expansive conception of the scientific enterprise, and of the place of "philosophical" concerns within the sciences, which would incorporate serious discussion of what is at issue and at stake in competing conceptions of a scientific discipline. Such a conception might be encouraged by recognition of the unappreciated affinities between Heidegger's ontological conception of science and philosophical naturalism. Rethinking Heidegger's philosophy of science in this way must, however, be left to another occasion.

Notes

1   All translations from *Sein und Zeit* and *Holzwege* are my own. Thanks to William Blattner and Taylor Carman for reviewing and suggesting changes in the *Holzwege* translations.

2   The word "possibility" can be misleading here. What Heidegger has in mind are not possible actualities (definite sets of objects, properties, and relations that might have obtained, but in fact do not), but actual possibilities (we are already situated within the world as oriented toward definite but not fully determinate ways for us to be). We can comport ourselves toward possibilities without needing to represent them to ourselves, even implicitly.

3   Note that Heidegger's terms for Dasein's most basic relations to itself and to other entities, such as "care," "concern," or "solicitude," refer not to mental states, but to whole ways of comporting oneself.

4   In his lectures on *History of the Concept of Time*, Heidegger (1985) credited phenomenology with a recovery of "the original sense of the a priori." He contrasted this sense of the a priori to post-Cartesian conceptions, for which "the term a priori has been attributed first and foremost to knowing," such that "knowledge is a priori when it is not based upon empirical inductive experience," and is thus "accessible first and only in the subject as such" (p. 73). Heidegger instead used the term "a priori" to designate what is ontologically prior, the conditions of possibility of entities themselves (as the kind of entities they are) rather than merely as the conditions of possibility of our *knowledge* of entities. Heidegger's call for reflection upon the a priori conditions of possibility of entities was thus directly opposed to armchair philosophy and a quest for a priori knowledge in the familiar sense. The a priori in Heidegger's sense is "nothing immanent, [nothing] belonging primarily to the sphere of the subject" (p. 74). Indeed, Heidegger's claim in *Being and Time* that all interpretation presupposes an understanding of being has the consequence that there can be no a priori knowledge in the usual philosophical sense.

5   Michael Friedman's (1992) interpretation in *Kant and the Exact Sciences* thus closely parallels Heidegger's understanding of how philosophy as ontological inquiry might

constructively relate to the positive sciences, even though Heidegger and Friedman differ significantly in their reading of Kant as a philosopher of science and of nature.

6   This issue has been central to John Haugeland's (1998: ch. 13; 2000) most recent work on Heidegger.

7   In this respect, Heidegger's work bears surprising affinities to McDowell 1994.

8   In this respect, Otto Neurath was an exception among the logical positivists. Rouse (2002: ch. 1) discusses the ironic parallels between Heidegger's and Neurath's turns toward naturalism.

9   Brandom (2002: ch. 9) gives a lucid account of what such a formalization of availability would look like.

10   "Circumspection" (*Umsicht*) denoted Dasein's situated attentiveness to relevant cues within its immersion in a concrete task-at-hand, in contrast to a disengaged observation of entities as occurrent.

11   Heidegger used the term "Lichtung" (literally a "clearing" in the woods, but etymologically related to "Licht," light) as an analogue to what he had earlier called an "understanding of being." He retains the ontological difference between entities and the clearing as the intelligibility (being) of the entities disclosed within it.

12   It is surely no coincidence that in Book X, Plato develops this point about an artifact (a bed), and that Heidegger begins the artwork essay with the claim that the equipmental character of shoes is more adequately revealed in a painting than by the shoes themselves or the concept of shoes.

13   Even here, however, it does so only nostalgically, since Heidegger strongly suggested that artworks can no longer work in this way in modernity.

14   The usual English translation of Heidegger's "Betrieb" as "ongoing activity" or "continuing activity" misses its overtones of business enterprise and factory works.

15   Throughout his career, Heidegger often addressed the prospect of philosophical governance of the various scientific disciplines in terms of the need for university reform. Crowell (1997) provides insightful discussion of this issue. One might well ask whether the "World-Picture" essay's account of the sciences as overriding any wider normative accountability partially responds to Heidegger's own disastrous attempt five years earlier to give philosophical direction to the University of Freiburg as its National Socialist rector.

16   Heidegger 1967: 67. In *Being and Time* (1953: 9–10) Heidegger had cited relativity theory as exemplary of an ontological reawakening in physics. The omission of Einstein's name alongside Heisenberg and Bohr 10 years later inevitably invites questions about Heidegger's possible deference to Nazi campaigns against "Jewish physics."

References

Brandom, Robert. 2002. *Tales of the Mighty Dead: Historical Essays in the Metaphysics of Intentionality.* Cambridge, MA: Harvard University Press.

Crowell, Steven. 1997. "Philosophy as a Vocation." *History of Philosophy Quarterly* 14: 255–76.

Friedman, Michael. 1992. *Kant and the Exact Sciences.* Cambridge, MA: Harvard University Press.

Haugeland, John. 1998. *Having Thought.* Cambridge, MA: Harvard University Press.

——. 2000. "Truth and Finitude: Heidegger's Transcendental Existentialism." In M. Wrathall and J. Malpas, eds., *Heidegger, Authenticity, and Modernity.* Cambridge, MA: MIT Press, 43–77.

Heidegger, Martin. 1950. *Holzwege.* Frankfurt am Main: Vittorio Klostermann.

——. 1953. *Sein und Zeit*. Tübingen: Max Niemeyer.

——. 1967. *What Is a Thing?*, trans. W. Barton and V. Deutsch. Chicago: Regnery.

——. 1969. *Zur Sache des Denkens*. Tübingen: Max Niemeyer.

——. 1982. *The Basic Problems of Phenomenology*, trans. A. Hofstadter. Bloomington: Indiana University Press.

——. 1985. *History of the Concept of Time: Prolegomena*, trans. T. Kisiel. Bloomington: Indiana University Press.

Husserl, Edmund. 1981. "Philosophy as Rigorous Science." In *Shorter Works*, eds. P. McCormick and F. Elliston. Notre Dame: University of Notre Dame Press, 166–97.

Kuhn, Thomas. 1970. *The Structure of Scientific Revolutions*, 2nd ed. Chicago: University of Chicago Press.

McDowell, John. 1994. *Mind and World*. Cambridge, MA: Harvard University Press.

Rouse, Joseph. 2002. *How Scientific Practices Matter: Reclaiming Philosophical Naturalism*. Chicago: University of Chicago Press.

Todes, Samuel. 2001. *Body and World*. Cambridge, MA: MIT Press.

10

From *ON "TIME AND BEING"*

Martin Heidegger

The End of Philosophy and the Task of Thinking

The title designates the attempt at a reflection which persists in questioning. The questions are paths to an answer. If the answer could be given, the answer would consist in a transformation of thinking, not in a propositional statement about a matter at stake.

The following text belongs to a larger context. It is the attempt undertaken again and again ever since 1930 to shape the question of Being and Time in a more primal way. This means: to subject the point of departure of the question in *Being and Time* to an immanent criticism. Thus it must become clear to what extent the *critical* question of what the matter of thinking is, necessarily and continually belongs to thinking. Accordingly, the name of the task of *Being and Time* will change.

We are asking:

1. What does it mean that philosophy in the present age has entered its final stage?
2. What task is reserved for thinking at the end of philosophy?

1. What does it mean that philosophy in the present age has entered its final stage?

Philosophy is metaphysics. Metaphysics thinks being as a whole — the world, man, God — with respect to Being, with respect to the belonging together of beings in Being. Metaphysics thinks beings as being in the manner of representational thinking which gives reasons. For since the beginning of philosophy and with that beginning, the Being of beings has showed itself as the ground (*arche*, *aition*). The ground is from where beings as such are what they are in their becoming, perishing and persisting as something that can be known, handled and worked upon. As the ground, Being brings beings to their actual presencing. The ground shows itself as presence. The present of presence consists

Martin Heidegger, pp. 55–73 from *On "Time and Being"*, trans. Joan Stambaugh. New York: Harper & Row, 1972 (reissued 1977). English language © 1972 by Harper & Row Publishers Inc. Reprinted by permission of HarperCollins Publishers Inc.

in the fact that it brings what is present each in its own way to presence. In accordance with the actual kind of presence, the ground has the character of grounding as the ontic causation of the real, as the transcendental making possible of the objectivity of objects, as the dialectical mediation of the movement of the absolute Spirit, of the historical process of production, as the will to power positing values.

What characterizes metaphysical thinking which grounds the ground for beings is the fact that metaphysical thinking departs from what is present in its presence, and thus represents it in terms of its ground as something grounded.

What is meant by the talk about the end of philosophy? We understand the end of something all too easily in the negative sense as a mere stopping, as the lack of continuation, perhaps even as decline and impotence. In contrast, what we say about the end of philosophy means the completion of metaphysics. However, completion does not mean perfection as a consequence of which philosophy would have to have attained the highest perfection at its end. Not only do we lack any criterion which would permit us to evaluate the perfection of an epoch of metaphysics as compared with any other epoch. The right to this kind of evaluation does not exist. Plato's thinking is no more perfect than Parmenides'. Hegel's philosophy is no more perfect than Kant's. Each epoch of philosophy has its own necessity. We simply have to acknowledge the fact that a philosophy is the way it is. It is not our business to prefer one to the other, as can be the case with regard to various *Weltanschauungen*.

The old meaning of the word "end" means the same as place: "from one end to the other" means: from one place to the other. The end of philosophy is the place, that place in which the whole of philosophy's history is gathered in its most extreme possibility. End as completion means this gathering.

Throughout the whole history of philosophy, Plato's thinking remains decisive in changing forms. Metaphysics is Platonism. Nietszche characterizes his philosophy as reversed Platonism. With the reversal of metaphysics which was already accomplished by Karl Marx, the most extreme possibility of philosophy is attained. It has entered its final stage. To the extent that philosophical thinking is still attempted, it manages only to attain an epigonal renaissance and variations of that renaissance. Is not then the end of philosophy after all a cessation of its way of thinking? To conclude this would be premature.

As a completion, an end is the gathering into the most extreme possibilities. We think in too limited a fashion as long as we expect only a development of recent philosophies of the previous style. We forget that already in the age of Greek philosophy a decisive characteristic of philosophy appears: the development of sciences within the field which philosophy opened up. The development of the sciences is at the same time their separation from philosophy and the establishment of their independence. This process belongs to the completion of philosophy. Its development is in full swing today in all regions of beings. This development looks like the mere dissolution of philosophy, and is in truth its completion.

It suffices to refer to the independence of psychology, sociology, anthropology as cultural anthropology, to the role of logic as logistics and semantics. Philosophy turns into the empirical science of man, of all of what can become the experiential object of his technology for man, the technology by which he establishes himself in the world by working on it in the manifold modes of making and shaping. All of this happens everywhere on the basis and according to the criterion of the scientific discovery of the individual areas of beings.

No prophecy is necessary to recognize that the sciences now establishing them-selves will soon be determined and guided by the new fundamental science which is called cybernetics.

This science corresponds to the determination of man as an acting social being. For it is the theory of the steering of the possible planning and arrangement of human labor. Cybernetics transforms language into an exchange of news. The arts become regulated-regulating instruments of information.

The development of philosophy into the independent sciences which, however, interdependently communicate among themselves ever more markedly, is the legitim-ate completion of philosophy. Philosophy is ending in the present age. It has found its place in the scientific attitude of socially active humanity. But the fundamental char-acteristic of this scientific attitude is its cybernetic, that is, technological character. The need to ask about modern technology is presumably dying out to the same extent that technology more definitely characterizes and regulates the appearance of the totality of the world and the position of man in it.

The sciences will interpret everything in their structure that is still reminiscent of the origin from philosophy in accordance with the rules of science, that is, techno-logically. Every science understands the categories upon which it remains dependent for the articulation and delineation of its area of investigation as working hypotheses. Their truth is measured not only by the effect which their application brings about within the progress of research.

Scientific truth is equated with the efficiency of these effects.

The sciences are now taking over as their own task what philosophy in the course of its history tried to present in part, and even there only inadequately, that is, the ontologies of the various regions of beings (nature, history, law, art). The interest of the sciences is directed toward the theory of the necessary structural concepts of the coordinated areas of investigation. "Theory" means now: supposition of the categor-ies which are allowed only a cybernetical function, but denied any ontological mean-ing. The operational and model character of representational-calculative thinking becomes dominant.

However, the sciences still speak about the Being of beings in the unavoidable supposition of their regional categories. They just don't say so. They can deny their origin from philosophy, but never dispense with it. For in the scientific attitude of the sciences, the document of their birth from philosophy still speaks.

The end of philosophy proves to be the triumph of the manipulable arrangement of a scientific-technological world and of the social order proper to this world. The end of philosophy means: the beginning of the world civilization based upon Western European thinking.

But is the end of philosophy in the sense of its development to the sciences also already the complete realization of all the possibilities in which the thinking of philosophy was posited? Or is there a *first* possibility for thinking apart from the *last* possibility which we characterized (the dissolution of philosophy in the technologized sciences), a possibility from which the thinking of philosophy would have to start out, but which as philosophy it could nevertheless not experience and adopt?

If this were the case, then a task would still have to be reserved for thinking in a concealed way in the history of philosophy from its beginning to its end, a task accessible neither to philosophy as metaphysics nor, and even less so, to the sciences stemming from philosophy. Therefore we ask:

2. *What task is reserved for thinking at the end of philosophy?*

The mere thought of such a task of thinking must sound strange to us. A thinking which can be neither metaphysics nor science?

A task which has concealed itself from philosophy since its very beginning, even in virtue of that beginning, and thus has withdrawn itself continually and increasingly in the time to come?

A task of thinking which – so it seems – includes the assertion that philosophy has not been up to the matter of thinking and has thus become a history of mere decline?

Is there not an arrogance in these assertions which desires to put itself above the greatness of the thinkers of philosophy?

This suspicion easily suggests itself. But it can as easily be removed. For every attempt to gain insight into the supposed task of thinking finds itself moved to review the whole of the history of philosophy. Not only this, but it is even forced to think the historicity of that which grants a possible history to philosophy.

Because of this, that supposed thinking necessarily falls short of the greatness of the philosophers. It is less than philosophy. Less also because the direct or indirect effect of this thinking on the public in the industrial age, formed by technology and science, is decisively less possible to this thinking than it was in the case of philosophy.

But above all, the thinking in question remains slight because its task is only of a preparatory, not of a founding character. It is content with awakening a readiness in man for a possibility whose contour remains obscure, whose coming remains uncertain.

Thinking must first learn what remains reserved and in store for thinking to get involved in. It prepares its own transformation in this learning.

We are thinking of the possibility that the world civilization which is just now beginning might one day overcome the technological-scientific-industrial character as the sole criterion of man's world sojourn. This may happen not of and through itself, but in virtue of the readiness of man for a determination which, whether listened to or not, always speaks in the destiny of man which has not yet been decided. It is just as uncertain whether world civilization will soon be abruptly destroyed or whether it will be stabilized for a long time, in a stabilization, however, which will not rest in something enduring, but rather establish itself in a sequence of changes, each of which presents the latest fashion.

The preparatory thinking in question does not wish and is not able to predict the future. It only attempts to say something to the present which was already said a long time ago precisely at the beginning of philosophy and for that beginning, but has not been explicitly thought. For the time being, it must be sufficient to refer to this with the brevity required. We shall take a directive which philosophy offers as an aid in our undertaking.

When we ask about the task of thinking, this means in the scope of philosophy: to determine that which concerns thinking, which is still controversial for thinking, which is the controversy. This is what the word "matter" means in the German language. It designates that with which thinking has to do in the case at hand, in Plato's language *to pragma auto* (cf. "The Seventh Letter" 341 C7).

In recent times, philosophy has of its own accord expressly called thinking "to the things themselves." Let us mention two cases which receive particular attention today. We hear this call "to the things themselves" in the "Preface" which Hegel has placed before his work which was published in 1807, *System of Science*,[1] first part: "The Phenomenology of Spirit." This preface is not the preface to the *Phenomenology*, but to the *System of Science*, to the whole of philosophy. The call "to the things themselves" refers ultimately – and that means: according to the matter, primarily – to the *Science of Logic*.

In the call "to the things themselves," the emphasis lies on the "themselves." Heard superficially, the call has the sense of a rejection. The inadequate relations to the matter of philosophy are rejected. Mere talk about the purpose of philosophy belongs to these relations, but so does mere reporting about the results of philosophical thinking. Both are never the real totality of philosophy. The totality shows itself only in its becoming. This occurs in the developmental presentation of the matter. In the presentation, theme and method coincide. For Hegel, this identity is called: the idea. With the idea, the matter of philosophy "itself" comes to appear. However, this matter is historically determined: subjectivity. With Descartes' *ego cogito*, says Hegel, philosophy steps on firm ground for the first time where it can be at home. If the *fundamentum absolutum* is attained with the *ego cogito* as the distinctive *subiectum*, this means: The subject is the *hypokeimenon* which is transferred to consciousness, what is truly present, what is unclearly enough called "substance" in traditional language.

When Hegel explains in the Preface (ed. Hoffmeister, p. 19), "The true (in philoso-phy) is to be understood and expressed not as substance, but just as much as subject," then this means: The Being of beings, the presence of what is present, is only manifest and thus complete presence when it becomes present as such for itself in the absolute Idea. But since Descartes, *idea* means: *perceptio*. Being's coming to itself occurs in speculative dialectic. Only the movement of the idea, the method, is the matter itself. The call "to the thing itself" requires a philosophical method appropriate in it.

However, what the matter of philosophy should be is presumed to be decided from the outset. The matter of philosophy as metaphysics is the Being of beings, their presence in the form of substantiality and subjectivity.

A hundred years later, the call "to the thing itself" again is uttered in Husserl's treatise *Philosophy as Exact Science*. It was published in the first volume of the journal *Logos* in 1910–11 (pp. 289 ff.). Again, the call has at first the sense of a rejection. But here it aims in another direction than Hegel's. It concerns naturalistic psychology which claims to be the genuine scientific method of investigating consciousness. For this method blocks access to the phenomena of intentional consciousness from the very beginning. But the call "to the thing itself" is at the same time directed against historicism which gets lost in treatises about the standpoints of philosophy and in the ordering of types of philosophical *Weltanschauungen*. About this Husserl says in italics (*ibid.*, p.340): *"The stimulus for investigation must start not with philosophies, but with issues and problems."*

And what is at stake in philosophical investigation? In accordance with the same tradition, it is for Husserl as for Hegel the subjectivity of consciousness. For Husserl, the *Cartesian Meditations* were not only the topic of the Parisian lectures in February, 1920. Rather, since the time following the *Logical Investigations*, their spirit accompanied the impassioned course of his philosophical investigations to the end. In its negative and also in its positive sense, the call "to the thing itself" determines the securing and development of method. It also determines the procedure of philosophy by means of which the matter itself can be demonstrated as a datum. For Husserl, "the principle of all principles" is first of all not a principle of content, but one of method. In his work published in 1913,[2] Husserl devoted a special section (section 24) to the determination of "the principle of all principles." "No conceivable theory can upset this principle," says Husserl (*ibid.*, p. 44).

"The principle of all principles" reads:

> that very primordial dator Intuition is a source of authority (Rechtsquelle) for knowledge, that whatever presents itself in "Intuition" in primordial form (as it were in its bodily reality), is simply to be accepted as it gives itself out to be, though only within the limits in which it then presents itself.

"The principle of all principles" contains the thesis of the precedence of method. This principle decides what matter alone can suffice for the method. "The principle of principles" requires reduction to absolute subjectivity as the matter of philosophy. The transcendental reduction to absolute subjectivity gives and secures the possibility of grounding the objectivity of all objects (the Being of this being) in its valid structure and consistency, that is, in its constitution, in and through subjectivity. Thus transcendental subjectivity proves to be "the sole absolute being" (*Formal and Transcendental Logic*, 1929, p. 240). At the same time, transcendental reduction as the method of "universal science" of the constitution of the Being of beings has the same mode of being as this absolute being, that is, the manner of the matter most native to philosophy. The method is not only directed toward the matter of philosophy. It does not just belong to the matter as a key belongs to a lock. Rather, it belongs to the matter because it is "the matter itself." If one wanted to ask: Where does "the principle of all principles" get its unshakable right, the answer would have to be: from transcendental subjectivity which is already presupposed as the matter of philosophy.

We have chosen a discussion of the call 'to the thing itself' as our guideline. It was to bring us to the path which leads us to a determination of the task of thinking at the end of philosophy. Where are we now? We have arrived at the insight that for the call "to the thing itself," what concerns philosophy as its matter is established from the outset. From the perspective of Hegel and Husserl – and not only from their perspective – the matter of philosophy is subjectivity. It is not the matter as such that is controversial for the call, but rather its presentation by which the matter itself becomes present. Hegel's speculative dialectic is the movement in which the matter as such comes to itself, comes to its own presence. Husserl's method is supposed to bring the matter of philosophy to its ultimately originary givenness, that means: to its own presence.

The two methods are as different as they could possibly be. But the matter as such which they are to present is the same, although it is experienced in different ways.

But of what help are these discoveries to us in our attempt to bring the task of thinking to view? They don't help us at all as long as we do not go beyond a mere discussion of the call and ask what remains unthought in the call "to the thing itself."

Questioning in this way, we can become aware how something which it is no longer the matter of philosophy to think conceals itself precisely where philosophy has brought its matter to absolute knowledge and to ultimate evidence.

But what remains unthought in the matter of philosophy as well as in its method? Speculative dialectic is a mode in which the matter of philosophy comes to appeal of itself and for itself, and thus becomes presence. Such appearance necessarily occurs in some light. Only by virtue of light, i.e., through brightness, can what shines show itself, that is, radiate. But brightness in its turn rests upon something open, something free which might illuminate it here and there, now and then. Brightness plays in the open and wars there with darkness. Wherever a present being encounters another present being or even only lingers near it – but also where, as with Hegel, one being mirrors itself in another speculatively – there openness already rules, open region is in play. Only this openness grants to the movement of speculative thinking the passage through that which it thinks.

We call this openness which grants a possible letting-appear and show "opening." In the history of language, the German word "opening" is a borrowed translation of the French *clairière*. It is formed in accordance with the older words *Waldung* (foresting) and *Feldung* (fielding).

The forest clearing (opening) is experienced in contrast to dense forest, called "density" (*Dickung*) in older language. The substantive "opening" goes back to the verb "to open." The adjective *licht* "open" is the same word as "light." To open something means: To make something light, free and open, e.g., to make the forest free of trees at one place. The openness thus originating is the clearing. What is light in the sense of being free and open has nothing in common with the adjective "light," meaning "bright" – neither linguistically nor factually.[3] This is to be observed for the difference between openness and light. Still, it is possible that a factual relation between the two exists. Light can stream into the clearing, into its openness, and let brightness play with darkness in it. But light never first creates openness. Rather, light presupposes openness. However, the clearing, the opening, is not only free for brightness and darkness, but also for resonance and echo, for sounding and diminishing of sound. The clearing is the open for everything that is present and absent.

It is necessary for thinking to become explicitly aware of the matter called opening here. We are not extracting mere notions from mere words, e.g., "opening," as it might easily appear on the surface. Rather, we must observe the unique matter which is adequately named with the name "opening." What the word designates in the connection we are now thinking, free openness, is a "primal phenomenon," to use a word of Goethe's. We would have to say a primal matter. Goethe notes (*Maxims and Reflections*, n. 993): "Look for nothing behind phenomena: they themselves are what is to be learned." This means: The phenomenon itself, in the present case the opening, sets us the task of learning from it while questioning it, that is, of letting it say something to us.

Accordingly, we may suggest that the day will come when we will not shun the question whether the opening, the free open, may not be that within which alone pure space and ecstatic time and everything present and absent in them have the place which gathers and protects everything.

In the same way as speculative dialectical thinking, originary intuition and its evidence remain dependent upon openness which already dominates, upon the opening. What is evident is what can be immediately intuited. *Evidentia* is the word which Cicero uses to translate the Greek *enargeia*, that is, to transform it into the Roman. *Enargeia*, which has the same root as *argentum* (silver), means that which in itself and of itself radiates and brings itself to light. In the Greek language, one is not speaking about the action of seeing, about *videre*, but about that which gleams and radiates. But it can only radiate if openness has already been granted. The beam of light does not first create the opening, openness, it only traverses it. It is only such openness that grants to giving and receiving at all what is free, that in which they can remain and must move.

All philosophical thinking which explicitly or inexplicitly follows the call "to the thing itself" is already admitted to the free space of the opening in its movement and with its method. But philosophy knows nothing of the opening. Philosophy does speak about the light of reason, but does not heed the opening of Being. The *lumen naturale*, the light of reason, throws light only on openness. It does concern the opening, but so little does it form it that it needs it in order to be able to illuminate what is present in the opening. This is true not only of philosophy's *method*, but also and primarily of its *matter*, that is, of the presence of what is present. To what extent the *subiectum*, the *hypokeimenon*, that which already lies present, thus what is present in its presence is constantly thought also in subjectivity cannot be shown here in detail. Refer to Heidegger, *Nietzsche*, vol. 2 (1961), pages 429 ff.

We are concerned now with something else. Whether or not what is present is experienced, comprehended or presented, presence as lingering in openness always remains dependent upon the prevalent opening. What is absent, too, cannot be as such unless it presences in the *free space of the opening*.

All metaphysics including its opponent positivism speaks the language of Plato. The basic word of its thinking, that is, of his presentation of the Being of beings, is *eidos, idea:* the outward appearance in which beings as such show themselves. Outward appearance, however, is a manner of presence. No outward appearance without light – Plato already knew this. But there is no light and no brightness without the opening. Even darkness needs it. How else could we happen into darkness and wander through it? Still, the opening as such as it prevails through Being, through presence, remains unthought in philosophy, although the opening is spoken about in philosophy's beginning. How does this occur and with which names? Answer:

In Parmenides' reflective poem which, as far as we know, was the first to reflect explicitly upon the Being of beings, which still today, although unheard, speaks in the sciences into which philosophy dissolves. Parmenides listens to the claim:

> ... *kreo de se panta puthestha*
> *emen aletheies eukukleos atremes etor*
> *ede broton doxas, tais ouk emi pistis alethes.*
> Fragment I, 28 ff.

> ... but you should learn all:
> the untrembling heart of unconcealment, well-rounded
> and also the opinions of mortals,
> lacking the ability to trust what is unconcealed.[4]

Aletheia, unconcealment, is named here. It is called well-rounded because it is turned in the pure sphere of the circle in which beginning and end are everywhere the same. In this turning, there is no possibility of twisting, deceit and closure. The meditative man is to experience the untrembling heart of unconcealment. What does the word about the untrembling heart of unconcealment mean? It means unconcealment itself in what is most its own, means the place of stillness which gathers in itself what grants unconcealment to begin with. That is the opening of what is open. We ask: openness for what? We have already reflected upon the fact that the path of thinking, speculative and intuitive, needs the traversable opening. But in that opening rests possible radiance, that is, the possible presencing of presence itself.

What prior to everything else first grants unconcealment in the path on which thinking pursues one thing and perceives it: *hotos estin* ... *einai*: that presence presences. The opening grants first of all the possibility of the path to presence, and grants the possible presencing of that presence itself. We must think *aletheia*, unconcealment, as the opening which first grants Being and thinking and their presencing to and for each other. The quiet heart of the opening is the place of stillness from which alone the possibility of the belonging together of Being and thinking, that is, presence and perceiving, can arise at all.

The possible claim to a binding character or commitment of thinking is grounded in this bond. Without the preceding experience of *aletheia* as the opening, all talk about committed and non-committed thinking remains without foundation. Where does Plato's determination of presence as *idea* have its binding character from? With regard to what is Aristotle's interpretation of presencing as *energeia* binding?

Strangely enough, we cannot even ask these questions always neglected in philosophy as long as we have not experienced what Parmenides had to experience: *aletheia*, unconcealment. The path to it is distinguished from the street on which the opinion of mortals must wander around. *Aletheia* is nothing mortal, just as little as death itself.

It is not for the sake of etymology that I stubbornly translate the name *aletheia* as unconcealment, but for the matter which must be considered when we think that which is called Being and thinking adequately. Unconcealment is, so to speak, the element in which Being and thinking and their belonging together exist. *Aletheia* is named at the beginning of philosophy, but afterward it is not explicitly thought as such by philosophy. For since Aristotle it became the task of philosophy as metaphysics to think beings as such onto-theologically.

If this is so, we have no right to sit in judgment over philosophy, as though it left something unheeded, neglected it and was thus marred by some essential deficiency. The reference to what is unthought in philosophy is not a criticism of philosophy. If a criticism is necessary now, then it rather concerns the attempt which is becoming more and more urgent ever since *Being and Time* to ask about a possible task of thinking at the end of philosophy. For the question now arises, late enough: Why is *aletheia* not translated with the usual name, with the word "truth"? The answer must be:

Insofar as truth is understood in the traditional "natural" sense as the correspondence of knowledge with beings demonstrated in beings, but also insofar as truth is interpreted as the certainty of the knowledge of Being, *aletheia*, unconcealment in the sense of the opening may not be equated with truth. Rather, *aletheia*, unconcealment thought as opening, first grants the possibility of truth. For truth itself, just as Being and thinking, can only be what it is in the element of the opening. Evidence, certainty in every degree, every kind of verification of *veritas* already move *with* that *veritas* in the realm of the prevalent opening.

Aletheia, unconcealment thought as the opening of presence, is not yet truth. Is *aletheia* then less than truth? Or is it more because it first grants truth as *adequatio* and *certitudo*, because there can be no presence and presenting outside of the realm of the opening?

This question we leave to thinking as a task. Thinking must consider whether it can even raise this question at all as long as it thinks philosophically, that is, in the strict sense of metaphysics which questions what is present only with regard to its presence.

In any case, one thing becomes clear: To raise the question of *aletheia*, of unconcealment as such, is not the same as raising the question of truth. For this reason, it was inadequate and misleading to call *aletheia* in the sense of opening, truth.[5] The talk about the "truth of Being" has a justified meaning in Hegel's *Science of Logic*, because here truth means the certainty of absolute knowledge. But Hegel also, as little as Husserl, as little as all metaphysics, does not ask about Being as Being, that is, does not raise the question how there can be presence as such. There is presence only when opening is dominant. Opening is named with *aletheia*, unconcealment, but not thought as such.

The natural concept of truth does not mean unconcealment, not in the philosophy of the Greeks either. It is often and justifiably pointed out that the word *alethes* is already used by Homer only in the *verba dicendi*, in statement and thus in the sense of correctness and reliability, not in the sense of unconcealment. But this reference means only that neither the poets nor everyday language usage, not even philosophy see themselves confronted with the task of asking how truth, that is, the correctness of statements, is granted only in the element of the opening of presence.

In the scope of this question, we must acknowledge the fact that *aletheia*, unconcealment in the sense of the opening of presence, was originally only experienced as *orthotes*, as the correctness of representations and statements. But then the assertion about the essential transformation of truth, that is, from unconcealment to correctness, is also untenable.[6]

Instead we must say: *Aletheia*, as opening of presence and presenting in thinking and saying, originally comes under the perspective of *homoiosis and adaequatio*, that is, the perspective of adequation in the sense of the correspondence of representing with what is present.

But this process inevitably provokes another question: How is it that *aletheia*, unconcealment, appears to man's natural experience and speaking *only* as correctness and dependability? Is it because man's ecstatic sojourn in the openness of presencing is turned only toward what is present and the existent presenting of what is present? But what else does this mean than that presence as such, and together with it the opening granting it, remain unheeded? Only what *aletheia* as opening grants is experienced and thought, not what it is as such.

This remains concealed. Does this happen by chance? Does it happen only as a consequence of the carelessness of human thinking? Or does it happen because self-concealing, concealment, *lethe* belongs to *a-letheia*, not just as an addition, not as shadow to light, but rather as the heart of *aletheia*? And does not even a keeping and preserving rule in this self-concealing of the opening of presence from which unconcealment can be granted to begin with, and thus what is present can appear in its presence?

If this were so, then the opening would not be the mere opening of presence, but the opening of presence concealing itself, the opening of a self-concealing sheltering.

If this were so, then with these questions we would reach the path to the task of thinking at the end of philosophy.

But isn't this all unfounded mysticism or even bad mythology, in any case a ruinous irrationalism, the denial of *ratio*?

I return to the question: What does *ratio*, *nous*, *noein*, perceiving (*Vernunft–Vernehmen*) mean? What does ground and principle and especially principle of all principles mean? Can this ever be sufficiently determined unless we experience *aletheia* in a Greek manner as unconcealment and then, above and beyond the Greek, think it as the opening of self-concealing? As long as *ratio* and the rational still remain questionable in what is their own, talk about irrationalism is unfounded. The technological scientific rationalization ruling the present age justifies itself every day more surprisingly by its immense results. But these results say nothing about what the possibility of the rational and the irrational first grants. The effect proves the correctness of technological scientific rationalization. But is the manifest character of what-is exhausted by what is demonstrable? Doesn't the insistence on what is demonstrable block the way to what-is?

Perhaps there is a thinking which is more sober than the irresistible race of rationalization and the sweeping character of cybernetics. Presumably it is precisely this sweeping quality which is extremely irrational.

Perhaps there is a thinking outside of the distinction of rational and irrational still more sober than scientific technology, more sober and thus removed, without effect and yet having its own necessity. When we ask about the task of this thinking, then not only this thinking, but also the question about it is first made questionable. In view of the whole philosophical tradition, this means:

We all still need an education in thinking, and before that first a knowledge of what being educated and uneducated in thinking means. In this respect, Aristotle gives us a hint in Book IV of his *Metaphysics* (1006a ff.). It reads: *esti gar apaideusia to me gignoskein tinon dei zetein apodeixin kai tinon ou dei*. "For it is uneducated not to have an eye for when it is necessary to look for a proof, and when this is not necessary."

This sentence demands careful reflection. For it is not yet decided in what way that which needs no proof in order to become accessible to thinking is to be experienced. Is it dialectical mediation or originary intuition or neither of the two? Only the peculiar quality of that which demands of us above all else to be admitted can decide about that. But how is this to make the decision possible for us before we have not admitted it? In what circle are we moving here, inevitably?

Is it the *eukukleos alethein*, well-founded unconcealment itself, thought as the opening?

Does the name for the task of thinking then read instead of *Being and Time*: Opening and Presence?

But where does the opening come from and how is it given? What speaks in the "It gives"?

The task of thinking would then be the surrender of previous thinking to the determination of the matter of thinking.

Notes

1 *Wissenschaft, scientia,* body of knowledge, not "science" in the present use of that word. For German Idealism, science is the name for philosophy. (Tr.)
2 English edition: *Ideas* (New York: Collier Books, 1962). (Tr.)
3 Both meanings exist in English for light. The meaning Heidegger intends is related to lever (i.e., alleviate, lighten a burden). (Tr.)
4 Standard translation: "It is needful that you should learn of all matters – both the unshaken heart of well-rounded truth and the opinions of mortals which lack true belief." (Tr.)
5 How the attempt to think a matter can at times stray from that which a decisive insight has already shown, is demonstrated by a passage from *Being and Time* (1927) (p. 262, New York: Harper & Row, 1962). To translate this word *(aletheia)* as "truth," and, above all, to define this expression conceptually in theoretical ways, is to cover up the meaning of what the Greeks made "self- evidently" basic for the terminological use of aletheia as a prephilosophical way of understanding it.
6 This statement has profound implications for Heidegger's book *Platons Lehre von der Wahrheit.* (Tr.)

BACHELARD

11

TECHNOLOGY, SCIENCE, AND INEXACT KNOWLEDGE: BACHELARD'S NON-CARTESIAN EPISTEMOLOGY

Mary Tiles

In order to constitute a well-defined scientific fact, it is necessary to put a coherent technique to work. The action of science is essentially complex. It is on the basis of truths which are factitious and complex, rather than on the basis of those which are adventitious and clear, that the active empiricism of science is developed.

–Bachelard 1934: 176

Objectivity cannot be detached from the social characteristics of proof … Naturally, once one goes from observation to experimentation, the polemical character of knowledge becomes even more clear. Then it is necessary for phenomena to be selected, filtered, purified, poured into the mold of instrumentation, produced according to the design of instruments. Instruments are only materialized theories. From them come phenomena which bear the marks of theory.

–Bachelard 1934: 16

It must be stated clearly: substances studied by informed materialism are no longer, properly speaking, *natural givens*. Their social label is hereafter profoundly marked. Informed materialism is inseparable from its social status.

–Bachelard 1953: 31

This sounds like social constructivism; the substances and phenomena studied by modern science are artificial constructs, and modern science is an inherently social activity, its knowledge is socially constituted. Yet we should remember that these statements were made at a time when other philosophers of science were talking about foundations, simplicity, and observation protocol statements. Bachelard was not writing in the environment of late twentieth-century science studies or sociology of knowledge, nor was he a precursor for that movement. Although one can trace lines of intellectual influence from Bachelard to Kuhn via Koyré, Bachelard is no sociologist of knowledge. In fact, much to Latour's disgust, Bachelard persists in talking about epistemology, about the knower and the object of his or her knowledge.[1] As

Hacking says, "no other philosopher or historian so assiduously studied the realities of experimental life, nor was anyone less inclined than he to suppose the mind unimportant (his applied rationalism)" (Hacking 1984: 119).

Bachelard is no ordinary epistemologist; his emphases on technology and technique, on doing and making, as opposed to observing and theorizing, change the problematic within which epistemological discussion takes place. He simply isn't looking in the same place that most philosophers of science look, he isn't asking the same questions. So it is perhaps not surprising that he doesn't come up with answers to their problems, which would enable us to categorize him neatly as being on one or other side in the so-called "science wars" of the late twentieth century. He is concerned with the dynamics of scientific thought, not its statics, and as a result his portrayal of science is of something dynamic, something in process, not finalized, but where progress has been made in large part because the direction of further progress has been indicated.

> Concepts and methods, all are a function of the experimental domain; all scientific thought must change when confronted with a new experience; a discourse on scientific method will always be a circumstantial discourse, it will not describe a definitive constitution of the scientific mind. (Bachelard 1934: 139)

Here we are well outside the framework of Descartes's *Discourse on Method*, and this non-Cartesian theme is made explicit in what is perhaps the most widely known of Bachelard's works on science – *Le nouvel esprit scientifique*, written in 1934. However, much of the groundwork for that position is laid down in his first book, published in 1928, *Essai sur la connaissance approchée*. This is a work which has received relatively little attention, but which is important if we want to know exactly what is the direction in which Bachelard turns in his philosophy of science and to appreciate his reasons for going in this direction. Although "approximation" and "inexactness" do not appear as central themes in Bachelard's later writings on science, the positions developed in *Essai sur la connaissance approchée* are taken for granted and recast in slightly different language (for example, in discussions of *phenomenotechnique*, *rationalisme ouverte*, and the repeated references to the importance of "perturbations" in the verification and correction of laws).

Pragmatism Redirected

In the foreword to his *Essai*, Bachelard expresses his admiration for the American pragmatists. The work of Dewey is explicitly referenced twice and James 14 times.[2] What could Bachelard have found in pragmatism to build upon and redirect? Most obviously we find a view of the nature of the acquisition of scientific knowledge as a *process*, and as a process of *approximation*, both of which are reflected in the title "*connaissance approchée.*"

We find James saying, for example:

> But as the sciences have developed farther, the notion has gained ground that most, perhaps, of our laws are only approximations ... investigators have become accustomed

to the notion that no theory is absolutely a transcript of reality, but that any one of them may from some point of view be useful. (James 1975: 33)

and

The truth of an idea is not a stagnant property inherent in it. Truth *happens* to an idea. It *becomes* true, is *made* true by events. Its verity *is* in fact an event, a process: the process namely of its verifying itself, its veri-*fication*. Its validity is the process of its valid-*ation*. (James 1975: 169–70)

Equally significantly, in the light of what was to become a distinctive theme in Bachelard's own philosophy, we find James linking the theme of the dynamic and approximative character of scientific knowledge to the idea that scientific understanding requires a break with common sense.

Science and critical philosophy thus burst the bounds of common sense. With science *naïf* realism ceases: "Secondary" qualities become unreal: primary ones alone remain. With critical philosophy, havoc is made of everything. The commonsense categories one and all cease to represent anything in the way of *being*; they are but sublime tricks of human thought, our ways of escaping bewilderment in the midst of sensation's irremediable flow. (James 1975: 91)[3]

Here we see James rejecting both the basic sensory categories of traditional empiricism and the primary qualities of traditional rationalism. Practical functionality, as a basis for acceptance, is emphasized in another remark.

Theories become instruments, not answers to enigmas, in which we can rest. We don't lie back on them, we move forward, and, on occasion, make nature over again by their aid. Pragmatism unstiffens all our theories, limbers then up and sets each one at work.... pragmatic method means: *The attitude of looking away from first things, principles, "categories," supposed necessities; and of looking toward last things, fruits, consequences, facts.* (James 1975: 32)

Thus James's pragmatism both takes the emphasis of epistemology in the opposite direction to that of foundationalism (away from first principles) and takes it beyond the framework of language, of the psychology of belief, and of knowledge representation to the real world. Included in consequences are practical effects, the powers conferred by advances in scientific understanding.

Truth for us is simply a collective name for verification processes, just as health, wealth, strength, etc., are names for other processes connected with life, and also pursued because it pays to pursue them. Truth is *made*, just as health, wealth and strength are made, in the course of experience. (James 1975: 104)

Not only do the pragmatists in this way break with a Cartesian framework for the epistemology of science, they do so in such a way that they begin to take scientific knowledge out into the world. The truth "made," although a truth *for* an individual, is not and cannot be "made" *by* an individual. The making process includes humans interacting with each other and with the world they inhabit. In other words, this is already social epistemology (as distinct from sociology of knowledge). These themes are taken up by Bachelard, but he takes them in a direction not fully anticipated by James, because there is a crucial difference between their basic positions.

Bachelard's criticism of pragmatism amounts to the accusation that it has made an insufficient break with the intellectualism and dualism of the Cartesian cognitive tradition. His proposed revision of our conception of scientific knowledge is much more radical. This revision centers on outlining a dialectic between the two seemingly contrary drives in modern science. One of the first moves he makes in *Essai sur la connaissance approchée* is to note, as did others such as Whitehead (1925: 3), that modern science contains seemingly contradictory epistemological imperatives: perfect knowledge would have to be both minutely detailed and universal. These contrary tendencies, while grounding the dynamism of modern science, cannot be simultaneously met in any supposedly complete and perfect description of reality. Universality requires abstraction – recognition of the same in the different requires stripping away detail – while increased detail helps to particularize our understanding of the distinctive nature of a specific locale or specimen. Bachelard concludes that scientific knowledge can never be complete, reality can never be fully known: "The concepts of reality and truth have to be given a new meaning in a philosophy of the inexact" (Bachelard 1928: 8). The being of reality resides in its resistance to knowledge, and the impossibility of complete, exact knowledge is taken as a founding epistemological postulate (Bachelard 1928: 13).

Bachelard faults pragmatism[4] for having concentrated only on science as the quest for knowledge of general laws, and for failing to see that the verification of such knowledge is intimately connected to the search for detailed understanding of particular conditions and to the use of technology in conducting experiments. He proposes a much more complex, multidimensional portrayal of the dynamics of scientific progress than that of continuous approximation toward *the truth*. The pragmatist proposes a picture of science as a sequence of approximations gradually converging on, but never actually fully attaining, complete or "true" knowledge of the material world. This however depends for its plausibility on being able to sustain the image of a continuously convergent sequence of approximations. But as Hacking notes ((1984: 119), Bachelard, like Kuhn, allows that there are radical discontinuities in the development of scientific concepts, theories, and experimental practices. But this creates a problem. Once there are discontinuities, how can our current position be any indicator either of where we will be in the future or of where the "truth" might lie? How can we even sustain talk of approximation or of progress?

Bachelard's response in his *Essai* is that this is justified and possible to the extent that we succeed in making more accurate measurements with new kinds of instruments, using them to reveal previously undetectable structures, forcing us to recognize complexity within what we had treated as simple. The conjunction of mathematics with instrumentation which is effected by measurement determines dimensions along which progress can be assessed even as that progress requires the (discontinuous) correction of concepts and the (discontinuous) introduction of new experimental procedures to accommodate the demands of new instrumental technology, technology which is itself the material realization of theoretical constructs (see also Bachelard 1934: 176–7). Thus Bachelard dives into the complexity and multidi-mensionality of scientific knowledge acquisition (as of the early twentieth century) to get to grips with the dimensions along which we can talk about getting better approximations. This takes us into the actual interfaces between theory and material reality, mediated by measurement practices, laboratory apparatus, and mathematical theory. Arguments for the position at which he arrives – one which is indebted to, but which wishes to move beyond, pragmatism – center on his treatment of the technology–science relation, his acknowledgments of the role of technology in sci-ence and his distinction between scientific and technological imperatives.

In the technological sphere, goals *are* achieved; devices are constructed and perform the function for which they were designed (although there are of course also failures). A remarkable feature of modern technological innovation is the way in which general principles are put to work in the realization of wholly new processes and products. It is here that the detailed and the general meet in the construction of complex devices, whether in the factory or the laboratory; in the process science acquires confidence in its claims to have advanced its understanding of (and mastery over) material reality. But such success is not to be mistaken for complete knowledge or understanding; it is achievable because (i) technological goals are never specified completely or with complete mathematical precision, so there is always a margin of imprecision built into criteria of success; (ii) the contexts of action within which success is deemed to have been achieved are closed and/or limited in kind. In other words, finite bounds are set on the demands for both generality and detail.

This makes the contexts of science (and scientists) in action *internal* to the episte-mology of science rather than relegating them to the *external* status of applied science, or mere laboratory technique. In order to develop his account of science as the ongoing process of refining approximate (inexact, incomplete) knowledge, Bachelard deploys contrasts between the demands of action and of understanding, insisting that the epistemic values of science are neither coincident with nor wholly separable from pragmatic values. Here it is clear that acquisition of scientific knowledge is not just an affair of the mind; the epistemology of science must therefore be concerned not just with mental operations but with thought, action, and their interrelation.

From Knowledge Approximated to Approximate Knowledge

Many key debates about the development and deployment of new technologies (GM foods, nuclear waste disposal, climate change) are about how to act in the face of incomplete knowledge. It is increasingly being recognized that we desperately need better ways of communicating what can and cannot be demanded of science, and of the economic and social costs of thinking that it can deliver "certainties" on the basis of which actions with assured outcomes can be planned. But in its focus on the truth of units of knowledge (whether statements or theories), epistemology has, for the most part, not admitted inexactness into the discussion of knowledge. There has been discussion of approximation to knowledge, or correct description, but not of the idea that in the context of science we should be talking about approximate knowledge as a form of knowing, a knowledge which internally acknowledges its own incompleteness and avenues of extension. There has been much discussion of uncertainty within the framework of probability and statistics, but the classical versions of the theory behind these is still a theory focused on the ideal of the (exact) truth of an individual statement.

This is all the more surprising because it is generally acknowledged that supposedly observational concepts, those derived from perception of sensory qualities, cannot be applied with mathematical precision. Hume (*A Treatise of Human Nature*, 1.4.2, 1.4.6; and *Two Essays* II.16) already argued from the nontransitivity of perceptual judgments of identity (A and B may be judged identical in color, and so may B and C, yet A may clearly be seen not to be of the same color as C) to the conclusion that strict identity, like necessary connection, is an idea without foundation in experience and hence without objective content. Even if our mathematized theories support a notion of quantitative precision, is this ground for presuming that items in the natural world instantiate those particular mathematical structures? Bachelard thinks not, quoting a remark by Borel: "The idea of an exact physical magnitude is a mathematical abstraction to which there is no corresponding reality" (Borel 1914: 73). A physical magnitude thus carries with it an air of imprecision which is part of its reality. No measure entitles us to abstract from this. So even though Bachelard accords mathematics a very significant role in science, he insists,

> it is a pure impossibility to fall even by chance on an exact knowledge of a reality, since a coincidence between thought and reality is a veritable epistemological monster, it is always necessary for the mind to be moved to reflect the diverse multiplicities which qualify the phenomenon studied by hiding its boundaries. (Bachelard 1928: 43)

For this reason we should always be conscious of the imprecision at the qualitative and quantitative borders of phenomena. A central plank in his philosophy of approximation is that this negative judgment should be incorporated in any positive affirmation made. It is a remarkable testament to the persistence of the Cartesian/ Newtonian mathematized mechanistic ontology that although other philosophers of science recognized the inevitable inexactness of all empirical detection and measurement, they did not build this into their representation of scientific knowledge but continued with representations (centering on classical first-order predicate logic) in which concepts have sharp boundaries and sentences are either true or false. Even Russell (1923) – who acknowledged that "All traditional logic assumes that precise symbols are being employed. It is therefore not applicable to this terrestrial life, but only to an imaginary celestial existence" (quoted in Sangalli 1998: 14) – continued to use it in attempting his foundational and reductionist accounts of our knowledge of the material world. These representations of knowledge as mathematically and/or logically precise have formed the backdrop against which the majority of epistemological discussion has taken place. Bachelard's move to count acknowledgment of its approximate or inexact character as part of what is known has, as we shall see, quite a profound effect on his account of science and its epistemology. It is also, as we need to remember, an integral part of his insistence that our epistemological focus must be shifted to the dynamics of the development of scientific knowledge. This means moving away from any theory of knowledge which first defines scientific knowledge in terms of the static, logical organization of theories and then moves to a discussion of how we can attain such knowledge.

Indeed, if all approximation were modeled on that of continuous approximation to the limit of an infinite series – a model derived from mathematics – something like a classical framework could be retained, for in that case there still is sense assigned to talk of exact knowledge: it is the infinite limit toward which our approximations tend but which they never attain. But what Bachelard does in the second and third books of the *Essai* is to *contrast* the finite approximations available within the empirical natural sciences with the infinite series of approximations available in mathematics. He challenges our right to ontologize mathematical exactness (which has no demonstrable empirical instantiation), urging rather a reexamined "naive" (nonreductionist) realism where reality is revealed as a hierarchy of orders of magnitude. In the first book he argues that neither logic nor a concentration on statements, judgments, or truths can provide the framework within which to discuss the epistemological dynamics of science. Instead we should be discussing the dynamics of conceptual development (the rectification of concepts).

Technology, Standardization, and Experimental Science

One of the crucial insights which Bachelard underscores using his discussion of the intimate relation between modern science and modern industrial technology is that induction can give neither complete affirmation nor complete refutation. Where it works well it gives us predictive power, but it does so only where it cannot give anything but approximate knowledge.

Modern industrial production is mass production, and the products which roll off a particular production line on a particular run are, to all intents and purposes, identical and are intended to be so. They are "concrete universals." The imperative to standardize and to make uniform in this way is utilitarian and pragmatic, and antedates modern experimental science. The importance of standardization of measurement units, of military equipment with interchangeable parts, and of legal procedures for effective governance and control was appreciated by ancient imperial powers, such and Rome and China. The connection between uniformity and the control which derives from reliability and predictability is underscored in Bachelard's discussion of induction.

He begins by noting that the principle of induction is not itself something learned from experience, rather it is the condition without which we could not talk of experience. (Bachelard 1928: 127). It is both a principle of knowledge acquisition and conservation. For him, conceptualization and induction represent the same operation; to think that there are re-identifiable objects with stable characteristics already requires induction (as indeed Hume pointed out). Where there can be problems is in the application of concepts to reality because here approximations are involved. Bachelard thus shifts the locus of uncertainty from the principle of induction to the problems of concept application and the imprecision of all empirical judgments of identity. Given the same conditions, the same thing will happen, but how can we know for sure that we have identical conditions? Exact identity is possible only in logic and mathematics; it cannot be empirically established. But one can intervene in the production of artefacts to make them as identical as possible, specifying the dimensions, materials, and production methods to be used. To the extent that uniformity is assured, inductive inference becomes more reliable.

Of course exact identity cannot be attained, and especially in periods before the development of modern science, and when reliance had to be placed on natural materials, there would always be differences that might make a difference to how an artefact might function but which might be undetectable or simply not be recognized as important, at the time of manufacture. The seventeenth-century developments in the theory of mechanics, so frequently heralded as marking the scientific revolution, were preceded by a period in which mechanical devices (cranks, pumps, siege engines, clocks, musical boxes, clockwork imitations of birds and animals) proliferated. In this context one can appreciate the importance of the distinction between accuracy and precision, between fine craftsmanship and mass production. Many of the novelty clockwork pieces demanded very great ingenuity and extraordinarily fine manual skill in the accurate production of small gears and other mechanical parts if they were actually to function in the coordinated way intended. But these were characteristically made as one-off pieces. Pumps for mines, cannon for ships might be a little less demanding in terms of the accuracy with which each part needed to be made, but uniformity, being able to replace one pump or pump part by another which would do the same job starts to become a priority as efficiency is linked to rationality and becomes a regulative concept. Precision relates to uniformity and ability to generalize by induction, to knowing that any of a whole stack of parts will fit, or that a measurement taken repeatedly will give the same result. A measuring instrument which is capable of being highly accurate but only under specially prepared conditions, is much less convenient and more costly to use than one which will be slightly less accurate but which will give stable readings through a wider variety of conditions. The emphasis on reliable function, crucial to control and to "efficiency," is intimately bound up with the felt need for precision as a route to ensuring uniformity.

It is for this reason that Bachelard associates this nest of concepts with (James's) pragmatism and insists that pragmatic values alone do not drive modern science. From a scientific point of view, every induction is provisional and subject to revision in the light of more detailed knowledge. But since one can be more confident of relatively imprecise predictions, it may not, from a practical point of view, always be reasonable to seek greater precision. Our question will be, "Have we left anything important out of the concept under which we recognize this object, or the present conditions?" The inclusion of "important" indicates both the context relativity and the pragmatic component in any such judgment.

Realization that the goals of effective, controlled action and scientific understanding, though interrelated, are not the same, emerges again in the 1990s in the work of René Thom and Lofti Zadeh (originator of the term "fuzzy logic"). René Thom says,

> If we only seek to gain power over things, we can resign ourselves to incomprehension, for we can act effectively without understanding the reasons for our success. (Thom 1994, quoted in Sangalli 1998: 29)

The deployment of fuzzy logic in a variety of electronic devices is based on the idea that a "fuzzy" controller doesn't require an exact theoretical model. Zadeh notes that complexity and precision bear an inverse relation to one another; as the complexity of a problem increases, the possibility of analyzing it in precise terms diminishes. Moreover, in many situations precision may be too costly, take too much time, or be otherwise inappropriate. He uses the example of parking a car.

> Usually, a driver can park a car without too much difficulty, because the final position of the car is not specified exactly. If it were specified with high precision, it would take days and perhaps months to park the car. (Zadeh 1992, quoted in Sangalli 1998: 15)

Of course we could imagine making the task easier by somehow linking the vehicle's controls to a GPS system, but what would be the point of introducing the extra technology if the normal standards of parking accuracy can be met without it?

Even if there were a point, we have no right to assume that ever more precise degrees of control will always be possible. There might be cases where a very small neglected detail has a very significant effect (what would now be called sensitive dependence on initial conditions). Induction works on the continuity assumption that small differences in initial conditions produce small differences in outcome; or the more precisely initial conditions are known, the more precisely outcomes can be predicted. But it cannot assure us that this is always the case, and Bachelard cites (1928: 133) an example given by Borel to show why this might be an unjustified assumption. In an idealized simple pendulum the sum of the kinetic and potential energy is constant. This is expressed in the equation

$$(dx/dt)^2 + x^2 = a^2,$$

where a is the amplitude of the oscillation, which means that the phase space representation of such motions is a circle. But if the system is under the influence of an external cause, however minimal, so that the energy of the system increases over time, as represented in the equation

$$(dx/dt)^2 + x^2 = a^2 + 2\lambda t$$

(where λ can be as small as one pleases), the motion ceases to be periodic; instead of a swinging pendulum we get a bob rotating always in the same direction. In phase space this is a matter of crossing the boundary between the region (a) where the curves of possible motions are closed and that where they are open.[5] In figure 11.1, V is velocity, q the angle of deflection, the shaded regions correspond to oscillations; the region outside corresponds to rotations.

This line of thinking runs counter to the idea that a more detailed theoretical model will always yield increased predictive precision and hence increasingly reliable control. That is, it problematizes the idea that increased control must rely on being able to make a complete deductive map of the processes involved, and so requires advances in scientific theory and computing power. Instead it suggests that control can be the product of a more qualitative understanding together with development of feedback mechanisms (the cybernetic approach).

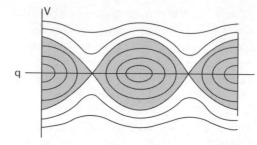

Figure 11.1: Pendulum "phase portrait" (the action of a pendulum).

In this context it is important to remember that some technological development occurred without the benefit of modern science and did not on its own give rise to modern science. Bachelard sees science as wanting a knowledge of the detailed working of things, of wanting to understand why normally reliable devices fail in particular circumstances, wanting to know what more evidence we should have had before making a prediction or a diagnosis which turned out to be incorrect. And, initially guided by experience with mechanical devices, the answers to such questions have typically been sought by trying to determine the microstructures of macroscopic objects and processes and again forming inductive generalizations about their behavior, finding their limitations, looking for more structure, and so on, in a dialectical movement between the need to make generalizations and the need, which arises from trying to apply them in the real world, to determine and understand their limitations. This is a dialectic of analysis and synthesis, but one which synchronizes conceptual operations with experimental procedures for analyzing and constructing.

On the other hand, Bachelard is also acutely aware that modern experimental science is possible only because of developments in instrumental technology. This is important on two counts: (i) there are similarities between the laboratory and the factory – laboratory phenomena are, like factory products, artefacts, i.e. concrete universals, (ii) the level of detail experimentally accessible is a function of the instrumental technology available. This first point indicates why Bachelard insisted that the objects of laboratory study are not natural givens. The laboratory is a carefully controlled environment in which the object of study is carefully prepared and shielded from interference or contamination, in an effort to analyze by isolating the particular aspect to be studied. Procedures and measuring apparatus are standardized. An important part of establishing the credibility of a report on experimental results (the reality of the phenomena claimed to have been observed) is its reliable reproducibility. Thus, in response to Descartes's discussion of the piece of wax, Bachelard says:

> The physicist doesn't take wax straight from the hive, but wax which is as pure as possible, chemically well defined, isolated as the result of a long series of methodical manipulations. The wax chosen is thus in a way a precise *moment* of the method of objectivation. It retains none of the odor of the flowers from which it was gathered, but it bears proof of the care with which it was purified. In other words it is realized in its experimental manufacture. Without this experimental manufacture, such a wax – in the pure form which is not its natural form – would not have come into existence. (Bachelard 1934: 173)

Measurement and Orders of Magnitude

Just as important, and more intimately related to the theme of empirical approximation, is the realization that experimental instrumentation and methods are a function
of what is already known and that they change as the knowledge base changes
(Bachelard 1928: 63). Thus experimental and measurement practices leading to improved precision may involve discontinuous change, rather than being a matter of
continuous improvement, as might be suggested by the mathematical picture of getting successively improved approximations (Bachelard 1928: 61). To get greatly improved precision (orders of magnitude better) requires not just better instruments but
new instrumental technologies such as the shift from optical to electron scanning
microscopes, or from sextant and chronometer to GPS. Experimental methods which
may be excellent at a given level of precision end up having their value diminished
after the introduction of new instrumentation capable of much greater precision.

Moreover, an object (or process) as revealed at a new order of "magnification" may
at first be quite unrecognizable. If one is to be convinced that what is revealed is an
objective feature of the object to be measured or studied rather than an artefact of the
instrumental device, there has to be a process of integrating different instrumental
approaches, and a development of associated practices of preparation and interpretation, coupled with some theoretical account of how the instrumentation works. But
the judgment of instrumental reliability is frequently more a technological than a
theoretically scientific judgment – that is, one ruled by pragmatic values. Does the
new piece of apparatus do the job required reliably in the conditions under which we
want to use it? That answer has to be given affirmatively if the instrumentation is to
be used in experimental investigation; experimental science is built on inductive,
practical trust in its technology, but it has to build into that trust a recognition of
limits of reliability both in terms of precision and in terms of conditions of use.[6]

However many refinements are made, each instrument and each technique has its
limits, and the precision of measures is sufficient to characterize the scientific methods
of a period. Bachelard suggests that while theories may be reborn after being eclipsed
for several centuries, a conquest of the more minute irremediably raises credibility
issues for the experimental knowledge of a previous epoch. At the same time it has to
be recognized that objects at different orders of magnitude have to be studied by
different methods and that at every level of experimentation there arises the question
"What can we neglect?" "That which we can neglect, we should neglect" (because
otherwise we won't get any generalization). Since absolute precision is impossible, we
have to be satisfied with a level of indetermination and have to know what that
appropriate level is for the domain in which we are working. Error is similarly
relative; there is a difference between being out by a milligram when working at the
level of grams, and when working at the level of kilograms (Bachelard 1928: 77).

Once you introduce a measure at a new level of minuteness, new qualities will
appear which entirely alter the object under consideration. Each order of magnitude
has its own realities, but these don't necessarily make sense at high or lower orders.
Practical and theoretical coordination of macro- and micro-views of an object or
process present what Bachelard sees as the most difficult and challenging scientific

problem; only once progress is made on these can there be claim to have any detailed understanding of an object (or process) identified at a relatively macro level.[7] The red blood we see with our ordinary eyes loses most of its color when revealed to our microscopical eyes. How do we know we are really seeing "the same thing": how can we be reassured the features we see, not detectable by unaided sight, are not an optical illusion? How are those microscopically revealed features going to help us explain what we have thought to be the functions and properties of blood? Integrating these levels of knowledge leads to a new understanding of blood, a new and "rectified" concept. It doesn't make blood, the red liquid we see when we cut ourselves, cease to be real, but it does cease to be a simple homogenous substance and is now something complex containing white and red corpuscles which themselves become objects of study.

If we put together recognition that when we talk about moving from one scale/ level of magnification/order of physical magnitude to another with the fact that physical magnitudes do not all stand in linear relationships to one another, we realize why there is a kind of natural hierarchy and why induction may not in general be expected to carry us smoothly from one level to another. Things at one scale will not be able to "work" in the same way as things at another. When length dimensions are multiplied by 10, surface area increases by a factor of 100 and volume by a factor of 1,000. Weight depends on volume; ability to exchange heat with the surrounding environment depends on surface area. In addition, objects defined at one scale usually cannot be sensibly measured at another. A coastline may theoretically be viewed as having a fractal dimension because its length is infinite, but for practical purposes (comparing the coastline of one country with that of another, say) there is an implicit limit to the accuracy that makes sense, in part because there is an inevitable degree of convention involved in deciding, for example, how much of a river estuary counts as part of the coast. Just because there are no definitive, objective answers to be given to such a question does not mean that coastlines aren't objectively real; as we know, the boundary between land and ocean is very important to us and changes significantly as ocean levels rise only by relatively small amounts. What differences are going to make a difference depends on the context of measurement and the uses to which its results are to be put. Bachelard emphasizes that there is no appearance of determinism without an exercise of choice, i.e. without putting to one side interfering or insignificant phenomena. These choices and abstractions, however gradually, become realized in simple mechanisms, which in turn inspire scientific visions of the world. Thus he traces the interconnections between mechanism, determinism, and reductionism. For everything to be determined everything would have to be reducible to mechanical properties (see Bachelard 1934: 107–11).

Non-Reductionism, Hierarchy, and Complexity

The position to which Bachelard is thus led is nonreductionist; it is one which views reality as hierarchically ordered by orders of physical magnitudes. It is a nested order of complex systems. Already in 1928 he was arguing for the central importance to our scientific understanding of the world of learning how to think about complex systems, and was aware of at least some of the difficulties involved; difficulties which before the advent of the electronic computer seemed to be almost insurmountable. His criticism of pragmatists for going along with other nominalist philosophers of science in limiting their conception of science to that of the quest for knowledge of the general laws of nature is that it leads them to think that epistemology of science needs to focus primarily on problems of induction, on how to to arrive at knowledge of general laws from observation of particulars. Bachelard indicates how this problem is transformed once we take it out of the context of logic and into the context of technological action, but he also insists that knowledge of general principles is only a first step. The pragmatists and nominalists give us only a very superficial view of scientific knowledge and hence of the epistemology of science (Bachelard 1928: 274). They give a "flat" depiction of the scientific endeavor (Bachelard 1928: 275). Induction is not the process that will help us to understand macro-level objects as complex systems of micro-objects; this requires the use of models and analogies together with methods such as those used by statistical mechanics to understand a gas as a system of molecules in motion. Epistemology is cut in two depending on whether one is going for generality or for details, organizing a system of marks or examining the prodigality of detail underlying the general, the fluctuations under the law (Bachelard 1928: 154). One cannot assimilate the effort which reveals to us the general characteristics of things with that which undertakes to appreciate the fluctuations. It will not lead us to realize that the simple is always the simplified, the product of a process of simplification (Bachelard 1934: 142–3).

Again, partly drawing on statistical mechanics and partly (one may suspect) on Poincaré's work on the n-body problem in Newtonian mechanics, Bachelard argues that it is not possible to have fully precise (complete) micro-state descriptions of macro-objects and/or processes. He uses language very reminiscent of Poincaré's shift to phase space to describe the role of mathematics as providing the spaces within which scientists can dream of possibilities.

Complexity is a theme which Bachelard also picks up in connection with his discussion of modern industrial technology, and it is again instructive to see how he draws the lines of comparison between the scientific and technological implications of complexity. He notes that in earlier periods complex devices were likely to be highly individualized, containing custom-made parts, whereas in modern industry increased complexity is accompanied by increased demands for generality in the form of interchangeability of parts. This is extended up to the level of factory workers – who now function as components of the assembly-line. The drive toward machines and organizations of increasing complexity, coupled with a governing idea of rational assembly is accompanied by a corresponding drive for precision and precise uniformity at the level of components. Bachelard notes that here we find in industry the same ideals of exactitude, precision, and generality as in speculative science. So have scientific and technological imperatives now become identical?

Bachelard thinks not. He notes that the driving forces are not the same in both cases. Science, he says, faces up to the fundamental irrationality of the given – there is always more to be known, always something we don't quite understand. Industry, on the other hand, wants to construct a reality which is fully understood and fully controllable. It imposes its "rationality": values dominate over knowledge in this situation. There is a normative element here. There is an end in view and a plan developed to achieve that end; the goal is to reorder part of the world, to make it conform to the demands of the plan and so realize the desired end. This means that to understand a technological device we need to know the end for which its was produced in addition to its physical description. If that end is the maximization of profits, then that will be reflected in the technology produced.

One of the reasons why complexity has come to drive requests for precision and interchangeability of parts is that there is another dialectic at work. Once some standardization of parts and quality control for materials is achieved, the work of design can increasingly be separated from that of fabrication. The more the design process is taken to govern and determine (in advance) the construction details, the more complex a deductive process it becomes. This in turn starts to facilitate mass production, which further entails "designing" the construction process. Theoretical scientific knowledge becomes important to the mathematical modeling and complex sets of calculations involved in the design process, which must take many variables into account. To the extent that everything is done according to plan, the result is an imposition of an intellectually conceived form on a part of the physical world. When the production process itself is governed with a view to efficiency and profitability, workers themselves become nothing more than interchangeable parts of the production machine, required repeatedly to perform exactly the same actions and thus turn themselves into efficiently operating machines (eventually to be replaced by robots in many cases). Scientific knowledge used in this fashion becomes part of the design, the plan, the idea, but appears separated from the production process by the division of labor into managers and workers. This whole structure can seem to be a validation of the dualism inherent in the intellectualist conception of scientific knowledge in which pure theoretical science comes up with the ideas, hypotheses which somehow have to be held up to the world they aim to describe.

But what is missing from the picture is the fact that managers alone can neither build a production line nor keep it going. There must also be the engineers and technicians who actually get things to work and keep them running. Bachelard is analogously urging the importance of the hands-on, technical work of experimental science to the process of generating scientific knowledge. If concepts had only a logical and theoretical life they would be exact. Approximation and inexactness arise in the bridge between theory and practice, between the idealized conceptual schema and its practical application. Logic lulls us into thinking that application of general predicates or of general principles should be unproblematic. But it is not – this is the arena of judgment, of skill, of expert knowledge, where a scientific concept must prove its worth. The mere theoretical schema is not yet a scientific concept; the inexactness involved in bringing it to application (to realization) isn't an externality but is internal to all concepts used in our cognitive depictions of the material world.

Technological artefacts are designed as far as possible as closed systems able to function in a (limited) variety of conditions. This means that perturbations must be automatically compensated for, or absorbed internally by components of the system without disrupting the intended functioning of the device. This limited independence from fluctuations in systems' immediate environment makes it possible to treat them as closed, and hence predictable systems. But as Bachelard points out, modern science also uses technology to construct its phenomena. It uses technology to create closed laboratory systems, to isolate the variables under study and thus to simplify phenom-ena. So laboratory science deals with purified, simplified phenomena. Natural systems generally are not closed systems and are susceptible to disruption by external factors. We treat them as if closed in order to model them mathematically, but cannot be assured of resulting predictions because we do not know what possible perturbing factors might arise.

The similarities between modern science and technology, and the experience of technological success may tempt us into thinking that complete knowledge and com-plete predictability are possible. Bachelard insists that even in the technological case, knowledge and predictability are at best assured only to a certain degree of precision and within certain limits; they are far from scientifically complete.

Action, Progress, and Moving Beyond Cartesian Intellectualism

The world which forms, for Bachelard, the object of our possible knowledge is endlessly rich and diverse. We encounter the world in multiple ways as we interact with it, seek to modify it, and are frustrated by its resistances. All of these multiple modes of empirical interaction afford pathways by which we can gain scientific knowledge. This knowledge has the character of being progressively acquired and always incomplete, always aware of possibilities for further refinement, of details and aspects ignored. Bachelard is critical of the pragmatists for postponing our right to claim knowledge and for failing to pay enough attention to the way ideas are applied and put in to action.

> Certainly the means of verification proposed by James are ingenious. He has accorded action the predominant role which is its due and which intellectualist theories have neglected. But this action only appears at the end.... pragmatism places its sanction at too distant a due date ... A truly active knowledge is one which is verified in each of its acquisitions. (Bachelard 1928: 265–6)
>
> It is the progress of knowledge which determines discriminations and which suggests more refined hypotheses requiring more careful verification. But it can no longer be a question of block verification, by a yes or a no, like that considered by James. At this level we accept approximative validations. (Bachelard 1928: 266)

Bachelard argues that in recognizing and setting bounds of the imprecision of our predictions (so we can say when we will count them as not being fulfilled), we thereby already delineate the path of progress. When we succeed in finding ways to increase precision, to reduce error, that in itself is an increase in our knowledge (1928: 266). All of this involves action; successful action has cognitive value, and without action the intellect on its own can make no progress in coming to know the material world. Bachelard thinks of all knowledge as incorporating within it a sense of success and of progress, an awareness of having achieved something as well as of what remains to be achieved. It is only this "vector" character of knowledge which under-writes a sense of continuity and of progress, for as Bachelard repeatedly points out, the history of science is fractured by discontinuities at the level of both theoretical and experimental practice. Thus he insists that verification occurs at all levels and is not a once-and-for-all affair. It is through verification that presentation becomes representation, and it is through use that representations get revised and refined. Revision in the direction of being able to include more detail, greater precision, is what Bachelard takes as progress. This is in part because it is not easy to obtain; it requires mobilization of other knowledge, new instruments, all of which work to establish links between the more detailed knowledge and other areas of scientific investigation. It is for this reason that detail is a mark of objectivity. Thus correction of concepts and theories in the acquisition of more detailed knowledge is for Bachelard where the real focus of epistemology should lie. It is again in the recognition of a change as a correction that cognitive standards are in play, not in making judgments of absolute truth or falsity, but of better and worse understandings or representations.

> Correction (*rectification*) ... is the true epistemological reality, since it is thought in action, in its profound dynamism. Thought is not to be explained by making an inventory of its acquisitions; a force permeates it which must be taken into account. Besides, a force is only well explained by its direction, its end. The terminus toward which experimental determinations tend can be affirmed, since the determinations are made following the schema of an approximation. Approximation is unachieved objectivation, but it is prudent, fruitful, and truly rational since it is conscious both of its insufficiency and its progress. (Bachelard 1928: 300)

There are here no foundations and reductionism is firmly rejected. More profound in its implications is Bachelard's rejection of what he calls the Cartesian intellectualist picture of science. By denying us the right to assume that the exact abstract structures of logic or mathematics are anywhere realized in reality, Bachelard thoroughly disrupts the concept of (scientific) "rationality," while also cutting away the structures which provided the rigid connections required to give sense to strict determinism. Simultaneously this undercuts the assumption that the only way to achieve increased control to achieve desired outcomes is through being able deductively to generate predictions on the basis of a theoretical model (which models causal connections as logical connections). Bachelard's critique of pragmatism amounts to a critique of models of instrumental or practical rationality according to which the intellectual work (reasoning) is separated from the practical work – putting the reasoning into effect. He is arguing that knowledge and its application are intertwined, whether we are talking about science or technology, reason or action. His conception of science is of an ongoing cognitive enterprise kept on course by continual corrective adjustments, making scientific reason more analogous to a fuzzy control system keeping a pole balanced upright on a moving trolley by making frequently repeated adjustments to counter the destabilizing effect of the trolley's motion, than to the natural light designed to reveal the blueprints of the created universe.[8]

Notes

1   Latour expresses his hostility to epistemology in several places. One of the more explicit is Latour 1984: 5–6.

2   No text by Dewey is cited in the bibliography. Two texts by William James are listed: 1913 and 1915.

3   James continues, expressing unease about the practical consequences of the gulf:

> The scope of the practical control of nature newly put into our hand by scientific ways of thinking vastly exceeds the scope of the old control grounded on common sense. Its rate of increase accelerates so that no one can trace the limit; one may even fear that the *being* of man may be crushed by his own powers, that his fixed nature as an organism may not prove adequate to stand the strain of the ever increasingly tremendous functions, almost divine creative functions, which his intellect will more and more enable him to wield. He may drown in his wealth like a child in a bath tub, who has turned on the water and who cannot turn it off. (James 1975: 91)

4   Although one should emphasize that he makes no reference to Peirce, whose work he might have evaluated differently.

5   This is a simple example of a phenomenal transition much discussed in the literature on chaos and self-organization. (See for example Prigogine and Stengers 1984.) What is interesting is how close Bachelard already is to aspects of their position, even though the detailed work on chaotic systems had to await the technological development of computing capacity. For example, Prigogine and Stengers say that phase space was introduced into physics to account for the fact that we do not "know" the initial state of the systems formed by a large number of particles. "Originally this was thought of as an expression of our ignorance of something that was in fact well determined." They then say, "the entire problem takes on new dimension once it can be shown that *for certain types of systems* an infinitely precise determination of initial conditions leads to a self-contradictory procedure. Once this is so, the fact that we never know a single trajectory but a group, an ensemble of trajectories in phase space, is not a mere expression of the limits of our knowledge. It becomes the starting point of a new way of investigating dynamics" (1984: 261).

Bachelard is already advocating that move: immediately after the Borel example he suggests developing constructive analysis along the lines suggesting by Klein (1898), where the curve expressing a law carries with is a halo of incertitude. "One is never sure of retrieving the same effects because one is not certain that the same conditions have been completely reproduced. The least divergence distances us from the symbolic point of an exact induction" (Bachelard 1928: 135). In fact, although Prigogine and Stengers do not mention it, one of the types of systems that fall under their description is the forced damped pendulum (Hubbard 1999). This draws on a phenomenon originally discovered by the intuitionist/constructivist mathematician Brouwer (1910), but often attributed to the Japanese mathematician Wada (see Yoneyama 1917), who constructed an example of three disjoint, connected open subsets of the unit disc in the real plane such that every point in the boundary of one is in the boundary of the other two. Essentially what Hubbard does is to adapt methods developed by Poincaré in his work on the three-body problem to show that the pendulum creates lakes of Wada in its phase space, basically meaning that for any sequence of gyrations one might choose, there is a solution for its equations of motion on which it will execute them. This might seem to suggest that perfectly accurate control is possible. The problem is that the motions are extremely unstable, so that the slightest error in the initial conditions will destroy them, making any actual device (such as a flywheel polishing a telescope mirror) essentially unpredictable. But Hubbard then points out that the very instability which creates this unpredictability is in fact useful if we get away from the idea that we have to have the whole task figured out ahead of time. If we have laser sensors that can react to the polishing as it proceeds, allowing very small adjustments to be made, the very instability means that the adjustments need only be small to achieve the result you want. Continuous small corrections can produce the desired result. He gives the analogy of the difference between an expert and an inexpert skier. The beginner plants his skis well apart, seeking stability, but when he tries to turn, finds he cannot. The expert skier places his skis close together and parallel. This is a highly unstable position but a slight wiggle of the hips allows him to negotiate irregularities and turns in the course; he doesn't plot his entire course at the top of the mountain. Although Bachelard does not have the full mathematical detail, he is drawing on the very mathematical work which would yield that detail, to project his argument that precise prediction is impossible and cannot be assumed even as an idealized goal. In taking on the implications of this conclusion he is moving us toward thinking of rationally guided action in terms of continual adjustment, rather than working to a fixed, antecedently given, fully worked-out plan.

6   See Hacking 1983 for a much more through discussion of these points in connection with the development of microscopes.

7   Newton and Locke were aware of this problem. Newton tried to base the claim of the universality of mechanics on scale invariance (the analogy of nature). "The extension, hardness, impenetrability, mobility, and inertia of the whole result from the extension, hardness, impenetrability, mobility and inertia of the parts; and hence we can conclude the least of all bodies to be also extended, and hard and impenetrable, and movable, and endowed with their proper inertia. And this is the foundation of all philosophy" (Newton 1966: 398–9).

8   When all is said and done Latour seems really not to have moved very far beyond Bachelard's position, although he has wrapped it up in different language and continues to rail against epistemology (but always as Cartesian epistemology; see Latour 1999: 296). He says, for example, that when Descartes asked how an isolated mind could be *absolutely* sure of anything about the outside world, "he framed his question in a way that made it impossible to give the only reasonable answer which we in science studies have slowly rediscovered three centuries later: that we are *relatively* sure of the many things with which are are daily engaged through the practice of our laboratories" (Latour 1999: 4). Bachelard already arrived at this position before the science studies people moved in.

References

Bachelard, Gaston. 1928. *La Essai sur la connaissance approchée*. Paris: J. Vrin

——. 1934. *Le nouvel esprit scientifique*. Paris: Presses Universitaires de France.

——. 1953. *Le matérialisme rationnel*. Paris: Presses Universitaires de France.

Borel, Emile. 1914. *Introduction géometrique à quelques théories physiques*. Paris: Gauthier-Villars.

Brouwer, L. E. J. 1910. "Zur Analysis Situs." *Mathematische Annalen* 68: 422–34.

Hacking, Ian. 1983. *Representing and Intervening*. Cambridge: Cambridge University Press.

——. 1984. "Five Parables." In Richard Rorty, J. B. Schneewind, and Quentin Skinner, eds., *Philosophy in History*. Cambridge: Cambridge University Press.

Hubbard, John H. 1999 "The Forced Damped Pendulum: Chaos, Complication and Control." *American Mathematical Monthly* 106(8) (Oct.): 741–58.

James, William. 1913. *L'idée de la vérité*, trans. Veil and David. Paris: Alcan.

——. 1915. *Precis de psychologie*, trans. Baudin-Bertier. Paris: Rivière.

———. 1975. *Pragmatism and The Meaning of Truth*, intro. by A. J. Ayer. Cambridge, MA: Harvard University Press. [*Pragmatism* was first published in 1907. *The Meaning of Truth* was first published in 1909.]

Klein, Félix. 1898. *Conférences sur les mathematiques*, trans. L. Laugel. Paris: Libr. Scien. A. Hermann.

Latour, Bruno. 1984. *Les microbes: guerre et paix suivi de irréductions*. Paris: Editions A. M. Métailié. English translation: *The Pasteurization of France*, trans. Alan Sheridan and John Law, Cambridge, MA: Harvard University Press, 1988.

———. 1999. *Pandora's Hope*. Cambridge, MA: Harvard University Press.

Newton, Isaac. 1966. *Principia*. Berkeley and Los Angeles: University of California Press. Republication of 1729 English translation by Andrew Motte, revised by Florian Cajori.

Prigogine, I. and I. Stengers. 1984. *Order Out of Chaos: Man's New Dialogue with Nature*. London: Heinemann.

Russell, Bertrand. 1923. "Vagueness." *Australasian Journal of Philosophy* 1: 84–92.

Sangalli, Arturo. 1998. *The Importance of Being Fuzzy and Other Insights from Between the Borders of Math and Computers*. Princeton: Princeton University Press.

Thom, René. 1994. "La magie contemporaine." In Y. Johannisse, ed., *La magie contemporaine – l'échec du savoir moderne*. Montréal: Québec/Amérique.

Whitehead, Alfred North. 1925. *Science and the Modern World*. New York: Macmillan.

Yoneyama, K. 1917. "Theory of Continuous Set of Points." *Tohoku Mathematics Journal* 11–12: 43.

Zadeh, L. A. 1992. "The Calculus of Fuzzy If–Then Rules." *AI Expert* 7(3) (March): 36–9.

12

From *ESSAI SUR LA CONNAISSANCE APPROCHÉE*

Gaston Bachelard

Knowledge and Technology – Approximate Realization

We thus come to this unanticipated paradox which shows knowledge to be fleeting and mobile, and the action which it brings to light to be solid and assured. Science is an enigma reborn, a solution leads to a problem. The real, for the researcher, is a cloud of possibilities and the study of the possible is a temptation against which the scientist, no matter how positivist he may be, is ill defended. Nothing is more difficult than constantly equating the mind with present reality. Technology, on the contrary, fully realizes its object, and this object, to be born, has had to satisfy so many heterogeneous conditions, that it avoids all skeptical objections. It is a factual proof in all senses of the term. Better, it is the decisive element in scientific confidence. Poincaré [*La Valeur de la science*] wrote: "If industrial development makes me happy, it is not only because it provides an easy argument to the advocates of science; it is above all because it gives the scientist faith in himself, and also because it offers an immense field of experience, where he is opposed by forces too colossal for him to have a way of thumbing his nose at it. Without this ballast, who would know that he had not left the earth, seduced by the mirage of some scholastic novelty, or that he would not despair believing that he has done nothing but dream."

Over and above this role of anchoring scientific culture, technology has a quality of evidence recognized by all. It is one of the reasons why mechanism became a criterion of clarity and took on an explanatory value. It is often imagined that this general reference to mechanism derives from the fact that we are a productive center of forces and that we are thus able to vivify and dynamize the geometry of movements which would otherwise be for us an empty spectacle. We believe moreover that humans take *all* their instruction from the external world and that they understand themselves as a function of the pure kinematics of nature which surrounds them. Whence their

Gaston Bachelard, pp. 156–67 (ch. IX) from *Essai sur la connaissance approchée*. Paris: J. Vrin, 1973 (first published 1928). French text © 1973 by Librarie Philosophique J. Vrin – Paris. Reproduced by permission of the publisher, http://www.vrin.fr. New English translation by Mary Tiles.

timidity in their initial employment of natural forces. This is the opinion of Reuleaux (*Cinématique*), among others: "These are not forces which are at first manifested to the human intellect, on their debut, but only well after the movements that they produce. The child is vividly struck by the sight of windmills, water wheels, stamping machines and, in general, all machines which execute regular motions, easy to take in at a glance; but he hasn't the least idea of the forces utilized to obtain these movements. Abstraction, indispensable for conceiving of force separate from movement, constitutes a sufficiently complicated mental operation that it needs a long time to achieve complete development. It is for this reason that, in the first machines to come from the, still little exercised, hands of man, force plays only a secondary role in relation to that which allows for the efforts of his limbs, working in a somewhat unconscious manner." Similarly Geiger believes that the motion of rotation was employed before the lever, because the lever is above all a transformer of force, not a transformer of movement. It is an essentially indirect way of overcoming large resistances. If clarity has a double root in the antiquity of experience in Humanity and in the individual, one could then perhaps understand why mechanism, or better kinematicism, is for the human mind the domain of security. Kinematics is a formal science with an undeniable purity. Its technical realizations, by following the plan, thus offer examples, paradigms susceptible of awakening and regulating speculative thought. Language has won its richness and its precision more by the hand than by the brain.

This solid, stable technical realization has, besides, a character which must detain the philosopher. Modern industry does not individualize the object it creates. Strange creation when the general takes precedence over the particular! In certain respects, the production-line is an application of the Aristotelian formal cause. It is there that the form is really at work, that it organizes matter. It translates its act with a singular clarity, with such an economy of traits, of means, of matter, that the general is visible at first sight without needing to be disengaged by a progressive elimination of detail.

The museum of molds of a foundry or a glassworks is a veritable collection of platonic ideas. It is a reservoir of kinds, the aesthetic history of fabrication. Types tend in modern industry towards being schemas. Doubtless the ornamentation often preserves traditional motifs, but these motifs are worked in following a program of stylization which utilizes detail as a function of the whole. In the stylized object the mind recognizes its traditional mark. The object is not opposed to the copy because the idea is not dispersed in diverse samples, but remains manifest and whole in each with its harmony and its elegance.

This schematic grace which cloaks manufactured objects is of the same order as the Bergsonian grace which in following curved lines and avoiding angles finds a sentiment of ease in perception, the easy anticipation of movement, "the pleasure of arresting in some way the march of time and of holding the future in the present." In the industrial product actually achieved, convenience is evident, palpable, as the temptation to less effort. It is bereft of ornaments which amuse with useless movement; but with the economy of forethought, its clarity appears. Here again future action is prefigured clearly in the present as an invitation to work, but it is a productive, beneficial, rapid action.

To the progressive mathematization of technology thus corresponds an aesthetic of occasions whose true force resides in forms which are more and more rationally adapted to matter and to action. The explanation is that this rational conquest of the useful follows the obscure and slow method of geometry and mechanics and that it does not know the individual successes of art. In the beginning, matter, manual labor, and time itself don't count ([cf.] Choisy). One can securely grasp these riches. But the time comes when everything is taken into account, even the water of the river whose energy the dynamo rationalizes. In the form of electricity, energy demands an economy which becomes the driving factor in technology. It is no longer a question of succeeding but of succeeding economically. The individual utility of a mechanism cannot be judged in the absolute, the useful has a "weight" as the determination of a physical magnitude, and it is with this weight that it must enter into the judgment of general utility.

At first sight it may seem strange to talk of an approximation to the useful. The useful, as exact, seems to answer a simple question: yes or no. What misleads us in this regard is that the technology of everyday life, so apt to inculcate principles in us, only offers examples of simple machines which one understands and can sum up at a glance. Thus the unskilled laborer knows little more than how to use the lever, the crank, and the hammer. Similarly, common utensils have a clearly dominant utility factor which entails a judgment as to their utility value. For example, a receptacle for liquid is useful to the extent that it is impermeable. But the concept of convenience must be related to the concept of utility; it is, in modern life, technically inseparable from it. To considerations of convenience are finally joined economic conditions. At a stroke utility is submitted to comparison, to coordination, to a veritable measure. Equations will fix the degree of finish compatible with costs of purchase, production, and distribution, a whole slew of criteria will intervene to justify the methods of economic work, to legitimate its imposition of uniformity

This uniformity is sought in the most complex machines. Here again the organization, as in Nature, is not oriented toward individuality. Not content with manufacturing items on a production line, groups are organized according to a standard, the gestures of the worker are regulated in space and time, making the whole factory into an *active generality* which will be applied elsewhere without alteration, gesture for gesture, time for time. This mechanization of the *will* which is achieved in the Taylorization of human initiative seems to us to conform to the philosophy of approximation. In effect, it works essentially to make detail irrelevant, to relieve it of all its picturesque value, all its circumstantial force, to reduce it to a level where it would be known not to have any action on the whole. Detail thus disappears from the order of magnitude where one works just as it disappears from the order of knowledge where one measures.

In the past, adjustment was made by a particular action "on demand," parts only had a provisional generality; on their incorporation in the machine, they had to submit to the stamp and individuality of the whole. It seemed that complexity and generality were practically contraries. In modern industry, the interchangeability of parts must be total. It is done to a limit of precision appropriately adapted to the machines in which it must be used. It is responsible for the success of American agricultural machinery, very large-scale constructions. Actually the ball bearings are checked, with a view to interchangeability, with a fully scientific precision. The specifications of the Danish constructor Johansson call for tolerances on the order of a micron.[1] Once again industry demands perfect generality.

This generality can go so far as to play the role of identity. It is thus that by a strange confidence in the constancy of commercial products, the candle of pure stearic acid produced by prewar industry was once taken as a unit of luminosity. Of course an electrical lamp standard (the international candle) was promptly substituted for it. But originally, the industrial product was supposed sufficiently identical in its diverse samples to serve as the basis for physical measurements. The carbon of sugar is similarly used in the practice of chemical analysis.

Whether in the determination of calibration and finish of parts, of tolerance admitted in the fluctuation of their characteristics, of the quality control in the course of their manufacture, or even in the theory of rational assembly, one finds the same ideal of exactitude, of precision, and of generality as in speculative science.

However, the sense of the internal dialectic is not the same in the two cases. Science faces up to the fundamental irrationality of the given. This irrationality ceaselessly provokes science, pushing it to ever-renewed efforts. Industry, on the contrary, seeks to inscribe in matter a rationality which, since it is desired, is clearly recognized. One searches for the rational, the other imposes it. In technology, the end is really integrated into the being which realizes it, it is the principle element in it and this time the "this must be so" does not resonate as a logical presumption, but as an order. The descriptive element must thus cede place to the normative element. The cognitive judgment appears in second place; the value judgment dominates and, in a way, prepares it.

In many other respects we will have to establish a divergence between the orientations of speculation and practice, and the matter of knowledge must react to the modes and forms of this knowledge. Thus to know the machine or the object created by the technician, we will have to follow a different method from that of pure science. The formal cause which we have already adjoined to considerations of habit is not itself sufficient. It is necessary to consider in addition the final cause. We will only understand a mechanism well if to its pure and simple description we join an examination of the harmony between means and end which gives way to endlessly repeated judgments. A technology is developed in the realm of ends.

But this is only an apparent complication, since the study of reality organized by technology is simplified by the very fact of the generality which we have shown to be its dominating characteristic. "Each type of concrete realization," wrote Louis Basso,[2] "represents as it were a little universe that it (technology) isolates from the whole of reality, in order to rearrange it for its proper usage, by applying the method which science employs for its disinterested ends. Only the complexity of this little universe being infinitely less, technology can undertake to compose it, at least by approximation, starting from abstract elements, while science can only confine its research to the discrimination of these elements themselves." But these abstract elements which, as we have said, prefigure the general in the "constructed nature (*natura structa*) of technology, do not remain in an isolated state. They form a system, and it is their whole organization which is transparent to the technician who understands the realized machine, who reads it as a purely kinematic device. The *pre-project*, the more or less detailed *programs*, seem to us to play epistemologically a role similar to diverse hypotheses in the work of scientific research. With always the difference that these are hypotheses verified a priori since the technological effort precisely consists in imposing them on the concrete. There is there an element of certitude which is lacking in a more passive knowledge.

But there is more. Matter itself is, so to say, disentangled from its irrational character, for it is no longer necessary to advance knowledge of it beyond the limits assigned by the end pursued. Matter needs only to fulfill the conditions of resistance which fix very clear limits of freedom for the technician and which permit him, as in the technical hypothesis, to attain security. For security seems indeed to be a form of rationality.

Automatically the solids employed in technological construction develop as much of a reaction as is required in order for them to resist external perturbations, so long as one remains, naturally, within the zone fixed by the calculations of resistance. It is here that Reuleaux sees the essential difference between cosmic and mechanical systems. If a satellite describes a circle around a planet, a force exerted on the satellite perpendicular to the orbit will change its motion if another exterior force doesn't come along to compensate for the perturbating force. We realize the same motion in the rotation of a wheel joined to an axle. If a force parallel to the axis of rotation suddenly appears, it will this time be *automatically* and *totally* compensated for by molecular reactions. Are we not here on the way to rationalization since we are bordering on the limitation of conditions? A cosmic system is open to prodigal contingency, it is by an arbitrary abstraction of unanalyzable external conditions that one can regain rationality. And it is the machine which is truly a world entirely closed and summarized in its schema, in its idea, freed from external possibilities and, consequently, fully analyzable.[3] "The difference between the two systems, cosmic and mechanical, thus essentially consists in the fact that in the former, sensible forces find themselves opposed by other independent sensible forces, while in the latter, on the contrary, one has on the one hand sensible forces and on the other latent forces which depend on the first." Further, p. 36: "It is a principle inherent in the very essence of the machine that the perturbatory movements must be hindered by latent forces." In other words, in principle, the machine does not waver; in its domain of security it is absolutely and *automatically* exact.

So, one can in other words say that the progress of instruments helps to reduce irrationality and to push it beyond the limits of precision bearing on experience. In this respect the conception of perfect exactitude is a seduction which could lead us to veritable errors.[4] "Though Newton, who formulated a theory of dispersion declared achromatism impossible in virtue of this theory, a London optician made a system of lenses which was practically achromatic, and opticians have successfully imitated him ever since." This theoretization of matter is truly the work of technology. It is only for technical ends that one can find this autocorrection of matter sufficient. But if it is sufficient, it is practically rational. We say further, it would be irrational to search further for a gratuitous precision. A badly balanced practice which was mistaken in the "weight" attached to factors of precision would be, in technology more than elsewhere, a disastrous failure of principle. It has been remarked that "too much adjustability" can entail worse distortions.[5]

It is no longer a question of a matter conceived as indifferent to form but of a matter wholly subsumed under its geometrical aspect and its mass. The matter employed by the technician answers to the same title as the directing schema of a provisional study and, in this way the testing of materials is not, as one might think, a submission to empiricism. The goal of this testing is to arrive at a generalization, which is legitimate, for the margin of error of the test is always sufficient to cover particular cases. Louis Basso has indicated how one is misled with regard to the real relation between theory and practice.[6] "The bearing of theory is far from being limited to the conception of a general functional schema. The machine is a reduced world which is materialized around this schema: the point becomes a metallic mass which cannot be displaced without friction, the forces of inertia engender vibrations, many other phenomena come to enter into consideration, each one of which must necessarily be the object of more or less complex theoretical predictions. Nothing is less empirical in itself than such an enterprise."

Or at least technical empiricism has such clear rules that it becomes a veritable method supported through and through by a theoretical effort. Thus the same words don't say the same thing in technology and in everyday life. Everyday experience does not often lead further than itself. It is at the end of its progress. Technical experience must be engaged in other ways. "For a long time," says Reuleaux,[7] "the efforts made with a view to creating the sewing machine remained fruitless, because people had obstinately wanted to reproduce hand sewing; but from the moment when it was decided to introduce a new way of sewing which was in better accord with mechanical exigencies, the spell was broken and the sewing machine passed into the domain of practice without delay."

This technical experience has, at bottom, the same requirements as a purely scientific experience. It is a mistake to believe that the fact of advancing between inequalities gives the technician a certain liberty. However, such is the opinion of Sorel.[8] "This liberty of the industrial mechanic is one of the fundamental elements which pragmatism must take into account, because it is one of the conditions of the progress of modern technology." The role of the engineer will moreover be to try to eliminate this liberty, to reduce the play of formulae of inequality. Even from the didactic point of view, one does not consider examining a machine beyond the limits of its normal functioning. These limits are, however, very confined. The modern machine is too complicated for a partial disequilibrium not to reverberate deleteriously throughout the assembly. At the smallest disharmony the machine no longer speaks. Grimshaw, analyzing the "language" of the steam engine, highlights the numerous sounds predictive of breakdown. These sounds indicate to the attentive mechanic "that there is play in the main gasket or that the crank is returned at each stroke of the piston: or even that the piston rod has too much play in the big end, or that there is water in the cylinder, or again that the steam pipe is set at too steep an angle … or even that the valves have been badly adjusted." Faced with such a sensibility and such a complexity, what remains of the margin of tolerance for variations in regimes? From inequalities in equalities, the technician finds himself conducted to a veritable necessity. The bonds by which he is surrounded are a little slack, but they are so numerous, that they end up immobilizing him. Finally the engineer is not an artist who creates and signs a work full of personality, he is a geometer, guardian of rational methods, veritable representative of the technical society of his epoch. He is, like the physician, going down the narrow road of approximate realizations. He sees a precise end.

The apparent indetermination which seems to be attached to industrial assembly solutions is perhaps due to the fact that technical variables are, when all is considered, greater in number than the unknowns of ordinary scientific research. One must not forget that in effect the scientist is himself before an artificial nature. He does not study the mixed global phenomenon offered by general casual representation. He dissects reality down to the interior of its categories, he studies them always under conditions of separation. In short, the scientist evicts innumerable parasites in order to follow the relation and evolution of several variables only. Industry does not have the same freedom. The most heterogeneous variables are imposed on it. An apparently insignificant neglect can thwart success. Thus Taylor has shown that in the work of turning metals, it is necessary to take 12 variables into account. Once the variables have been made the object of separate study, their rational organization can itself be extremely complicated. Industry, which is a search for the optimum, can require long efforts of elaboration. "In reuniting the 12 elements in question," says Taylor,[9] "one ends up establishing 12 long mathematical formulae; in order to determine the two points that the mechanic needs to know – that is to say, the speed and transverse action – one must solve an equation containing 12 unknowns. This requires 6 hours of continuous work by a mathematician. For 18 years, we had mathematicians occupied solely in looking for simplifications of this problem, and the result was that the calculation can be made in 20 seconds and by any worker."

Other research has discerned even more numerous variables. The simple work of a shovel required observation of 30 to 40 factors. In the organization of a mechanized factory, Taylor took into account and examined 70,000 elements. He adds with his robust confidence: "when the work is finished, it will be possible to formulate all the laws governing the movements of the workers."

Will one find a general formula to completely relate these variables which empiricism leaves dispersed, a sort of industrial potential function from which one will derive all the factors on which the energy, time, and pain of men is expended? One can imagine so. Financial success has peremptorily consecrated the work of Taylor. The coordination of factors studied reveals, after all, a single technological will. Without doubt it is a question of a construction where the empirical method plays an important role, but following rules that one would hope to enumerate faultlessly. Industrial harmony must achieve its mathematization.

Sometimes, however, the equilibrium of such heterogeneous factors can be found to be impossible; one can irremediably tip the balance. This is what happens, for example, when a cognitive judgment is supplanted by a judgment of another kind; two orders of practice can then in effect interfere and bring two degrees of precision, having no correspondence with each other, to the determination of the same object. Surprising anomalies ensue. With great acuity, Borel points out[10] that certain monetary evaluations surpass in precision the practical evaluation of the objects of which they indicate the value. "For example, the area of a district of Paris, of thousands of square meters, is often indicated to the nearest hundredth of a square meter, that is to say to five decimal places, this evaluation presupposing that the dimensions of a plot of land (which is, for example, 30 meters by 30 meters) are known with an error of less than one millimeter. If the price of land is calculated at about 1,000 francs per square meter, it would be necessary, in order to evaluate the price to the nearest franc, to measure the dimensions with an error of less than a tenth of a millimeter; yet these dimensions are evaluated more or less to the nearest centimeter. However, cases will occur where this price will be calculated to the nearest franc and will not be *rounded*; it is necessary to seek the explanation of this contradiction in the opposition between the scientific spirit and the juridical spirit." The juridical mind finds an approximate judgment repugnant. The price of an object can, with tolerance and after the fact, be rounded; but legally, and to be properly accounted for, it is first fixed with all the precision afforded by the monetary system. Thus a plot of land whose area is not known exactly is valued economically with the most precise exactitude. There is here an affirmative element which gives knowledge a definitive character, but which profoundly troubles the implicit pragmatism which guides everyday life.

Even in its most minutely regulated applications, human practice is thus traversed by inexplicable inconsistencies. It is not that practice will be required to develop purely logically: the technician will be more willing than the scientist to accept a philosophy which establishes the truth by a judgment of finality. Again it is necessary that this finality be rational and rationally pursued. In short, practice must preserve a solid homogeneity, the full coherence of a program of actions. It is in this sense that it must search for the philosophic conditions of its progress. It must rejoin a pragmatism which is active, desired, constructed, and gradual, a veritable system of schematic will. This is once again the problem of approximate knowledge.

Notes

Editor's note: the imprecise referencing and annotation are features of the original.

1   D. Bellet, *Evolution de l'industrie*, p. 192.
2   *Revue philosophique*, Jan.–Feb. 1925, p. 72
3   Reuleaux, *Cinématique: Principes fondamenteaux d'une théorie générale des machines*, p. 34
4   Weber, *Le rythme du progrès*, p. 234.
5   See Grimshaw, *Procédés méchanique spéciaux et tours de mains*, p. 314.
6   *Revue philosophique*, p. 72.
7   Cited by Sorel, *De l'utilité du pragmatisme*, p. 340.
8   Ibid., p. 321
9   Taylor, *Œvres postumes*, trans. Schwers, p. 36.
10 *Revue philosophique*, Dec. 1924, p. 329

CANGUILHEM

13

REASSESSING THE HISTORICAL EPISTEMOLOGY OF GEORGES CANGUILHEM

Hans-Jörg Rheinberger

L'histoire des sciences c'est la prise de conscience explicite, exposée comme théorie, du fait que les sciences sont des discours critiques et progressifs pour la détermination de ce qui, dans l'expérience, doit être tenu pour réel.[1]

Canguilhem's ideal was to understand science as a complex family of disciplined human efforts to approach the truth about something in the real world.[2]

There is no doubt that the history of science which Georges Canguilhem (1904–95) has left to us is a history highly laden with epistemology. But with what kind of epistemology do we have to do? With Dominique Lecourt, Jean Gayon contends that "'epistemological history' corresponds better to what Canguilhem did," whereas 'historical epistemology' would characterize the work of Canguilhem's predecessor Georges Bachelard.[3] The relation between these two terms is the main question which I would like to pursue in this chapter.[4] I will thereby restrict myself essentially to Canguilhem's writings after 1960 and thus not enter into a discussion of *Le normal et le pathologique*, his medical thesis first published in 1943, nor *La connaissance de la vie* of 1952, nor *La formation du concept de reflexe aux XVIIe et XVIIIe siècle*, his doctoral thesis in philosophy of 1955. The aspect of Canguilhem's work grounded especially in the first of these three books has received considerable attention in the recent past and has won Canguilhem the label of a 'vital rationalist.'[5] It should nevertheless briefly be mentioned that a certain solidarity may be perceived between Canguilhem's historical assessment and vision of health, disease, and life as exposed in his first book, on the one hand, and his vision of science on the other, at least inasmuch as he conceives of both life and science as, first, genuinely autocorrective processes and, second, as existing in an essential precarity. Although a thorough attempt at a periodization of Canguilhem's thought has not yet been undertaken, commentators and students like Etienne Balibar have suggested that there occurred "if not a rupture, then at least a rekindling" in the work of Canguilhem around 1960.[6] This was after he had been installed as the successor of Gaston Bachelard in the chair for the history of science and as director of the Institute for the History of Science and Technology at the Sorbonne, and after he encountered Michel Foucault, who asked him to report on his philosophical thesis *Folie et déraison: Histoire de la folie à l'âge classique*. Foucault's second book, *La naissance de la clinique: Une archéologie du regard médical*, was then published in 1962 in a series on the history and philosophy of biology and medicine directed by Canguilhem. It is to Foucault that we owe one of the most telling characterizations of Canguilhem: "This man whose oeuvre is sincere, deliberately limited, and carefully devoted to a particular domain in a history of the sciences which, in all events, does not represent a spectacular discipline, this man has found himself in a certain manner present in those debates in which he took care never to figure. . . . In the whole debate of ideas that preceded or followed the movement of 1968, it is easy to find the place of those who, from a distance or through close contact, have been formed by Canguilhem."[7]

In his assessment of Canguilhem's philosophy, Balibar quotes from a television interview with Canguilhem in 1964–5 by Alain Badiou:

> *Question*: "Do you mean to say that the expression 'scientific knowledge' is a pleonasm?"
> *Answer*: "You have understood me perfectly. That is what I want to say. A knowledge which is not scientific is no knowledge. I maintain that 'true knowledge' is a pleonasm; 'scientific knowledge' as well; 'science and truth' as well; that all this is the same. That does not mean that for the human spirit there is no aim or no value outside truth, but it does mean that you cannot claim that to be knowledge which is no knowledge, that you cannot give that name to whatever way of living has nothing to do with truth, that is, with rigor."

A few years later, in February 1968, he declared at the Sorbonne before a student audience assembled by the journal *Raison présente* and the *Union rationaliste*:

> One day it seems I scandalized all the students of philosophy who saw a television broadcast. The students, and many of their professors, because I said this: there is no truth other than scientific truth, there is no philosophical truth. I am perfectly willing to take upon me here what I have said elsewhere. But saying that there is no truth other than scientific truth, or that there is no objectivity other than scientific knowledge, does not mean that philosophy has no object. . . . There is no philosophical object in the sense that there is a scientific object which is precisely the one that science constitutes theoretically and experimentally . . . but then I don't mean to say that there is no object of philosophy.[8]

At first glance, these quotations would seem to indicate the exact opposite of something to be appreciated as an historically informed epistemology. The second quote displaces the question from knowledge/truth to the object of science, and it is indeed the question of the object of science and the question of the object of the history of the sciences, the latter being registered in the same space as the object of philosophy mentioned in the quotation, that might give us an entry point to an understanding of Canguilhem's epistemology. My reflections will thus start with an analysis of *L'objet de l'histoire des sciences*, a paper delivered by Canguilhem in October 1966 in Montréal, upon the invitation of the Canadian Society for History and Philosophy of Science and later chosen as the opening paper for the collection of *Etudes d'histoire et de philosophie des sciences* published in 1968. In a footnote to this paper Canguilhem expresses his debt to seminar discussions with his students at the Institut d'Histoire des Sciences et des Techniques in the years 1964–5 and 1965–6. In the course of the following exposition, I will also consider the Introduction to the collection of essays on *Idéologie et rationalité dans l'histoire dees sciences de la vie* published about 10 years later in 1977.

Canguilhem opens his paper on the object of the history of the sciences with a critique: The relation between history of science and science as its object is not unproblematic and direct, as has frequently been assumed; it is not possible to map the relationship of the sciences with their objects directly onto the history of science's own relation to the sciences. Canguilhem distances himself vigorously from such a deliberate historical positivism, or "pure history," as he calls it. "The pure history of botany in the eighteenth century can comprise nothing else under the name of botany than what the botanists of the epoch assigned to themselves as their domain of exploration. Pure history reduces the science it studies to the field of investigation which is assigned to it by the researchers of the epoch, and to the kind of view they cast on the field. But is this science of the past of a science of today?"[9] History of science, understood as the past of a science of today, is not a blueprinted paper version of the chronological progress of a particular science in a particular time period read backwards. "Between the chemistry of oxidation and the biochemistry of enzymatic oxidations, vegetable physiology had first to become cellular physiology – and one knows well what kind of resistances the cellular theory of organisms encoun-tered – and then to detach itself from the first conceptions of the cell and the protoplasm to approach the study of metabolism at the molecular level... . One realizes therefore why the past of a science of today does not confound itself with that same science in its past."[10]

We encounter here a distinction that became central for Canguilhem's thinking about the sciences, which even could be claimed to be the pivotal point around which, for him, the necessity for an epistemological foundation of the history of the sciences turns: "Without epistemology," he claims, "it would thus be impossible to distinguish two kinds of history of science, that of superseded knowledge and that of sanctioned, that is, still actual because acting, knowledge."[11] Giving credit to Gaston Bachelard for having insisted on this distinction, Canguilhem appears to push the door wide open to what has for a long time been so severely criticized by general historians and historians of science alike as "whiggish history,"[12] and bluntly as "nor-mative epistemology" more recently by representatives of sociological and anthropo-logical science studies.[13] Why do we need, as historians of science, to sort out real knowledge from discarded knowledge, and why, for practicing our trade as historians of science, do we need to be in touch with recent science and with its epistemo-logical groundwork? It will require some further exposition in order to show that the implications of the distinction between sanctioned and superseded knowledge are decidedly more complicated. They carry not only the whole burden of the notion of "historical recurrence" developed by Bachelard in his late epistemological writings in order to think history of science in a nonteleological manner, and yet stick to the idea that science carries along with it the most irreversible of all histories.[14] Canguilhem also takes the concept of historical recurrence a step further and makes it an inherent part of historical methodology itself.

Taking issue with Emile Meyerson and putting himself in line with Alexandre Koyré and Gaston Bachelard, Canguilhem states: "The history of the sciences is not the reversed progress of the sciences, that is, the perspectivization of past stages for which the truth of today is the vanishing point. It is an effort to research and to make understandable to what extent the superseded notions or attitudes or methods have themselves, in their time, been a supersession and, consequently, to what extent the superseded past remains the past of an activity for which one has to preserve the qualification of being 'scientific'."[15]

This leads Canguilhem to make a first fundamental distinction with respect to the objects of the sciences. The objects of the sciences themselves are eminently historical entities. The "truth of today" is not perennial, it is deliberately qualified as "of today," and it is not perennially related to an unchanging "nature." The "natural object, outside of any discourse held about it, is not of course the scientific object. Nature is not of and by itself partitioned into scientific objects and phenomena. It is science which constitutes its objects from the moment when it has invented a method to conceive, by way of propositions capable to be integrally composed, of a theory which is controlled by care about the possibility of being wrong."[16] Science might even be said to be the most deeply historical enterprise of culture: For it lies in the very definition of its activity that in order to remain science it has to supersede itself. It *is* science only in the process of a permanent *becoming*. As Gilles Renard, in his study on Georges Canguilhem's epistemology, has recently stated: "To put it another way, its [science's] progress is not related to an ontology whose knowledge would escape that progress, or to a reality that one could apprehend without it, but to its technical and theoretical *autocorrection*."[17] We see therefore that Canguilhem attributes a deep temporality to scientific activity itself, to the extent that the products of science might be addressed as cultural-historical objects *par excellence*. This temporality itself can be and usually is a fragmented one. It is not homogenous, it has a different pace according to the domains where knowledge installs itself as scientific knowledge: "The time of the coming of the scientific truth, the time of veri-fication, has a liquidity or viscosity which is different for the different disciplines within the same periods of general history."[18]

What, then, can be said about the object of the history of the sciences? Writing the history of the objects of science, these cultural objects *par excellence*, consequently cannot be done in the fashion of a "natural history," as Canguilhem puts it.[19] A third level of object formation has to be introduced to account for the activity of the historian of science with its own epistemological dimension: "The object in the history of the sciences has nothing to do with the object of the science. The scientific object, constituted by a methodical discourse, is second to, although not derived from the natural, primary object which one could, by playing with the meaning of the term, call pre-text. The history of the sciences exerts itself on these secondary, non-natural, cultural objects, but it is not derived from them, just as they are not derived from the primary ones. The object of the historical discourse is, in fact, the historicity of the scientific discourse, inasmuch as this historicity represents the effectuation of a project which is normed from within, but traversed by accidents, retarded and rerouted by obstacles, interrupted by crises, that is, of moments of judgment, and of truth."[20] If the objects of science are pushed into the future, the objects of the history of sciences are pulled from past. There is a certain inevitable "anachronism" involved in this position.[21]

As a consequence, the historiography of science operates according to its own proper time regime. It creates its own temporal order in which it exposes that cultural object which is the scientific object. One of the perennial problems of this temporal order is the relation between continuity and break in the development of the sciences. Canguilhem is aware of the danger of oversimplifying Bachelard's notion of an epistemological rupture between a prescientific mode of thinking entrenched in the immediacy of everyday life with its lived illusions, and a scientific one that needs the surroundings of a research culture to flourish. The question is: how do we reflect our own historical periodizations? Canguilhem does not follow Thomas Kuhn with his essentially sociopsychological construction of an alternation between phases of normal science and revolutionary science. Science, according to Canguilhem, is revolutionary in its very core, but the pace and scope of its displacing dynamics can vary widely according to cultural circumstances and according to the state of the scientific object itself. I will come back to this problem when discussing the notion of scientific ideology.

A typical historical manifestation of the relation between continuity and break is the question of the 'precursor.' Canguilhem devotes considerable effort to denouncing the search for precursors in the history of the sciences. The very notion of the precursor for him is the paradigmatic expression of a historiography of science that confounds the object of the history of science with the object of science itself. "If one substitutes the logical time of the relations of truth for the historical time of their invention, one aligns the history of science with science itself, the object of the former with that of the latter, and one creates that artifact, that false historical object which is the precursor."[22] In doing so science itself would become deprived of its essential historical dimension. It would become the logical deployment of a truth conquest: "At the limit, if there would exist precursors, the history of the sciences would lose all meaning, because science itself would only apparently have a historical dimension."[23] The precursor marks a caricature of the history of science, it is the very counterimage of the notion of historical recurrence. In the critique of the notion of precursor, the Canguilhemian epistemology reaches its full sophistication and intellectual power, as exemplified in an early case study on the development of cell theory,[24] and in a late study on the history of the life sciences since Darwin, where Canguilhem touches upon the work of Mendel, whose name acquired iconic status, even becoming an "historical pleonasm" in advance.[25]

Let me mention another problem. It is connected to the fact that acting in the temporal order of the historiography of science needs epistemological coupling to the progress of the sciences and ideally even to the frontiers of research. The historian of science changes and has to change his own narrative with the advancement of science: "Thus the historiography of the sciences, the history of the progressive relation of intelligence to truth, secretes its own time, and it does so differently according to the moment of progress at which it assigns itself the task of reviving, in the earlier theoretical discourses, that which the language of the day can still understand."[26] That means, practicing the history of the sciences is itself a fundamentally historical enterprise, an enterprise which differentially takes account of the cultural embedding of would-be scientific objects in accordance with and depending upon its own cultural embedding. Just as science creates permanently superseded scientific objects, so the narratives about that movement by the historian of the sciences create constructs that become superseded. Their supersession, however, is not of the same intrinsic autocritical mode as that of the sciences, because the discourse on the history of the sciences is motivated not only by the surpassing of its own achieved standards, but first and foremost by its epistemological coupling to the scientific movement:

> Doubtless it goes without saying that scientific progress by epistemological rupture imposes the frequent rewriting of the history of a discipline that one cannot exactly call the same, because under the usual name, perpetuated by linguistic inertia, there is hidden a different object. Besides the personality of the authors, it is not only through the amount of accumulated knowledges that [e.g.] *La logique du vivant* (1970) by François Jacob differs from the second edition (1950) of the *History of Biology* by Charles Singer. It is from the fact of the discovery of the structure of DNA (1953) and the introduction of new concepts into biology, either under conservative terms such as organization, adaptation, heredity, or other hitherto unheard of terms such as message, program, teleonomy.[27]

The sense of historical ruptures and filiations can only come to the historian of the sciences from his contact with recent science, even if mediated through epistemology, which in turn ruptures his own discourse. This contact is productive "under the condition that it is vigilant, as Gaston Bachelard has taught it. Thus understood, the history of the sciences cannot but be precarious, involved in its own rectification."[28]

The trivial aspect of the historical embeddedness of the historiography of the sciences is that one has to be aware that historians of science, just like historians in general, are making time-dependent choices and connecting things according to particular historical conjunctures. But it is not the sheer contingency of the historical circumstances that induces them to do so; there is rather an inner reason for them to rewrite, from time to time, the history of their disciplines and the history of the transdisciplinary dynamics of scientific objects. The nucleus of this inner constraint and contrivance is the historicity of the scientific objects themselves with their developments and displacements. "A science is a discourse which receives its norm from its critical rectification. If this discourse has a history whose course the historian claims to reconstitute, it is because it *is* a history of which the epistemologist has to reactivate its meaning."[29] As a simple thought experiment we could ask ourselves how a history of evolution would look under the assumption that new findings would corroborate the view that acquired characteristics are not hereditarily transmitted. Our pictures of Lamarck and Darwin would change accordingly.

The main point remains that the historian's object is different from the object of the scientist. The historian's object is the scientific object in its contextualized movement, in the frame of an historico-epistemological discourse *sui generis*. The object of the history of the sciences is a cultural object that does not belong to the category of universals. "The history of the sciences, insofar as it applies itself to the object delimited above, has not only to do with a group of sciences without inner cohesion but also with the nonscientific, the ideological, with political and social practice. This object therefore has its natural theoretical place not in this or that science, where history would expose it; neither has it its place in politics or pedagogics. The theoretical place of this object is to be found nowhere but in the history of the sciences themselves, for it is this history, and it alone, that constitutes the specific domain where the theoretical questions which are posed by scientific practice in its continual becoming find their place."[30] And because the object of the history of the sciences is not a scientific object, the history of the sciences consequently "is not a science."[31] It is of a different epistemological order. It is not easy to convey an appropriate picture of an activity that understands itself working on its own proper terrain by constantly tracing and retracing historical connections whose possibility and whose plausibility derive from an assessment, at least in the last instance, of the actual state of scientific affairs in the domain covered by that activity. Practicing the history of science is thus a constant constructive effort: The history of the sciences has to be *made* and *remade*, and it cannot simply be described once and for all. With these reflections, the Canguilhem of the 1960s and 1970s is at the very least reflecting and rationalizing his own learning process.

The space thus carved out for an epistemological history of science is filled by Canguilhem on many occasions and by many examples taken from the history of the life sciences. His epistemology is inseparable from the concrete historical material on which it operates and with which it works. His preferred way of doing so is via conceptual order.[32] Canguilhem has a clear predilection for a history of concepts, as a history of problems meandering through the historical space of the sciences, the "restitution," as Renard calls it, "of a conceptual itinerary."[33] While he talks about scientific objects, it is mainly concepts which he addresses. In this respect, Canguilhem belongs to the broad movement of a qualified history of ideas that flourished around the middle of the twentieth century, including representatives such as Alexandre Koyré, Arthur Lovejoy, A. C. Crombie, and Karl Rothschuh. But there is another aspect to this emphasis on concepts. In his assessment of Canguilhem's conception of science, Balibar speaks of "the fundamental idea according to which the typical units of knowledge are not 'theories' but 'concepts'." And he explains: "It is in the circulation – that is, the translation, transposition, and generalization – of concepts that the application or their 'work' is effected and that they are examined."[34] We know that, among others, François Jacob has taken up this point and suggested that it appears to be a characteristic of the life sciences in particular to be organized less around theories than around concepts which, according to epoch and domain, display narrower or wider generalizations.[35]

Let us consider a brief example. On several occasions, Canguilhem points to the crucial function of Claude Bernard's concept of *milieu intérieur* for the development of a physiological order of thinking in its own right in the second half of the nineteenth century. For Canguilhem, the concept of internal environment marks a turning point in the history of the life sciences which is also connected to a new space of experimentation. The concept might even be said to have been provoked within that new space, in the sense in which Bachelard in his *Le nouvel esprit scientifique* had talked about conceptual circumstantiality in general; Canguilhem invokes Bachelard in this context: "The concepts and the methods, everything is a function of the domain of experimentation. Every scientific thought must change before a new experience; a discourse on the scientific method will always be a discourse of circumstance, it will never describe the definitive constitution of the scientific spirit."[36] Neither anatomy nor chemistry, Canguilhem contends in this context, are sufficient to resolve a genuinely physiological question; what is needed is a kind of experimentation on animals which, in that it allows one to follow the mechanism of a function in a living being, "leads to the discovery of phenomena which it alone can bring to light and which nothing else could have induced one to foresee."[37] But in Canguilhem's view, Claude Bernard's technique of vivisection, the inauguration of experimentation on living beings, is intimately connected to and in a sense also enabled by a profound conceptual relocation: "Let us be clear about this point," he says; "it is the *concept* of interior environment which is given as the theoretical foundation of the *technique* of physiological experimentation."[38] The *in vitro* experimentation of twentieth-century biochemistry and enzymology, then, will turn the *milieu intérieur* itself into an object of intense investigation, and eventually subject it to a profound reconfiguration. Thus although Canguilhem recognizes that "the history of the sciences can without doubt distinguish and admit several levels of objects within the specific theoretical domain which it constitutes: documents to catalogue; instruments and techniques to describe; methods and questions to interpret; concepts to analyze and to critique," it is and remains for him this latter task alone which confers the real "dignity of a history of the sciences" on all the other activities of the historian.[39]

Whereas we can rightly identify Claude Bernard's case as an instance of epistemological rupture, there are other instances where movements in conceptual space appear much more dispersed and of a more global order. Such is the case for Canguilhem's study of the concept of biological regulation throughout the eighteenth and the nineteenth centuries.[40] Here Canguilhem shows that "one cannot undertake the history of 'regulation' without beginning with the history of the 'regulator,' which is a history composed of theology, astronomy, technology, medicine, and even sociology at its birth."[41] Following a meandering but rigorous analytical course that takes him through Leibniz's idea of pre-established harmony, Linnaeus's economy of nature, Buffon's spontaneous organization of living matter, Watt's steam engine regulator, and Malthus's concept of human population regulation, Canguilhem works his way through the nineteenth century with its opposition – in France – between Comtean regulation from without and Bernardian regulation from within, where again the inner milieu plays a crucial role, ending up with *Entwicklungsmechanik* at the beginning of the twentieth century. It is only with Hans Driesch's book *Die organischen Regulationen* of 1901, claims Canguilhem, that the concept of biological regulation finally reaches the point at which an autocritical biological discourse finds its irreversible crux. This rupture is marked by the plural in Driesch's book and installs regulatory networks as the key for an understanding of biological organization. It installs them as an autocorrective scientific object, as a "particular problematic which inhabits [a science] and which defines a homogenous field of conceptual filiation."[42] Canguilhem was a master of conceptual landscape painting, both in the foreground and in the distance.

Here we have arrived at a point where another notion comes into play which Canguilhem developed in his later writings. It is the notion of "scientific ideology" already mentioned above. Canguilhem said he was inspired to use the term by contemporary work of Louis Althusser.[43] The basic idea of scientific ideology is triple. First, scientific ideologies are systems of thought whose objects, as compared to the standards of these systems, are hyperbolic; that is, they are not (or not yet) in the realm of, nor under the control of, that system. Second, a science never installs itself on virgin terrain, it never springs immediately from lived experience, but only from a field prepared by a scientific ideology. Conversely, such an ideology is usually the extension of a science into a field that comes into a lateral focus. Third, a scientific ideology is not to be confounded with a 'false science.' It derives its power and prestige from an institutionalized science whose principles it seeks to try out in another field.[44] To summarize with a suggestive formula to be found in one of Canguilhem's papers on the history of physiology: "The problems ... do not necessarily emerge on the terrain where they find their solution."[45] Whether the term 'scientific ideology' was a good choice may be questioned. Perhaps we would be just as well to talk about the orienting function of overgeneralizations.

Balibar comes to the conclusion that, above all, the notion of scientific ideology represents the vanishing point of a long series of tendencies dispersed in Canguilhem's work which converge toward the conviction that "there can be no history of truth which is *only* the history of truth, nor a history of science which is *only* the history of science."[46] The construction is such that "there is always already a dialectics of scientificity and ideologization, or better yet, of *ideologization and of de-ideologization* of the concept which is constitutive of knowledge."[47] To say it in the words of Canguilhem himself: "Wanting to do the history of truth alone makes one do an illusory history."[48]

We can qualify this statement as an attempt at broad recognition of the insight that the microbreaks of scientific autocorrection, which themselves permanently result in discarding the truths of today, are always accompanied by and embedded in macrobreaks and displacements whereby the dynamics of whole fields of knowledge become reoriented and reconfigured. If Canguilhem's famous expression of Galileo's "being in the truth"[49] (taken up by Foucault in *L'ordre du discours*) – already implies that "the veracity of the saying-the-truth of science does not consist in the true reproduction of some truth forever inscribed in the things or in the intellect,"[50] then these broader displacements need all the more to be perceived as more encompassing cultural shifts. After all, as Renard summarizes in an almost Husserlian but very appropriate manner, which congenially seizes Canguilhem's ultimate message: "it is humanity that produces science and not the other way round."[51] The sciences therefore remain embedded in the social and technological concerns of humanity from which they sprang. As Canguilhem declares with respect to one of his favorite subjects of analysis, Claude Bernard's experimental medicine: "Experimental medicine then is only one of the configurations of the demiurgic dream being dreamt by all industrial societies in the middle of the nineteenth century, in an age where, through the bias of their application, the sciences became a social power."[52] Canguilhem's way of doing history of science thus pays tribute to Bernard's own claim, in his *Introduction à l'étude de la médecine expérimentale*, that "with the help of these active experimental sciences man becomes an inventor of phenomena, a veritable counter-master of creation."[53]

Acknowledgments

I thank Jean Gayon for his generous hints and valuable suggestions and comments.

Notes

1 Georges Canguilhem, "L'objet de l'histoire des sciences." In *Etudes d'histoire et de philosophie des sciences* (Paris: Vrin, 3rd ed. 1975), pp. 9–23, 17.

2 Marjorie Grene, "The Philosophy of Science of Georges Canguilhem: A Transatlantic View." *Revue d'histoire des sciences* 53 (2000): 47–63.

3 Jean Gayon, "The Concept of Individuality in Canguilhem's Philosophy of Biology." *Journal of the History of Biology* 31 (1998): 307 n. 8; see also Dominique Lecourt, *L'épistémologie de Gaston Bachelard* (Paris: Vrin, 1969).

4 Dominique Lecourt was among the first to write on Canguilhem's "historical epistemology." See his "L'histoire épistémologique de Georges Canguilhem" in his *Pour une critique de l'épistémologie (Bachelard, Canguilhem, Foucault)* (Paris: Maspéro, 1972). See also the special volume on "Canguilhem en son temps/Georges Canguilhem in his time," *Revue d'histoire des sciences* 53(1) (2000).

5 François Delaporte, ed., *A Vital Rationalist: Selected Writings from Georges Canguilhem*, with an Introduction by Paul Rabinow and a critical bibliography by Camille Limoges, trans. Arthur Goldhammer (New York: Zone Books, 1994). Jean-François Braunstein et al., *Actualité de Georges Canguilhem: Le normal et le pathologique* (Paris: Synthélabo, 1998).

6 Etienne Balibar, "Science et vérité dans la philosophie de Georges Canguilhem." In *Georges Canguilhem, Philosophe, historien des sciences* (Paris: Albin Michel, 1993), p. 58.

7 Michel Foucault, "La vie, l'expérience et la science." In *Dits et écrits 1954–1988*, vol. IV (Paris: Gallimard, 1994), pp. 263–76.

8 Quoted in Balibar, *Georges Canguilhem*, pp. 58–59, 60.

9 Georges Canguilhem, "Le rôle de l'épistémologie dans l'historiographie scientifique contemporaine." In *Idéologie et rationalité dans l'histoire des sciences de la vie* (Paris: Vrin, 2nd ed. 1981), p. 13.

10 Ibid., p. 15.

11 Canguilhem, *Etudes*, p. 13.

12 Herbert Butterfield, *The Whig Interpretation of History* (New York: Charles Scribner's Sons, 1951).

13 See e.g. Andrew Pickering, ed., *Science as Practice and Culture* (Chicago: University of Chicago Press, 1992); Bruno Latour, *We Have Never Been Modern* (Cambridge, MA: Harvard University Press, 1993).

14 Georges Canguilhem, "L'histoire des sciences dans l'oeuvre épistémologique de Gaston Bachelard." In *Etudes*, pp. 181–4.

15 Canguilhem, *Etudes*, p. 14.

16 Ibid., pp. 16–17.

17 Gilles Renard, *L'épistémologie chez Georges Canguilhem* (Paris: Editions Nathan, 1996), p. 33.

18 Canguilhem, *Etudes*, p. 19.

19 Ibid., p. 18.
20 Ibid., p. 17.
21 Helge Kragh, *An Introduction to the Historiography of Science* (Cambridge: Cambridge University Press, 1987), p. 74.
22 Canguilhem, *Etudes*, p. 22.
23 Ibid., pp. 20–1.
24 Georges Canguilhem, "La théorie cellulaire." In *La connaissance de la vie* (Paris: Vrin, 2nd ed. 1975), pp. 43–80.
25 Georges Canguilhem, "Sur l'histoire des sciences de la vie depuis Darwin." In *Idéologie*, 99–119.
26 Canguilhem, *Etudes*, p. 20.
27 Canguilhem, *Idéologie*, p. 27.
28 Canguilhem, *Etudes*, p. 20. See also Gaston Bachelard, "L'actualité de l'histoire des sciences [1951]." In *L'engagement rationaliste* (Paris: Presses Universitaires de France, 1972), pp. 137–52.
29 Canguilhem, *Idéologie*, p. 21.
30 Canguilhem, *Etudes*, pp. 18–19.
31 Ibid., p. 23.
32 Lecourt, *Pour une critique*.
33 Renard, *L'épistémologie*, p. 7.
34 Balibar, *Georges Canguilhem*, p. 68.
35 François Jacob, *La logique du vivant* (Paris: Gallimard, 1970), p. 21.
36 Gaston Bachelard, *Le nouvel esprit scientifique* (Paris: Presses Universitaires de France, 1968), p. 135.
37 Georges Canguilhem, "Théorie et technique de l'expérimentation chez Claude Bernard." In *Etudes*, p. 148.
38 Ibid., p. 148.
39 Ibid., p. 19.
40 Georges Canguilhem, "La formation du concept de régulation biologique aux XVIIIe et XIXe siècle." In *Idéologie*, pp. 81–99.
41 Ibid., p. 83.
42 Renard, *L'épistémologie*, p. 77.
43 François Bing and Jean-François Braunstein, "Entretien avec Georges Canguilhem." In *Actualité de Georges Canguilhem*, p. 128.
44 Georges Canguilhem, "Qu'est-ce que une idéologie scientifique?" In *Idéologie*, pp. 33–45.
45 Georges Canguilhem, "La constitution de la physiologie comme science." In *Etudes*, p. 237.
46 Balibar, *Georges Canguilhem*, p. 66.
47 Ibid., p. 69.
48 Canguilhem, *Idéologie*, p. 45.
49 Georges Canguilhem, "Galilée: La signification de l'oeuvre et la leçon de l'homme." In *Etudes*, p. 46.
50 Canguilhem, *Idéologie*, p. 21.
51 Renard, *L'épistémologie*, p. 47.
52 Canguilhem, *Etudes*, p. 140.
53 Claude Bernard, *Introduction à l'étude de la médecine expérimentale* (Paris: Editions du cheval Ailé, 1945), p. 71.

14

THE OBJECT OF THE HISTORY
OF SCIENCES

Georges Canguilhem

Considered under an aspect such as that afforded by the proceedings of a conference, the history of sciences could pass as a section heading rather than as a discipline or a concept.[1] A section can be inflated or distended almost indefinitely since it is created only by a procecdural rule, whereas a concept, because it contains an operational or judgmental norm, cannot change in respect of its extension without there being some corresponding change in its intension. So it comes about that one can equally well include under the heading of the history of sciences the description of a recently discovered portulan or a thematic analysis of scientific theory construction. It is therefore worth asking those who are interested in the history of sciences about the point of doing it; what is the founding idea of the history of sciences? There are certainly several questions which have been and continue to be asked about this activity. These questions concern *What?*, *Why?*, and *How?* But it turns out that a principal question which ought to be asked, but hardly ever is, is the question, Of what? *Of what* is the history of sciences the history? That this question has not been asked is related to the fact that it is generally believed that the answer lies in the very expression history *of* sciences or of *a* science.

Recall briefly how the questions of *Who?*, *Why?*, and *How?* are most often formulated today.

The question *Who?* entails the question *Where?* In other words the need for research and teaching in the history of sciences leads to its residing here or there in the institutional space of universities, depending on the antecedently specified discipline in which the need is felt. Bernhard Sticker, director of the Institute for the History of Sciences in Hamburg, has underlined the contradiction between destination, knowledge, and method. Its destination necessarily localizes the history of sciences to Faculties of Science, its method puts it in the Faculty of Philosophy. If one holds it to be a species of a genus, the history of sciences ought to have its place

Georges Canguilhem, "The Object of the History of Sciences," from *Etudes d'histoire et de philosophie des sciences*. Paris: J. Vrin, 1983. French text © 1983 by Librarie Philosophique J. Vrin – Paris. Reproduced by permission of the publisher, http://www.vrin.fr. New English translation by Mary Tiles.

in a central institute for historical disciplines. In fact, the specific interests of historians on the one hand, and of scientists on the other, only leads them to the history of sciences as a byway. General history is predominantly political and social history, completed by a history of religious or philosophical ideas. The history of a society as a whole, whether of its legal institutions, its economics, or of its demography, does not necessarily require a history of scientific methods or theories as such, even though philosophic systems are related to popularized scientific theories, that is, scientific theories watered down into ideologies. On the other hand, scientists as such have no need for the history of science beyond a bare minimum of philosophy without which they could not talk to nonscientists about their science. It is indeed rare, above all in France (with the exception of Bourbaki), for them to incorporate any philosophy into the exposition of the results of their specialized studies. If they occasionally become historians of sciences it is for reasons derived from the intrinsic demands of their research.

It is not then unprecedented for their field of competence to guide them in the choice of questions thought to be of primary interest. This was the case with Pierre Duhem in the history of mechanics, with Karl Sudhoff and Harvey Cushing in the history of medicine. As to philosophers, they can be led into the history of sciences either traditionally and indirectly via the history of philosophy, to the extent that such and such a philosopher has, in his time, called on a triumphant science to clarify the ways and means of a recalcitrant knowledge; or more directly via epistemology, to the extent that its critical consciousness of contemporary methods of a science which is adequate to its object feels itself bound to celebrate their power by recalling the difficulties which delayed their conquest. For example, if it is of little interest to the biologist, and even less to the mathematical probability theorist, to enquire about what, in the nineteenth century, could have prevented Auguste Comte and Claude Bernard from admitting the validity of statistical calculations in biology, it is not the same for those who, in epistemology, deal with probabilistic causality in biology. But it remains to be shown – as will be attempted below – that if philosophy sustains a more direct relation to the history of sciences than does history or science, it is on the condition that, as a result, it accept a new status concerning its relation to science.

The answer to the question *Why?* is symmetrical to the answer to the question *Who?* There are three reasons for doing the history of sciences: historical, scientific, and philosophical. The historical reason, which is extrinsic to science understood as a verified discourse on a limited area of experience, resides in the practice of commemorations; in the fact of rivalries in research over intellectual paternity, in quarrels over priority, such as that evoked by Joseph Bertrand in his eulogy on Neils Henrich Abel, which concerns the discovery in 1827 of elliptical functions. This reason is an academic fact, related to the existence and function of Academies and the multiplicity of national Academies. There is a more explicitly scientific reason, expressed by scientists in their function as researchers rather than academics. He who happens on a theoretical or experimental result which had been up to that time inconceivable, which is disconcerting for his contemporaries, does not get any support for want of possible communication within the scientific community. And because the scientist believes in the objectivity of his discovery, he looks to see whether perchance his thought has not already been previously thought. It is in looking to the past for an accreditation for his discovery, due to a temporary lack of ability to find this in the present, that an inventor invents his predecessors. It is thus that Hugo de Vries rediscovered Mendelism and discovered Mendel. Finally, the strictly philosophical reason comes from this: without reference to an epistemology, a theory of knowledge would be a meditation on the void, and without relation to a history of sciences an epistemology would be a less important labor which was completely superfluous to the science of which it pretends to speak.

The relations between history of sciences and epistemology can be understood in two opposite ways. Dijksterhuis, the author of *The Mechanization of the World Picture*, thinks that the history of sciences is not only the memory of science but also the laboratory of epistemology. His words have often been cited, and his thesis has found favor with many specialists. This thesis has a lesser-known precedent. In his Eulogy to Cuvier, Flourens, referring to the *History of the Natural Sciences*, published by Magdalaine de Saint-Agy, declares that to do the history of sciences is to "put the human mind into experience. ... to make an experimental theory of the human mind." Such a conception turns back to copy the relation between the history of sciences and the sciences of which it is the history from the relation between that science and the objects of which it is the science. In fact the experimental relation is just one of these relations, and it does not go without saying that it is this relation which must be imported and transplanted from science into history. In other words, this thesis about historical methodology leads, in its recent defender, to the epistemological thesis that there exists an eternal scientific method, somnolent in some epochs, vigilant and active in others. Gerd Buchdahl holds this thesis to be naive, and one would agree, so long as the empiricism or the positivism which inspires it can pass as such. It is not unintended that positivism should be denounced here. Between Flourens and Dijksterhuis, Pierre Lafitte, a confirmed disciple of Auguste Comte, defined the role of the history of sciences as that of a "mental microscope," having the revelatory effect of introducing a delay and distance into the contemporary exposition of scientific knowledge, by mentioning the difficulties encountered in invention and in the propagation of knowledge of it. With the image of the microscope we remain within the laboratory, and we find positivist presuppositions in the idea that history is only an injection of duration into the exposition of scientific results. The microscope procures the enlargement of a development which is given independently, but which is visible only by its means. Here again the history of sciences is to the sciences as a scientific detection apparatus is to the already constituted objects which it detects.

In order to understand the function and significance of a history of sciences one can oppose to the model of the laboratory that of the school, or of the tribunal, of an institution and of a place where judgments are brought to bear on the past of knowledge, on the knowledge of the past. But here one needs a judge. It is the epistemologist who is called to furnish history with the principle of judgment by teaching it the most recent language spoken by some science, chemistry for example, and in thus permitting it to retreat into the past, back to the time when this language ceases to be intelligible or translatable into any more loose or more commonplace language which was spoken before. The language of nineteenth-century chemistry finds its semantic lacunae in the period before Lavoisier, because Lavoisier instituted a new nomenclature. Now as it happens it has not been sufficiently remarked upon that, in the preliminary discourse to his *Elementary Treatise on Chemistry*, Lavoisier at the same time assumed responsibility for two decisions about which one could have complained to him, that he had "changed the language that our masters have spoken," and that he had not given in his work "any history of the opinions of those who have preceded me." It is as if he had understood, in the Cartesian manner, that it is all one to found a new science and to wholly sever relations with that which had been improperly occupying its place. Without epistemology it would thus be impossible to discern two sorts of history, both of which are said to be history of sciences – that of lapsed knowledge and that of sanctioned knowledge, i.e. science which is still current because still being used. It was Gaston Bachelard who opposed lapsed history to sanctioned history, to the history of the facts of scientific experimentation or conceptualization appreciated in their relation to fresh scientific values.

The idea which Alexandre Koyré had of the history of sciences, the idea which his works have illustrated, was not fundamentally different. Although Koyre's epistemology was closer to that of Meyerson than to that of Bachelard, being more sensitive to the continuity of rational functions than to the dialectic of rationalizing activity, it is because of it that *Galilean Studies* and *The Astronomical Revolution* were written as they were. In addition, in order to remove from this difference on appreciation of epistemological ruptures any appearance of being a contingent or subjective fact, it is not without interest to note that in general Koyré and Bachelard were interested in exactly successive periods in the history of science, periods which were not equally equipped for giving a mathematical treatment of physical problems. Koyré begins with Copernicus and ends with Newton, which is where Bachelard begins. As a result, the epistemological orientation of history according to Koyré can serve as a verification of Bachelard's view that a continuist history of sciences is the history of a young science. The epistemological theses of the historian Koyré are, firstly, that science is theory and that theory is fundamentally mathematization – Galileo, for example, is Archimedean more than Platonist; and secondly that there is no way of avoiding error in the generation of scientific truth. To write the history of a theory is to write the history of the hesitations of the theoretician. "Copernicus ... was not Copernican." When invoking the image of the school or of the tribunal to characterize the function and significance of a history of sciences which is not forbidden to bring judgments of scientific value to bear, it is just as well to avoid a possible misunderstanding. A judgment on this matter is neither a purge nor an execution. The history of sciences is not the progress of sciences in reverse, i.e. the putting into perspective of outmoded stages whose truth is today on the point of disappearing. It is an effort to enquire into and give an understanding of the extent to which outmoded notions or attitudes or methods were, in their time, successful; and consequently of the respect in which the outmoded past remains the past of an activity for which it is necessary to retain the term "scientific." To understand what gave instruction in its time is as important as exposing the reasons for its destruction by what followed.

How is the history of sciences done, and how should it be done? This question brings us closer to the question to come: *of what* does one write the history in the history of sciences? In fact, it is most often presupposed that this question has been resolved simply, it seems, by not being asked. This is apparent in certain debates between opponents whom anglosaxon authors would designate "*externalists*" and "*internalists.*" Externalism is a way of writing the history of sciences by seeing certain events – which one continues to call scientific more by tradition than as a result of critical analysis – as being conditioned by their relation with economic and social interests, with technical demands and practices, with religious or political ideologies. It is, in sum, an enfeebled and impoverished Marxism, which has currency in rich societies. Internalism – held by the externalists to be idealism – consists in thinking that there is no history of sciences unless one places oneself in the very interior of the scientific discipline in order to analyze the methods by which it seeks to satisfy the specific norms which permit it to be defined as science rather than as technology or ideology. From this perspective, the historian of sciences must adopt a theoretical attitude with respect to what is retained as a fact of theory, consequently using hypotheses, paradigms, in the same way as scientists themselves.

It is evident that both positions come to assimilate the object of the history of sciences to the object of a science. The externalist sees the history of science as an explication of a cultural phenomenon via the conditioning of its global cultural milieu, and consequently assimilates it to a naturalistic sociology of institutions, neglecting entirely the interpretation of a discourse having pretensions to truth. The internalist sees in the facts of the history of sciences – for example, the facts of simultaneous discovery (infinitesimal calculus, conservation of energy) – facts of which one cannot write the history without a theory. Here, therefore, the fact of the history of sciences is treated as a scientific fact, on the basis of an epistemological position which consists in privileging theory relative to the empirically given.

Now what must be questioned is the attitude which one could call spontaneous, and which is in fact almost universal, which consists in aligning the history of science with science when it is a matter of the relation of knowledge to its object. We thus ask, "Of what exactly is the history of sciences the history?"

When one talks of the science of crystals, the relation between the science and the crystals is the genitive relation, as when one talks of the mother of a kitten. The science of crystals is a discourse on the nature of crystals, the nature of crystals being nothing other than crystals considered in their self-identity, independently of all uses to which man puts them but for which they were not naturally destined. From the time when crystallography, crystalline optics, and mineral chemistry are constituted as sciences, the nature of crystals is the content of the science of crystals, i.e. an account made up of objective propositions laid down by a process of hypothesis and verification which has been forgotten in favor of its results. When Helene Metzger wrote *La Génèse de la Science des Cristeaux* [The Genesis of the Science of Crystals] she composed a discourse on discourses held on the nature of crystals, discourses which were not at first good discourses, in terms of which crystals have become the object of which their science gives an account. Thus the history of sciences is the history of an object which is a history, which has a history, whereas science is the science of an object which is not history, which does not have a history.

Crystals are given objects. Even if it is necessary in the science of crystals to take account of a history of the earth and a history of minerals, the time of this history is itself an object already given in that science. Thus the object, crystal, has, relative to the science which takes it for an object of knowledge-to-be-obtained, the sort of independence from discourse, which leads one to call it a natural object.[2] This natural object, outside of all discourse held on it, is indeed not a scientific object. Nature is not of itself cut and partitioned into scientific objects and phenomena. It is science which constitutes its object from the time when it invents a method for forming, by propositions capable of being consistently combined, a theory controlled by the concern of finding itself to be mistaken. Crystallography is constituted from the time when crystalline species are defined by constant angles of faces, by systems of symmetry, by the regularity of truncations at summits as a function of systems of symmetry. "The essential point," said Hauy, "is that theory and crystallization ended up meeting each other and finding themselves in accord with one another."

The object in the history of sciences has nothing in common with the object of science. The scientific object, constituted by methodical discourse, is second in relation to, even though not derived from, the initial natural object which we can, playing on meanings, gladly call a pre-text. The history of sciences practices on these second, nonnatural, cultural objects but does not derive from them any more than these second objects derived from the first. The object of historical discourse is, in effect, the historicity of scientific discourse, inasmuch as this historicity represents the carrying out of an internally law-governed project, but one which is traversed by accidents, retarded or deflected by obstacles, interrupted by crises, i.e. moments of judgment and of truth. It has not been sufficiently noted that the birth, in the eighteenth century, of history of sciences as a literary genre presupposes, as a condition of its historical possibility, knowledge of two scientific revolutions; for it needs at least two. In mathematics, the algebraic geometry of Descartes, then the infinitesimal calculus of Leibniz–Newton; in mechanics and cosmology, the *Principles* of Descartes and the *Principia* of Newton. In philosophy, and more exactly in the theory of knowledge, i.e. the theory of the foundations of science, Cartesian innatism and the sensualism of Locke. Without Descartes, without the rending from tradition, a history of science could not begin. But, according to Descartes, knowledge is without history. It needed Newton, and the refutation of Cartesian cosmology, for history, ingratitude for the proclaimed birth of science which breaks with its origins, to appear as a dimension of science. The history of sciences is the explicit awareness, expressed as theory, of the fact that the sciences are critical and progressive discourses for the determination of that which, in experience, must be taken for real. The object of the history of sciences is thus not an object given there; it is an object to which incompleteness is essential. The history of sciences can in no way be the natural history of a cultural object. Too often it is done as natural history, because science is identified with scientists and scientists with their civil and academic biographies, or even because science is identified with its results and these results with their present pedagogical statement.

The object of the historian of sciences can only be delimited by a decision which assigns it its interest and importance. Moreover, this is always so at bottom, even in the case when this decision only obeys a tradition observed without critique. Take for example the history of the introduction and extension of probabilistic mathematics into biology and the human sciences in the nineteenth century; it did not correspond to any natural object, knowledge of which could be supposed to be a replica or a full description. Therefore, the historian himself constitutes an object starting from a current stage of the biological and human sciences, a stage which is neither the logical consequence nor the historical result of any previous stage of *one* distinctive science, neither of the mathematics of Laplace, nor the biology of Darwin, nor the psychophysics of Fechner, nor the ethnology of Taylor, nor the sociology of Durkheim. But on the contrary biometry and psychometry could only be constituted by Quetelet, Galton, Catell, and Binet once nonscientific practices had had the effect of furnishing observation with a homogeneous matter, one which was susceptible of being mathematically treated. The human form, object of Quetelet's studies, presupposes the institution of national armies, of conscription and the interest accorded to criteria of reform. Intellectual aptitudes, the object of Binet's study, presuppose the institution of compulsory primary schools and the interest accorded to criteria of mental retardation. Thus the history of sciences, to the extent that it is applied to the object delimited above, is related not only to a group of sciences without intrinsic cohesion but also to nonscience, to ideology, to political and social practice. So this object does not have a natural theoretical place in this or that science, where history will go to set it apart, any more than it has one in politics or pedagogy. The theoretical place of this object is not to be sought anywhere other than in the history of sciences itself, for it is that, and that alone, which constitutes the specific domain where theoretical questions posed by scientific practice and its growth find their place. Quetelet, Mendel, Binet, Simon invented unforeseen relations between mathematics and initially nonscientific practices: selection, hybridization, orientation. Their inventions were responses to questions which they asked themselves in a language which they had to forge. The critical study of these questions and these responses: this is the proper object of the history of sciences, which is sufficient for that history to avoid the possible objection of being externalist in conception.

The history of sciences can doubtless distinguish and admit several levels of objects in the specific theoretical domain which it constitutes: documents to catalogue, instruments and techniques to describe, methods and questions to interpret, concepts to analyze and criticize. This last task alone confers on the preceding ones the dignity of the history of science. To be ironical about the importance accorded to concepts is easier than to understand why without them there is no science. The history of instruments or of academies is only the history of sciences if one places them in the context of their use and the role they play in relation to theory. Descartes needed Ferrier to grind his optical glass, but it was he who produced the theory of the curves to be obtained by the grinder.

A history of scientific results can only be a chronological register. The history of science concerns an axiological activity, the search for truth. It is at the level of questions, of methods, of concepts that scientific activity appears as such. This is why the time of the history of sciences could not be an offshoot of the general course of time. The chronological history of instruments or of results can be cut up according to the periods of general history. The civil time in which the biographies of scientists are written is the same for all. The time of the advent of scientific truth, the time of verification, has a liquidity or a viscosity which is different for different disciplines in the same period of general history. The periodic classification of the elements by Dimitri Mendeleyev precipitated the progress of chemistry and hastened atomic physics; meanwhile other sciences retained a slow gait. So the history of sciences, history of the progressive rapport between intellect and truth, secretes its time itself, and it does it differently according to the moment of progress from which it starts when it gives itself the task of reviving that part of antecedent theoretical discourse which the language of the day still permits it to understand. A scientific invention promotes certain discourses which were not understood at the time when they were held, such as that of Gregor Mendel, and it annuls other discourses whose authors still thought they should form a school. The sense of historical rupture and filiation cannot come to the historian of science other than by his contact with fresh science. The contact is established by epistemology, on condition that it is vigilant, in the manner which Gaston Bachelard has taught. So understood, the history of sciences can only be uncertain, subject to rectification. For the modern mathematician, the relation of succession between Archimedes's method of exhaustion and infinitesimal calculus is not what it was for Montacla, the first great historian of mathematics. There is no definition of mathematics possible before mathematics, i.e. before the continuing succession of inventions and decisions which constitute mathematics. "Mathematics is a process," said Jean Cavaillès. Under these conditions the history of mathematics can only take from the mathematics of today a provisional definition of that which is mathematical. In the light of this, much interesting work loses its mathematical interest, becoming in the light of a new rigor, a matter of trivial applications.

One has the right to require of every theory that it furnish proof of its practical efficacy. What, therefore, in the eyes of the historian of science, is the practical effect of a theory which offers him recognition of the autonomy of his discipline as constituting the place where theoretical questions posed by scientific practice are studied? One of the most important practical effects is the elimination of what J. T. Clark has called "the virus of the precursor." Strictly speaking, if precursors exist, the history of science loses all significance, since science itself would only apparently have a historical dimension. If in antiquity, in the era of the closed world, someone had been able to be, in cosmology, the precursor of a thinker in the age of the infinite universe, a study in the history of sciences and ideas such as that by Alexandre Koyré would be impossible. A precursor would be a thinker, a seeker who had formerly come to the end of a road achieved more recently by another. Complaisance in seeking, finding, and celebrating precursors is the clearest symptom of an inaptitude for critical epistemology. Before putting two journeys end to end on a road, it would first be a good idea to be sure that it is indeed a question of the same road. In a coherent knowledge one concept has a relation to all others. Having supposed that the sun was the center of the universe does not make Aristarchus of Samos a precursor of Copernicus, nor does he lend authority to Copernicus. To change the center of reference of celestial motions is to relativize up and down, is to change the dimensions of the universe, in short, is to compose a system. Now Copernicus reproached all astronomical theories before his own for not being rational systems. A precursor would be a thinker in several times, in his own and in that, or those, which are assigned as continuing his thought, as executing his incomplete enterprise. The precursor is thus a thinker that the historian believes he can extract from his cultural embodiment in order to insert him in another; this comes back to considering concepts, discourses, and speculative or experimental gestures as being able to be displaced and replaced in an intellectual space where the reversibility of relations has been obtained by forgetting the historical aspect of the object of which it is treating. How many precursors have not thus been sought for Darwinian transformism by naturalists, or philosophers, or only by publicists for the eighteenth century! The list of precursors would be long. The extreme example will be rewritten in the manner of Dutens, the *Recherches sur l'origine des decouvertes attribuées aux modernes* [Investigation into the Origin of Discoveries Attributed to the Moderns] (1776). When Dutens wrote that Hippocrates had known of the circulation of the blood, and that the system of Copernicus belongs to the ancients, one smiles at the idea that he forgets that Harvey needed the anatomy of the Renaissance and mechanical models, and that the originality of Copernicus consisted in researching the mathematical possibility of the motion of the earth. One must then smile just as much at those more recent authors who salute Réaumur or Maupertius as precursors of Mendel, without having noticed that the problem that is posed by Mendel is unique to him and that he solved it by the invention of a concept which was without precedent, that of an independent hereditary characteristic. In short, inasmuch as a critical analysis of texts and work brought together by the telescoping of historical duration has not explicitly established that there is in one and another enquirer the same question and the same direction of research, identity of meaning of regulative concepts, identity of system of concepts from which precedents take their sense, it is artificial, arbitrary, and inadequate to an authentic project in the history of science to place two scientific authors in a logical succession of commencement and achievement, or of anticipation and realization. By substituting the logical time of relations of truth for the historical time of their invention, one aligns the history of science on science, the object of the first on that of the second, and one creates this artifact, this false historical object which is the precursor. Alexandre Koyré wrote: "The notion of a precursor is for the historian a very dangerous notion. It is true, doubtless, that ideas have a *quasi*-autonomous development, – i.e. born in a mind, they come to maturity and bear their fruit in another; and that it is, in virtue of this fact, possible to do the history of problems and their solutions; it is equally true that later generations are not interested in those who preceded them other than as they see in them their ancestors or their precursors. It is always evident – or as least should be – that no one is ever considered the precursor of someone else; and could not have been. Also, to portray them as such is the best means of prohibiting any understanding of them."

The precursor is the man of science of whom one knows well only in retrospect that he has run in front of all his contemporaries, and before he who is believed to be the victor on the course. Not to be conscious of the fact that the "precursor" is a creator of a certain history of sciences and not an agent of the progress of science, is to accept as real the condition of his possibility, the imaginary simultaneity of before and after in a sort of logical space.

In criticizing a false historical object, we have attempted to justify, by counter-example, the conception that we have proposed of a specific delimitation by the history of sciences of its object. The history of sciences is not a science and its object is not a scientific object. To do the history of sciences, in the most operative sense of the term, is one of the functions, and by no means the easiest, of philosophical epistemology.

Notes

1 It is common in French discussions not to presume that the various sciences are sufficiently uniform to license the use of the singular "science" in general philosophical discussions. Thus, although this sounds clumsy in English, I have stuck to the plural form when Canguilhem uses it. [Trans.]

2 Doubtless a natural object is not naturally natural, it is an object of everyday experience and of perception in a culture. For example, the mineral object and the crystalline object have no significant existence outside the activity of the quarry, mine, or foundry. To be detained here on this banality would be to digress.

FOUCAULT

15

FOUCAULT'S PHILOSOPHY OF SCIENCE: STRUCTURES OF TRUTH/STRUCTURES OF POWER

Linda Martín Alcoff

Michel Foucault's formative years included the study not only of history and philosophy but also of psychology: two years after he took a *license* in philosophy at the Sorbonne in 1948, he took another in psychology, and then obtained, in 1952, a Diplôme de Psycho-Pathologie. From his earliest years at the Ecole Normale Supérieure he had taken courses on general and social psychology with one of most influential psychologists of the time, Daniel Lagache, who was attempting to integrate psychoanalysis with clinical methods (Eribon 1991: 42). Foucault's studies included experimentation and clinical instruction in which patients were presented in an amphitheater just as in the days of Charcot. For several years after he had received his Diplôme, Foucault continued research in psychopathology, observing practices in mental hospitals, reportedly sometimes volunteering in experiments himself. He purchased the material required to administer Rorschach tests, subjecting numerous friends to the inkblots, and became, when it was founded, the honorary president of the French Rorschach organization. Foucault's first books were on the topic of the psychological characterization of madness and the development of clinical medicine, and his last books addressed the development of psychological norms within the establishment of modern prisons as well as within an increasingly scientific approach to sex. Thus, he worked at the intersections of philosophy and psychology throughout his life. Nonetheless, Foucault reported feeling an intense aversion to psychology no less than to philosophy (Sheridan 1980: 5).

Foucault's *oeuvre* is almost exclusively focused on particular knowledges in the human sciences, especially those aspects which pertained to psychological theories of human behavior, capacity, and normative functioning. His historical approach to these, and his emphasis on the role of discourse, is readily reminiscent of Thomas Kuhn's similarly historical approach to scientific method with its emphasis on the role of paradigms and the untranslatability of paradigm-dependent objects. Both Kuhn and Foucault are important figures in the development of postpositivist philosophies of science; neither trusted progressivist accounts of scientific development or saw the development of science as primarily a story about improved reference; both emphasized the social context of science but rejected an account of science as ideological;

and both insisted that the guiding organizations of scientific knowledge – discourse or paradigm – are not simply constraining of what scientists can see but, more significantly, constitutive and enabling for the production and solving of problems, the construction of data, and thus the production of new knowledge. Yet Foucault continues to be little read or discussed in the circles of analytic philosophy of science. This, I believe, has three main reasons: because his work addresses the human rather than the natural sciences – sciences, that is, that many consider immature, lacking methodological consensus or sufficient formalization; because he is associated with antirational French postmodernism (an association that is controversial among continental philosophers); and because he wrote a great deal about power. Whether Foucault should be classified as a rationalist, antirationalist, or arationalist is less philosophically important to assess, depending as these classifications do on prior commitments or understandings of reason and theories of knowledge. More important for the philosophical evaluation of Foucault's contribution is to assess his account of the knowledges in the human sciences and their specific implication in power.

Foucault did not offer a generalized and universal account of the epistemologies and methodologies of the human sciences, one addressing questions of objectivity, covering laws, or models of explanation, as was done by philosophers such as Ernest Nagel, Emile Durkheim, William Dray, or Maurice Natanson. Nor does he give a methodological critique or develop a clear normative alternative, like Max Horkheimer. Rather, Foucault gives us case studies of very particular *bodies* of knowledge, and occasional generalizations about the techniques of power in relation to these knowledges and the most effective way to analyze these techniques. In two books, *The Order of Things* and *The Archaeology of Knowledge*, he offers, respectively, a very broad account of major discursive shifts in the epistemic frameworks of the human sciences over the last four centuries and a philosophically descriptive account of the relationship between discourse and knowledge. This last work is indeed his most general and considered by many to be his weakest. Much more influential are his less broad and more specific case studies, from which he develops generalizations which have an unclear range of applicability.

Like all of the highly original philosophers, Foucault's work poses significant challenges of interpretation and assessment. How are we to take his occasional generalizations when he so often counsels us to analyze only locally and particularly? How can we understand his views on reference, or realism, or an assortment of other contemporary philosophical concerns, when he refuses to address these issues directly? Even more than other continental philosophers who were interested in epistemology and the sciences – such as Lyotard or Deleuze – Foucault's works read as histories and sociological studies of specific sciences rather than philosophical engagements with the major questions about knowledge, the real, or justification in science. His relationship to the history of European philosophy is no less difficult to draw out than his relationship to Anglo-American philosophy, since Foucault refers very little to the canonical philosophers but chooses rather to discuss such unfamiliar figures as Buffon, Pinel, Jacob Lenz, Charcot, Rusche and Kirchheimer, and Condillac. Yet it is clear that Foucault's work offers a paradigm shift of its own, with new articulations of problems and new objects of study. Any critical interpretation would be unfair that refuses to engage with Foucault on his own ground, that is, his case studies in the human

sciences, and chooses instead to focus solely on his occasional universal pronounce-
ments. At the very least, one should look at the context within which those pro-
nouncements arise. A surprising number of excellent philosophers, such as Habermas
(1987) and Rorty (1986), have seriously misunderstood Foucault because they neglect
to address his particular cases or to use those as a guide to interpret his generalizations.[1]

Beyond the particular depictions he gives of the rise of certain sciences, and of their
interrelationships and institutional genesis, Foucault's original contributions to the
study of science have mainly to do with how he conceptualized its co-constitutive
relationship to power. As in the Hegelian tradition, for Foucault the conceptual and
methodological approaches used in producing knowledge are historically contingent;
thus "one cannot speak of anything at any time" (Foucault 1972: 45). This affects not
just how objects are interpreted, but how phenomena are grouped into objects, as
well as in some cases how phenomena, such as stable sexual identities or criminal
personalities, come into existence. The conditions by which objects of study emerge
are, then, as he says, "many and imposing"(Foucault 1972: 45). Unlike the founda-
tionalists but like the coherentists, for Foucault perception has no causal primacy of
ontological preexistence, but neither does an imagined abstracted process of conceptualization. Foucault does not separate perception from conceptualization: the object,
the mode of perception, and the concept are produced simultaneously, after which
come competing explanatory theories (see e.g. Foucault 1975: 125).

> The object [of discourse] does not await in limbo the order that will free it and enable it
> to become embodied in a visible and prolix objectivity; it does not preexist itself, held
> back by some obstacle at the first edges of light. It exists under the positive conditions of
> a complex group of relations. (Foucault 1972: 45)

In this early period of his work Foucault develops analyses based on the multiple
patterns of relations between elements within a structure of statements, rather than
with something exterior to that structure, such as intentions or causal forces such as
profit motives, or the separate and specific interests of a state or ruling class. Instead
he argues that it is at the level of the internal relations of elements within a discourse
that conditions of possibility can be found (see e.g. Foucault 1972: 46).

> What properly belongs to a discursive formation and what makes it possible to delimit
> the group of concepts, disparate as they may be, that are specific to it, is the way in
> which these different elements are related to one another ... (Foucault 1972: 59–60)

Thus in *Madness and Civilization* he traces the variable characterizations and responses
to insanity in Europe from the end of the Middle Ages through the Classical and to
the Modern period, connecting the treatment of madness in each to other knowledges and approaches to knowledge at the time. In the initial work he did on this
topic, Foucault contrasted "Madness" as a socially contextualized positive category,
with a prediscursive, stable "madness," intending to show that there is no easy causal
or other correlation between the two: that the unofficial "madness" does not explain
officially recognized "Madness." However, by about 1960 he abandoned the idea that

such a contrast could be made. Instead, he began to theorize all of what is called madness as a social construction, since the "norm" of human behavior and functioning that is used to demarcate the sane from the insane is so implicated in other aspects of its contemporary cultural discourses that it is impossible to find even the outlines of an untouched "madness." All that can be said across these different representations is that differences do occur; nothing can be inferred about a sameness below the differences, however remotely it might be attached to them.

Nonetheless, what *can* be shown is how many relations exist between the given treatment of madness and its contemporary knowledges and orientations. In the late middle ages madness was schematized in terms of excess and irregularity, and those exhibiting mad behavior were grouped with other excessive and irregular types such as fools and simpletons, drunkards, and debauchers. The mad/sane distinction was neither medicalized nor psychologized but understood on the terms of a morality which counseled moderation and thus grouped the immoderate together. Thus in the Renaissance the mad were loaded onto ships and exiled from the city but left largely to their own devices. Madness later became construed through the binary of reason and madness, as the concept of secular reason began to dominate, in the classical age, the contrast of human and nature. Then in the modern period we have the more familiar medicalized psychological contrast of sanity/insanity, in which moral discourse is replaced by a discourse of impartial truth. But the move from excess to unreason to clinical insanity does not represent any sort of epistemic progress toward understanding a recurrent phenomenon, but merely reflects the alterations of intelligibility within discursive relations at a given time. These relations organize the formation of social constructions, rendering the mad consecutively as excessive (and grouped with drunkards, debauchers, etc.), as the unfortunate (and grouped with the poor and the homeless), and as the sick (and grouped apart from all others so that a specific kind of "cure" can be administered).

Foucault became very interested in the development during this latter period of the administrative oversight of those subject to confinement. From a period where lepers were confined to colonies and left to themselves, and fools were loaded onto ships to be delivered to "the uncertainty of fate" (Foucault 1965: 11), in later periods the inhabitants of such excluded spaces came to be looked after and cared for, and still later, came to be cured or rehabilitated. A new discourse of state obligation to the unfortunate emerged during the classical period, which aimed to take charge of them at the nation's expense. This resulted in significantly restricting their individual freedom (Foucault 1965: 48). Foucault quotes the French King's decree in 1656 at the founding of the Hôpital Genéral, created to oversee and combine poorhouses, madhouses, and rest homes for military veterans, which grants the directors, appointed for life,

> all the power of authority, of direction, of administration, of commerce, of police, of jurisdiction, of correction and punishment over all the poor of Paris, both within and without the Hôpital Genéral. (Foucault 1965: 40)

In the modern period this administrative function became much more associated with a science of psychological diagnosis and cure, expanding by far the degree of intervention and organization of an inmate's daily life. Each of these three periods, then, portraying madness as part of the excessive and irregular, as part of the irrational, or as part of the psychologically abnormal, represent alternative discursive regimes, alternative social valuations of science, alternative forms of power relations, and alternative legitimations of authority over both life and truth.

It is in Foucault's second major work, *The Birth of the Clinic*, that he begins to formulate more clearly then he did in *Madness and Civilization* his notion of "discourse," which is never more exactly defined, or less vague, than Kuhn's idea of a "paradigm," yet is nonetheless explanatorily powerful. (Those critical of Foucault's vagueness might recall that, in the first edition of *The Structure of Scientific Revolutions*, Margaret Masterman found 21 separate meanings of the term "paradigm": see Masterman 1970). Hacking perhaps defines discourse best as a system of possibility for a web of belief (Hacking 1984: 48). Discourses govern existing groups of statements and regulate the generation and distribution of new statements. In circular fashion, a statement is defined as meaningful if it is statable within a discourse. The rules of a discursive formation do not mandate specific truth-values for specific statements, but open up a delimited space in which some statements can be meaningfully expressed and understood. Foucault also at this point introduced his concept of the *episteme*, defined as the set of internal relations that unite the discursive practices that give rise to the sciences. The episteme is not a method or form of knowledge or type of rationality, but simply the totality of relations, e.g. the relations of similarity, analogy, and difference, that give rise to discursive regularities and thus give unity to a discursive formation. This includes the way in which the reasonable has been demarcated from the unreasonable, the true from the false, and the intelligible from the unintelligible.

Like the structuralists Foucault sees discourses as generative and not simply organizational. A fully articulated belief emerges from the prescriptions of a discourse which provides both resources and limitations for the possibilities of cognition. But this account is not necessarily inconsistent with realism: discourses do not determine the truth-value that any given belief has, but whether it *can* have a truth-value. Discourses govern the generation of statements whereas epistemes, at a greater remove, govern the ways in which objects can be constructed and justificatory procedures can be imagined.

For Foucault, statements cannot be adequately analyzed merely as the bearers of propositional content; they are also bearers of an "enunciative function" which

> is identified neither with grammatical acceptability nor logical correctness, and which requires if it is to operate: a referential (which is not exactly a fact, a state of things, or even an object, but a principle of differentiation); a subject (not the speaking consciousness, not the author of the formulation, but a position that may be filled in certain conditions by various individuals); an associated field (which is not the real context of the formulation, the situation in which it was articulated, but a domain of coexistence for other statements); a materiality (which is not only the substance or support of the articulation but a status, rules of transcription, possibilities of use and re-use). (Foucault 1972: 115)

A justified belief will require the proper kind of connection, then, between the statement and a referential, a subject, an associated field, and a materiality. Though the terms "referential" and "materiality" might imply an extradiscursive realm to which the statement must have some relations of correspondence, Foucault in the passage just quoted explicitly eliminates this inference by giving a discursive characterization of these terms. A referential is a principle of differentiation by which the object world comes to be constituted; a materiality is a set of rules of transcription that affect possibilities of use for specific statements in diverse contexts.

Especially important to note is the way in which for Foucault forms of subjectivity are connected to a discourse. Foucault argues that a discursive practice sets out the "legitimate perspective for a subject of knowledge" (Foucault 1997: 11) and involves rituals that determine "the individual properties and agreed roles of the speakers" (Foucault 1972: 225). In other words, who may speak about what to whom is determined at the level of the discursive formation. Foucault in several of his books develops detailed examples showing how this account can illuminate various moments in the human sciences, and showing how previously unconsidered connections appear between disparate discourses, such as language and medicine and economics, when we attend to the similarities in their enunciative functions.

In *The Birth of the Clinic*, Foucault applies his discursive approach to the development of medical science and challenges the prevailing view that saw the development of the medical clinic as the cornerstone of a move from superstition to science, or from practices primarily of classifications to empirical methods based on observation. Prior to the emergence of the medical clinic, diseases were generally conceptualized by Europeans as abstract essences, like Platonic "forms," that were in some fundamental sense independent of particular bodies. Diseases could become apparent through their visible symptoms, but the visible did not dominate the identification or construction of disease as an object of analysis. Conceptual a priori configurations were primary in identifying disease, and the visible individual bodily manifestations could not provide any independent information and could well be misleading. The primary question asked of patients was not "where does it hurt?" but "what is the matter with you?"

A series of epidemics gave rise to the development of the clinic, from which emerged a new construction of disease as phenomenon, or visible object, rather than concept, or Form. Perception thus came to take a much more important role. However, Foucault contested the usual explanation of this shift, which credits the progressivist development of pathological anatomy to revolutionary movements that began to displace the hegemony of religious and moral restrictions on dissection and anatomical inspection. This explanation assumes that the mere increased access to bodies in itself made possible the attainment of new perceptual information from which developed more empirically based theories. Foucault argues that not only the *amount* but also the *form* of perception changed: clinicians had to have reasons to give primacy to their perceptions over their classification schemas. They had to know both what to see and how to see it. The gaze could only function successfully when connected to a system of understanding that dictated its use and interpreted its results; a pre-conceptual gaze would be useless. Before the clinic, observable symptoms were viewed as only the distorted manifestations of disease in its pure, natural form. The physician had to look past the observable specifics to "see" disease. When the specific and comparative observations of bodily lesions, spots, sores, malformations, and loss of function were organized as the symptoms of living, alterable, and contingent processes, then disease came to be seen not as outside of life, existing independently in its timeless essence waiting to invade life, but as a form of life itself.

Foucault's treatment of perception, then, locates it not at the foundation of pathological identifications, nor as the mere byproduct of theoretical shifts. The disinterment of bodies would not be sufficient in itself to produce the innovations in pathology documented in this period, but in Foucault's view it is no coincidence that the changes occurred during the development of clinics. Clinics not only enabled comparative observations of disease manifestation; even more importantly, they reorganized the procedures of knowing, or what Foucault called enunciative functions. Clinics were set up for the operation of the gaze, with patients lined up and laid out for inspection in a context understood to be the site of research and teaching as much as of the delivery of treatment. The individuality of cases became more important under the increased attentiveness and authority of the gaze, and clinics authorized the gaze and authorized the physician to ask to see, away from the social norms of domestic spaces. Out of this, new classifications emerged for organizing patients, medical knowledge, and authority. New possibilities of analogy and new relations of similarity and of relevance were developed. The subject-positions of physician and patient were altered, with the individual physician gaining in authority. Disease as an object was reformulated, making possible the appearance of new theories of etiology.

To separate out the content of this new knowledge from its institutional context is to separate belief from justification. The development of scientific knowledge is always guided by a "body of anonymous, historical rules, always determined in the time and space that have defined a given period, and for a given social, economic, geographical, or linguistic area"(Foucault 1972: 117). The broad arena within which Foucault locates and analyzes specific knowledges is not transportable across different discursive domains or different enunciative functions. Justification, and theory-choice, cannot be adequately explained by a reductive account that would eclipse consideration of these contextual conditions.

With the current acceptance in the West of alternative medical approaches, from holism to herbalism to Chinese medicine, which have gained a following only because their real practical results are sometimes better than that garnered by "high-tech" orthodox medicine, we should be able to recognize without much intuitive strain that the knowledges developed in the clinics were not the only way to identify and effectively treat human disease. Foucault's account does not require that we repudiate Western medicine's claims of truth, but it certainly complicates how we may understand what truth is. In the story of medicine's contingent historical development, Foucault offers not a causal account, but an expanded explanatory account that refers to nothing outside the practices of medicine as exerting efficacious, independent causality, but he never questions that those practices are engaging with life and death, pain and suffering, or that they altered the elaboration of disease sometimes successfully in ways they had precisely intended.

After *The Birth of the Clinic* Foucault turned his own gaze to broader questions about the shifts in discursive formation that he had charted in the classical and modern era. He also began to raise more questions about the authorization processes by which knowledges produced in specific contexts grew and spread and metamorphosed through many diverse terrains. Overall, he became interested in the human sciences, those sciences that receive scant respect yet are favored by an enormous influence in institutions, from courts to classrooms to prisons. He refused the "why" question about theoretical developments and transformations and focused rather on the how, the what, and the who: how did knowledges proliferate, diversify, and expand? What new objects for study and analysis came into view? And:

> Who is speaking? Who, among the totality of speaking individuals, is accorded the right to use this sort of language? Who is qualified to do so? Who derives from it his own special quality, his prestige, and from whom, in return, does he receive if not the assurance, at least the presumption that what he says was true? (Foucault 1972: 50)

Given his account of how new knowledge emerged in very specific institutional arrangements with very specific actors as well as discursive formations, these how, what, and who questions are not epistemically tangential or "merely" political. Unless we hold that our current knowledge is absolute, final, and noncontingent (where contingent means that it might have evolved otherwise with no necessary sacrifice of truth or reference), unless we hold, in short, positivist accounts of contemporary knowledges, we must allow that such questions may be relevant for any adequate story of epistemic justification. Foucault makes the case for this conclusion most clearly in his two later works, *Discipline and Punish* and *History of Sexuality*.

The debate over whether and how political considerations affect science is similar to the old debate about how irrationality is involved. The solution for those who want to protect science from what they worry is a slippery slope of disauthorization is to erect partitions. So irrationality, admitted certainly to playing a role in the development of hypotheses, as well as successful but logically baseless leaps of inference or abduction, intuitions, and so forth, was safely sequestered to the context of discovery, securely away from the context of justification. Similarly, politics is thought by many to affect scientific applications or technologies, choice of research topics, certainly funding sources and who gains entry to the profession, but not, ultimately, theory-choice.

Foucault's originality is to deconstruct the problem of showing precisely where power enters science by conceptualizing science as a field of power relations where power is, in other words, always already there. He defines power as "the multiplicity of force relations immanent in the sphere in which they operate and which constitutes their own organization"(Foucault 1978: 92). Because power is without sovereignty or centrality, because it is not analogous to the law in the sense of exerting specific controls over a range of actions and expressions, but is rather contextually based, productive, and relational, we can begin to think of the power operative in science as occurring from the ground up, rather than entering illicitly through the back door. His is not a sense of power as having a uniform or coherent strategy that seeks to manifest itself the same in every instance, but more as a system of possibilities for relations that interact, reinforce, and strengthen other operations. "Relations of power," he says, "are not in a position of exteriority with respect to other types of relationships (economic processes, knowledge relationships, sexual relations), but are immanent in the latter"(Foucault 1978: 94). But most importantly, power's condition of possibility is the "moving substrate of force relations" which thus engender only "local and unstable" states of power (Foucault 1978: 93).

This approach eclipses the question of whether power's relation to science is intrinsic, as the Strong Program advocates and postmodernists are thought to believe, or extrinsic, as most philosophers of science, whether sympathetic to continental approaches or not, believe. The question no longer makes sense because power is not, strictly speaking, something that has a relation to science or the social, but is organic to them and emerges from them. It is neither ontologically nor conceptually separable, but more like the famous duck/rabbit picture drawing, where we have been trained to see only the duck but the rabbit has been there all along. The relations in the clinic that yielded new forms of knowledge emerged from specific social relations between persons, between domains of discourse, between institutions. These social relations, intelligible as a grid of power, are not extrinsic to the knowledges developed, but neither does their form exhaustively explain those knowledges, or eclipse in any way their truth-value.

Many of Foucault's critics have misconstrued his account as one that would replace rational processes of theory choice with ideological or politically strategic operations, as if he were arguing for power and against knowledge. But Foucault also held that there is "no power relation without the correlative constitution of a field of knowledge," that "power produces knowledge," and that "power and knowledge directly imply one another" (Foucault 1979: 27). In other words, not only does power provide the site for the elaboration of knowledge, but knowledge itself has a constituting effect on power relations, for example, in establishing and/or reinforcing hierarchies of epistemic authority.

In his study of the birth of the modern prison as both a universal method of punishment and one oriented toward rehabilitation rather than (simple) retribution, Foucault describes the emergence of a new approach to the prisoner and "a redistribution of the economy of punishment," from torture as a public spectacle to an increasingly minute set of controls and surveillance over the body of the prisoner. The object of rehabilitation was the prisoner's soul, but the prisoner's body was viewed as the means to reach the soul. Thus, there emerged a new economy of suspended rights, a controlled and limited, rather than spectacular, power over the body, and a new manner of judgment that categorized and rendered punishable not only the actions of the prisoners, but also their motivation, personality, spiritual alignment, and passions. In this way a new discourse of knowledge about the criminal was produced – a new field of the statable, a new "truth" of crime – which gave impetus to the subsequent development of psychology. Where before one was asked merely "did you do it?," now one was asked questions about motivation, intention, feelings, and so on. Thus new categories of criminals developed, involving what we now call first- and second-degree murder, manslaughter, and an analysis of the full complexities of causality. The goal of punishment changed from the public demonstration of the sovereign's absolute authority to the production of normatively functioning citizens. Prisons attempted to bring the desires, the feelings, and the personality of the prisoner into a range of normality. A legal judgment of guilt became a psychological assessment of normalization.

Foucault describes this shift as a new tactics of power which produced processes of individuation. Analogously to Nietzsche's argument that Christian guilt effected a more developed reflective interiority of the self, Foucault suggests that the disciplinary regimen resulted in an altered, expanded, and more intense subjective life of the individual, who began to approach himself as an object for attentive and vigilant observation. In the previous era dominated by torture, enforcement was intermittent and there was a realm of free space, both for the prisoners to shout out as they were being hauled off or tortured and for the public to respond. In the developing prison system, the discourse and practice of normalization became much more efficient and invasive, developing a microphysics of power directed on the body to produce minute self-imposed techniques of discipline, setting out the number of minutes spent on dressing, eating, praying, exercising, defecating, working, and sleeping. Actions were broken down into small segments for reorganization toward greater efficiency with the aim of creating totally useful time. To enforce these regimens prisons were built on Jeremy Bentham's model of the Panopticon, a circular spatial arrangement with a central guard tower assuring the possibility of permanent visibility and thus encouraging permanent self-monitoring.

Here, then, emerged a new object of study: "Man." The target of the prison apparatus – the body – was studied with the detailed precision relegated before to insects or specimens of plant-life: marked, invested, trained, evaluated, compared, and made the carrier of signs. New modes of expertise emerged and proliferated as the knowledges developed in the prison were transported to, and transformed in, hospitals, schools, factories, military camps, and insane asylums, spurring the development of Taylorism and techniques of mass production.

We are still within the epoch of discipline. Time management skills have filtered into our nonworking life, as the means to reduce stress, to find more quality time for our children or spouse, to increase enjoyment and productivity. Women's magazines especially are filled with helpful hints about how to cram every waking minute with useful activity, how to multitask, how to regulate our daily life into slotted periods for exercise, work, leisure, and romance.

The new disciplinary economy of power stimulated the human sciences toward pursuing knowledge about Man through statistical analyses, surveys, polls, and controlled experiments. Norms were developed for every possible activity: the normative amount of time it takes to learn arithmetic, to recover from an operation, to fall asleep, to be convinced to buy a brand of detergent. Persons became individuated through a meticulous measurement of their differences from a norm that imposed homogeneity. It makes perfect sense, then, to describe this as Foucault did: as a new regime for the circulation of power and of knowledge, involving a division and proliferation of forms of expertise, new types of epistemic relations, new institutionally constituted objects of knowledge, and new instrumentalities to direct operational determinations.

The representation of science as prone to ideological manipulation might picture this state of affairs as follows: that power relations confer authority on knowledges that are in the interests of the dominant. But everyone came to be more or less subject to these norms of behavior, and the new subject-positions of authority produced by this new domain were not imaginable before: the "trainer," the psychologist, the efficiency expert, the pollster. Power cannot be separated from justification processes when they occur within and are administered by authority figures operating with norms of model behavior. Nor can it be separated from truths which relate facts about new experiences, practices, techniques, and indeed objects that did not preexist the current power/knowledge era, such as the criminal personality, the delinquent adolescent, the socially maladjusted, or the pervert.

In *History of Sexuality* Foucault turns to the creation of knowledges and procedures for knowing sexuality. From the disciplined body he turns to the desiring body whose pleasures have become invested with power/knowledge. The confessional practices of the Church instituted a mediating relationship between priest and penitent for the extraction of truth on the basis of which to confer absolution and divine guidance. But the penitent had to confess in detail not only actions but thoughts, inclinations, and fantasies. In contemporary therapeutic relationships, the mediating relation between expert interpreter and the client who recounts the raw experiential data remains the same, and sex continues to be given pride of place in the individual's life, but here the operation aims at producing an objective diagnosis within a domain of science. Foucault calls into question whether the dramatic shift from the religious to the secular approach to sexuality spells a true change of content, given that there is not much change in form. Whether sex is the domain of the church or of science, it is the object of invasive scrutiny and of evaluation in comparison to a norm. What is new in the modern regime is the entrance of what he names "biopower," or the focus on populations and their reproduction as the proper business of science and the state, to be regulated, statistically tabulated, observed, and governed. The deployment of a national output requires a plethora of new discourses about sexuality in relation to reproduction, and the heterosexual couple's private conduct is now a matter of government reports, therapeutic guidance, medical intervention, and UN policy.

In the sphere of more private relations, Foucault is concerned about the shift that has occurred from practices to identities, from the engagement of acts, licit or illicit, to the articulation of birth-to-death stable categories of identity governing both reproductive and pleasurable activities. The confession/therapeutic procedures, enflamed by the pleasures of both listening and telling, work not only to reveal but to incite and spread ideas about new possibilities. Thus do the categories of perversion multiply, and the power of the norm increases since it marks not only what one has done but what one is, one's past as well as one's future.

The essential features of this new "sexuality", he tells us,

> are not the expression of a representation that is more or less distorted by ideology, or of a misunderstanding caused by taboos; they correspond to the functional requirements of a discourse that must produce its truth. (Foucault 1978: 68)

The truth in sex we seek today is not the poetic, lyrical, mystical, or spiritual truth, nor is it the practical knowledge or *ars erotica* having to do with the techniques and skills of pleasure (though these are minor discourses). Mainly the truth we seek is the propositional truth, the set of statements that accurately correspond and thus refer to the matter about human sexuality, its detailed diversity, its effects and causes. But for Foucault, this truth is not enough. All of the objects of scientific study may not be social constructions in the same degree, but the persuasiveness of his case studies suggests we need to ask questions about regimes of power in order to assess the claims to truth, not to reveal them as simple untruths, but to uncover the more complete story of how at least some truths have emerged.

So what then of science? Foucault's view makes possible the idea, as Mary Rawlinson puts it, of struggling against a system of truth (Rawlinson 1987). Not to divest it from power, but to reorient its functionality and organizational relations. And in this, the political and epistemic motives cannot be easily segregated. One wants relief from the disciplinary nightmare, certainly, but one also wants a fuller account of the truth about the human sciences themselves.

Note

1   Foucault's sympathetic interpreters range from the cautious (e.g. Gutting 1989), who want to reclaim Foucault for a broadly rationalist tradition, to the extreme (e.g. Allen 1993), who want to read Foucault as an epistemic nihilist. Ian Hacking (1984, 1986) and Joseph Rouse (1987, 1993, 1994) are probably somewhere in between. My own account is more in line with theirs and Gutting's; my critique of Allen's can be found in Alcoff 1999.

References

Alcoff, Linda Martín. 1999. "Becoming an Epistemologist." In Elizabeth Grosz, ed., *Making Futures: Explorations in Time, Memory, and Becomings*. Ithaca: Cornell University Press.

Allen, Barry. 1993. *Truth in Philosophy*. Cambridge, MA: Harvard University Press.

Dews, Peter. 1987. *Logics of Disintegration*. New York: Verso.

Eribon, Didier. 1991. *Michel Foucault*, trans. Betsy Wing. Cambridge, MA: Harvard University Press.

Foucault, Michel. 1965. *Madness and Civilization*, trans. R. Howard. New York: Pantheon.

——. 1970. *The Order of Things*, trans. A. Sheridan. New York: Random House.

——. 1972. *The Archaeology of Knowledge*, trans. A. M. Sheridan Smith. New York: Pantheon Books.

——. 1975. *The Birth of the Clinic: An Archaeology of Medical Perception*, trans. A. M. Sheridan Smith. New York: Random House.

——. 1978. *The History of Sexuality*, vol. 1, trans. Robert Hurley. New York: Random House.

——. 1979. *Discipline and Punish*, trans. Alan Sheridan. New York: Vintage Books.

——. 1997. *Essential Works of Foucault, Volume I*, ed. Paul Rabinow. New York: The New Press.

Gutting, Gary. 1989. *Michel Foucault's Archaeology of Scientific Reason*. Cambridge: Cambridge University Press.

Habermas, Jürgen. 1987. *The Philosophical Discourse of Modernity*, trans. Frederick Lawrence. Cambridge, MA: MIT Press.

Hacking, Ian. 1984. "Language, Truth and Reason." In Martin Hollis and Steven Lukes, eds., *Rationality and Relativism*. Cambridge, MA: MIT Press.

——. 1986. "The Archaeology of Foucault." In David Couzens Hoy, ed., *Foucault: A Critical Reader*. Oxford: Blackwell.

Masterman, Margaret. 1970. "The Nature of a Paradigm." In Imre Lakatos and Alan Mus-
    grave, eds., *Criticism and the Growth of Knowledge*. Cambridge: Cambridge University Press,
    pp. 58–89.
Rawlinson, Mary. 1987. "Foucault's Strategy: Knowledge, Power,and the Specificity of
    Truth." *The Journal of Medicine and Philosophy* 12: 371–95.
Rorty, Richard. 1986. "Foucault and Epistemology." In David Couzens Hoy, ed., *Foucault: A
    Critical Reader*. Oxford: Blackwell.
Rouse, Joseph. 1987. *Knowledge and Power: Toward a Political Philosophy of Science*. Ithaca:
    Cornell University Press.
——. 1993. "Foucault and the Natural Sciences." In John Caputo and Mark Yount, eds.,
    Foucault and the Critique of Institutions. Philadelphia: Pennsylvania State University Press,
    pp. 137–62.
——. 1994. "Power/Knowledge." In Gary Gutting, ed., *The Cambridge Companion to Foucault*.
    New York: Cambridge University Press.
Sheridan, Alan. 1980. *Michel Foucault and the Will to Truth*. New York: Routledge.

16

From *THE HISTORY OF SEXUALITY*, VOL. I: *AN INTRODUCTION*

Michel Foucault

1. Objective

Why these investigations? I am well aware that an uncertainty runs through the sketches I have drawn thus far, one that threatens to invalidate the more detailed inquiries that I have projected. I have repeatedly stressed that the history of the last centuries in Western societies did not manifest the movement of a power that was essentially repressive. I based my argument on the disqualification of that notion while feigning ignorance of the fact that a critique has been mounted from another quarter and doubtless in a more radical fashion: a critique conducted at the level of the theory of desire. In point of fact, the assertion that sex is not "repressed" is not altogether new. Psychoanalysts have been saying the same thing for some time. They have challenged the simple little machinery that comes to mind when one speaks of repression; the idea of a rebellious energy that must be throttled has appeared to them inadequate for deciphering the manner in which power and desire are joined to one another; they consider them to be linked in a more complex and primary way than through the interplay of a primitive, natural, and living energy welling up from below, and a higher order seeking to stand in its way; thus one should not think that desire is repressed, for the simple reason that the law is what constitutes both desire and the lack on which it is predicated. Where there is desire, the power relation is already present: an illusion, then, to denounce this relation for a repression exerted after the event; but vanity as well, to go questing after a desire that is beyond the reach of power.

But, in an obstinately confused way, I sometimes spoke, as though I were dealing with equivalent notions, of *repression*, and sometimes of *law*, of prohibition or censorship. Through stubbornness or neglect, I failed to consider everything that can

Michel Foucault, "Objectives" (pp. 81–91) and "Method" (pp. 92–102) from *History of Sexuality*, vol. I, trans. Robert Hurley. New York: Vintage, 1980. © 1978 by Random House Inc., New York. Originally published in French as *La Volonté du Savoir*. French text © 1976 by Editions Gallimard. Reprinted by permission of Georges Borchardt Inc., Éditions Gallimard, and Penguin Books Ltd.

distinguish their theoretical implications. And I grant that one might justifiably say to me: By constantly referring to positive technologies of power, you are playing a double game where you hope to win on all counts; you confuse your adversaries by appearing to take the weaker position, and, discussing repression alone, you would have us believe, wrongly, that you have rid yourself of the problem of law; and yet you keep the essential practical consequence of the principle of power-as-law, namely the fact that there is no escaping from power, that it is always-already present, constituting that very thing which one attempts to counter it with. As to the idea of a power-repression, you have retained its most fragile theoretical element, and this in order to criticize it; you have retained the most sterilizing political consequence of the idea of power-law, but only in order to preserve it for your own use.

The aim of the inquiries that will follow is to move less toward a "theory" of power than toward an "analytics" of power: that is, toward a definition of the specific domain formed by relations of power, and toward a determination of the instruments that will make possible its analysis. However, it seems to me that this analytics can be constituted only if it frees itself completely from a certain representation of power that I would term – it will be seen later why – "juridico-discursive." It is this conception that governs both the thematics of repression and the theory of the law as constitutive of desire. In other words, what distinguishes the analysis made in terms of the repression of instincts from that made in terms of the law of desire is clearly the way in which they each conceive of the nature and dynamics of the drives, not the way in which they conceive of power. They both rely on a common representation of power which, depending on the use made of it and the position it is accorded with respect to desire, leads to two contrary results: either to the promise of a "liberation," if power is seen as having only an external hold on desire, or, if it is constitutive of desire itself, to the affirmation: you are always-already trapped. Moreover, one must not imagine that this representation is peculiar to those who are concerned with the problem of the relations of power with sex. In fact it is much more general; one frequently encounters it in political analyses of power, and it is deeply rooted in the history of the West.

These are some of its principal features:

—*The negative relation.* It never establishes any connection between power and sex that is not negative: rejection, exclusion, refusal, blockage, concealment, or mask. Where sex and pleasure are concerned, power can "do" nothing but say no to them; what it produces, if anything, is absences and gaps; it overlooks elements, introduces discontinuities, separates what is joined, and marks off boundaries. Its effects take the general form of limit and lack.

—*The insistence of the rule.* Power is essentially what dictates its law to sex. Which means first of all that sex is placed by power in a binary system: licit and illicit, permitted and forbidden. Secondly, power prescribes an "order" for sex that operates at the same time as a form of intelligibility: sex is to be deciphered on the basis of its relation to the law. And finally, power acts by laying down the rule: power's hold on sex is maintained through language, or rather through the act of discourse that creates, from the very fact that it is articulated, a rule of law. It speaks, and that is the rule. The pure form of power resides in the function of the legislator; and its mode of action with regard to sex is of a juridico-discursive character.

—*The cycle of prohibition:* thou shalt not go near, thou shalt not touch, thou shalt not consume, thou shalt not experience pleasure, thou shalt not speak, thou shalt not show thyself; ultimately thou shalt not exist, except in darkness and secrecy. To deal with sex, power employs nothing more than a law of prohibition. Its objective: that sex renounce itself. Its instrument: the threat of a punishment that is nothing other than the suppression of sex. Renounce yourself or suffer the penalty of being suppressed; do not appear if you do not want to disappear. Your existence will be maintained only at the cost of your nullification. Power constrains sex only through a taboo that plays on the alternative between two nonexistences.

—*The logic of censorship.* This interdiction is thought to take three forms: affirming that such a thing is not permitted, preventing it from being said, denying that it exists. Forms that are difficult to reconcile. But it is here that one imagines a sort of logical sequence that characterizes censorship mechanisms: it links the inexistent, the illicit, and the inexpressible in such a way that each is at the same time the principle and the effect of the others: one must not talk about what is forbidden until it is annulled in reality; what is inexistent has no right to show itself, even in the order of speech where its inexistence is declared; and that which one must keep silent about is banished from reality as the thing that is tabooed above all else. The logic of power exerted on sex is the paradoxical logic of a law that might be expressed as an injunction of nonexistence, nonmanifestation, and silence.

—*The uniformity of the apparatus.* Power over sex is exercised in the same way at all levels. From top to bottom, in its over-all decisions and its capillary interventions alike, whatever the devices or institutions on which it relies, it acts in a uniform and comprehensive manner; it operates according to the simple and endlessly reproduced mechanisms of law, taboo, and censorship: from state to family, from prince to father, from the tribunal to the small change of everyday punishments, from the agencies of social domination to the structures that constitute the subject himself, one finds a general form of power, varying in scale alone. This form is the law of transgression and punishment, with its interplay of licit and illicit. Whether one attributes to it the form of the prince who formulates rights, of the father who forbids, of the censor who enforces silence, or of the master who states the law, in any case one schematizes power in a juridical form, and one defines its effects as obedience. Confronted by a power that is law, the subject who is constituted as subject – who is "subjected" – is he who obeys. To the formal homogeneity of power in these various instances corresponds the general form of submission in the one who is constrained by it – whether the individual in question is the subject opposite the monarch, the citizen opposite the state, the child opposite the parent, or the disciple opposite the master. A legislative power on one side, and an obedient subject on the other.

Underlying both the general theme that power represses sex and the idea that the law constitutes desire, one encounters the same putative mechanics of power. It is defined in a strangely restrictive way, in that, to begin with, this power is poor in resources, sparing of its methods, monotonous in the tactics it utilizes, incapable of invention, and seemingly doomed always to repeat itself. Further, it is a power that only has the force of the negative on its side, a power to say no; in no condition to produce, capable only of posting limits, it is basically anti-energy. This is the paradox of its effectiveness: it is incapable of doing anything, except to render what it dominates incapable of doing anything either, except for what this power allows it to do. And finally, it is a power whose model is essentially juridical, centered on nothing more than the statement of the law and the operation of taboos. All the modes of domination, submission, and subjugation are ultimately reduced to an effect of obedience.

Why is this juridical notion of power, involving as it does the neglect of everything that makes for its productive effectiveness, its strategic resourcefulness, its positivity, so readily accepted? In a society such as ours, where the devices of power are so numerous, its rituals so visible, and its instruments ultimately so reliable, in this society that has been more imaginative, probably, than any other in creating devious and supple mechanisms of power, what explains this tendency not to recognize the latter except in the negative and emaciated form of prohibition? Why are the deployments of power reduced simply to the procedure of the law of interdiction?

Let me offer a general and tactical reason that seems self-evident: power is tolerable only on condition that it mask a substantial part of itself. Its success is proportional to its ability to hide its own mechanisms. Would power be accepted if it were entirely cynical? For it, secrecy is not in the nature of an abuse; it is indispensable to its operation. Not only because power imposes secrecy on those whom it dominates, but because it is perhaps just as indispensable to the latter: would they accept it if they did not see it as a mere limit placed on their desire, leaving a measure of freedom – however slight – intact? Power as a pure limit set on freedom is, at least in our society, the general form of its acceptability.

There is, perhaps, a historical reason for this. The great institutions of power that developed in the Middle Ages – monarchy, the state with its apparatus – rose up on the basis of a multiplicity of prior powers, and to a certain extent in opposition to them: dense, entangled, conflicting powers, powers tied to the direct or indirect dominion over the land, to the possession of arms, to serfdom, to bonds of suzerainty and vassalage. If these institutions were able to implant themselves, if, by profiting from a whole series of tactical alliances, they were able to gain acceptance, this was because they presented themselves as agencies of regulation, arbitration, and demarcation, as a way of introducing order in the midst of these powers, of establishing a principle that would temper them and distribute them according to boundaries and a fixed hierarchy. Faced with a myriad of clashing forces, these great forms of power functioned as a principle of right that transcended all the heterogeneous claims, manifesting the triple distinction of forming a unitary regime, of identifying its will with the law, and of acting through mechanisms of interdiction and sanction. The slogan of this regime, *pax et justitia*, in keeping with the function it laid claim to, established peace as the prohibition of feudal or private wars, and justice as a way of suspending the private settling of lawsuits. Doubtless there was more to this development of great monarchic institutions than a pure and simple juridical edifice. But such was the language of power, the representation it gave of itself, and the entire theory of public law that was constructed in the Middle Ages, or reconstructed from Roman law, bears witness to the fact. Law was not simply a weapon skillfully wielded by monarchs; it was the monarchic system's mode of manifestation and the form of its acceptability. In Western societies since the Middle Ages, the exercise of power has always been formulated in terms of law.

A tradition dating back to the eighteenth or nineteenth century has accustomed us to place absolute monarchic power on the side of the unlawful: arbitrariness, abuse, caprice, willfulness, privileges and exceptions, the traditional continuance of accomplished facts. But this is to overlook a fundamental historical trait of Western monarchies: they were constructed as systems of law, they expressed themselves through theories of law, and they made their mechanisms of power work in the form of law. The old reproach that Boulainvilliers directed at the French monarchy – that it used the law and jurists to do away with rights and to bring down the aristocracy – was basically warranted by the facts. Through the development of the monarchy and its institutions this juridico-political dimension was established. It is by no means adequate to describe the manner in which power was and is exercised, but it is the code according to which power presents itself and prescribes that we conceive of it. The history of the monarchy went hand in hand with the covering up of the facts and procedures of power by juridico-political discourse.

Yet, despite the efforts that were made to disengage the juridical sphere from the monarchic institution and to free the political from the juridical, the representation of power remained caught within this system. Consider the two following examples. Criticism of the eighteenth-century monarchic institution in France was not directed against the juridico-monarchic sphere as such, but was made on behalf of a pure and rigorous juridical system to which all the mechanisms of power could conform, with no excesses or irregularities, as opposed to a monarchy which, notwithstanding its own assertions, continuously overstepped the legal framework and set itself above the laws. Political criticism availed itself, therefore, of all the juridical thinking that had accompanied the development of the monarchy, in order to condemn the latter; but it did not challenge the principle which held that law had to be the very form of power, and that power always had to be exercised in the form of law. Another type of criticism of political institutions appeared in the nineteenth century, a much more radical criticism in that it was concerned to show not only that real power escaped the rules of jurisprudence, but that the legal system itself was merely a way of exerting violence, of appropriating that violence for the benefit of the few, and of exploiting the dissymmetries and injustices of domination under cover of general law. But this critique of law is still carried out on the assumption that, ideally and by nature, power must be exercised in accordance with a fundamental lawfulness.

At bottom, despite the differences in epochs and objectives, the representation of power has remained under the spell of monarchy. In political thought and analysis, we still have not cut off the head of the king. Hence the importance that the theory of power gives to the problem of right and violence, law and illegality, freedom and will, and especially the state and sovereignty (even if the latter is questioned insofar as it is personified in a collective being and no longer a sovereign individual). To conceive of power on the basis of these problems is to conceive of it in terms of a historical form that is characteristic of our societies: the juridical monarchy. Characteristic yet transitory. For while many of its forms have persisted to the present, it has gradually been penetrated by quite new mechanisms of power that are probably irreducible to the representation of law. As we shall see, these power mechanisms are, at least in part, those that, beginning in the eighteenth century, took charge of men's existence, men as living bodies. And if it is true that the juridical system was useful for representing, albeit in a nonexhaustive way, a power that was centered primarily around deduction (*prélèvement*) and death, it is utterly incongruous with the new methods of power whose operation is not ensured by right but by technique, not by law but by normalization, not by punishment but by control, methods that are employed on all levels and in forms that go beyond the state and its apparatus. We have been engaged for centuries in a type of society in which the juridical is increasingly incapable of coding power, of serving as its system of representation. Our historical gradient carries us further and further away from a reign of law that had already begun to recede into the past at a time when the French Revolution and the accompanying age of constitutions and codes seemed to destine it for a future that was at hand.

It is this juridical representation that is still at work in recent analyses concerning the relationships of power to sex. But the problem is not to know whether desire is alien to power, whether it is prior to the law as is often thought to be the case, when it is not rather the law that is perceived as constituting it. This question is beside the point. Whether desire is this or that, in any case one continues to conceive of it in relation to a power that is always juridical and discursive, a power that has its central point in the enunciation of the law. One remains attached to a certain image of power-law, of power-sovereignty, which was traced out by the theoreticians of right and the monarchic institution. It is this image that we must break free of, that is, of the theoretical privilege of law and sovereignty, if we wish to analyze power within the concrete and historical framework of its operation. We must construct an analytics of power that no longer takes law as a model and a code.

This history of sexuality, or rather this series of studies concerning the historical relationships of power and the discourse on sex, is, I realize, a circular project in the sense that it involves two endeavors that refer back to one another. We shall try to rid ourselves of a juridical and negative representation of power, and cease to conceive of it in terms of law, prohibition, liberty, and sovereignty. But how then do we analyze what has occurred in recent history with regard to this thing – seemingly one of the most forbidden areas of our lives and bodies – that is sex? How, if not by way of prohibition and blockage, does power gain access to it? Through which mechanisms, or tactics, or devices? But let us assume in turn that a somewhat careful scrutiny will show that power in modern societies has not in fact governed sexuality through law and sovereignty; let us suppose that historical analysis has revealed the presence of a veritable "technology" of sex, one that is much more complex and above all much more positive than the mere effect of a "defense" could be; this being the case, does this example – which can only be considered a privileged one, since power seemed in this instance, more than anywhere else, to function as prohibition – not compel one to discover principles for analyzing power which do not derive from the system of right and the form of law? Hence it is a question of forming a different grid of historical decipherment by starting from a different theory of power; and, at the same time, of advancing little by little toward a different conception of power through a closer examination of an entire historical material. We must at the same time conceive of sex without the law, and power without the king.

2. Method

Hence the objective is to analyze a certain form of knowledge regarding sex, not in terms of repression or law, but in terms of power. But the word *power* is apt to lead to a number of misunderstandings – misunderstandings with respect to its nature, its form, and its unity. By power, I do not mean "Power" as a group of institutions and mechanisms that ensure the subservience of the citizens of a given state. By power, I do not mean, either, a mode of subjugation which, in contrast to violence, has the form of the rule. Finally, I do not have in mind a general system of domination exerted by one group over another, a system whose effects, through successive derivations, pervade the entire social body. The analysis, made in terms of power, must not assume that the sovereignty of the state, the form of the law, or the over-all unity of a domination are given at the outset; rather, these are only the terminal forms power takes. It seems to me that power must be understood in the first instance as the multiplicity of force relations immanent in the sphere in which they operate and which constitute their own organization; as the process which, through ceaseless struggles and confrontations, transforms, strengthens, or reverses them; as the support which these force relations find in one another, thus forming a chain or a system, or on the contrary, the disjunctions and contradictions which isolate them from one another; and lastly, as the strategies in which they take effect, whose general design or institutional crystallization is embodied in the state apparatus, in the formulation of the law, in the various social hegemonies. Power's condition of possibility, or in any case the viewpoint which permits one to understand its exercise, even in its more "peripheral" effects, and which also makes it possible to use its mechanisms as a grid of intelligibility of the social order, must not be sought in the primary existence of a central point, in a unique source of sovereignty from which secondary and descendent forms would emanate; it is the moving substrate of force relations which, by virtue of their inequality, constantly engender states of power, but the latter are always local and unstable. The omnipresence of power: not because it has the privilege of consolidating everything under its invincible unity, but because it is produced from one moment to the next, at every point, or rather in every relation from one point to another. Power is everywhere; not because it embraces everything, but because it comes from everywhere. And "Power," insofar as it is permanent, repetitious, inert, and self-reproducing, is simply the over-all effect that emerges from all these mobilities, the concatenation that rests on each of them and seeks in turn to arrest their movement. One needs to be nominalistic, no doubt: power is not an institution, and not a structure; neither is it a certain strength we are endowed with; it is the name that one attributes to a complex strategical situation in a particular society.

Should we turn the expression around, then, and say that politics is war pursued by other means? If we still wish to maintain a separation between war and politics, perhaps we should postulate rather that this multiplicity of force relations can be coded – in part but never totally – either in the form of "war," or in the form of "politics"; this would imply two different strategies (but the one always liable to switch into the other) for integrating these unbalanced, heterogeneous, unstable, and tense force relations.

Continuing this line of discussion, we can advance a certain number of propositions:

—Power is not something that is acquired, seized, or shared, something that one holds on to or allows to slip away; power is exercised from innumerable points, in the interplay of nonegalitarian and mobile relations.

—Relations of power are not in a position of exteriority with respect to other types of relationships (economic processes, knowledge relationships, sexual relations), but are immanent in the latter; they are the immediate effects of the divisions, inequalities, and disequilibriums which occur in the latter, and conversely they are the internal conditions of these differentiations; relations of power are not in superstructural positions, with merely a role of prohibition or accompaniment; they have a directly productive role, wherever they come into play.

—Power comes from below; that is, there is no binary and all-encompassing opposition between rulers and ruled at the root of power relations, and serving as a general matrix – no such duality extending from the top down and reacting on more and more limited groups to the very depths of the social body. One must suppose rather that the manifold relationships of force that take shape and come into play in the machinery of production, in families, limited groups, and institutions, are the basis for wide-ranging effects of cleavage that run through the social body as a whole. These then form a general line of force that traverses the local oppositions and links them together; to be sure, they also bring about redistributions, realignments, homogenizations, serial arrangements, and convergences of the force relations. Major dominations are the hegemonic effects that are sustained by all these confrontations.

—Power relations are both intentional and nonsubjective. If in fact they are intelligible, this is not because they are the effect of another instance that "explains" them, but rather because they are imbued, through and through, with calculation: there is no power that is exercised without a series of aims and objectives. But this does not mean that it results from the choice or decision of an individual subject; let us not look for the headquarters that presides over its rationality; neither the caste which governs, nor the groups which control the state apparatus, nor those who make the most important economic decisions direct the entire network of power that functions in a society (and makes *it* function); the rationality of power is characterized by tactics that are often quite explicit at the restricted level where they are inscribed (the local cynicism of power), tactics which, becoming connected to one another, attracting and propagating one another, but finding their base of support and their condition elsewhere, end by forming comprehensive systems: the logic is perfectly clear, the aims decipherable, and yet it is often the case that no one is there to have invented them, and few who can be said to have formulated them: an implicit characteristic of the great anonymous, almost unspoken strategies which coordinate the loquacious tactics whose "inventors" or decisionmakers are often without hypocrisy.

—Where there is power, there is resistance, and yet, or rather consequently, this resistance is never in a position of exteriority in relation to power. Should it be said that one is always "inside" power, there is no "escaping" it, there is no absolute outside where it is concerned, because one is subject to the law in any case? Or that, history being the ruse of reason, power is the ruse of history, always emerging the winner? This would be to misunderstand the strictly relational character of power relationships. Their existence depends on a multiplicity of points of resistance: these play the role of adversary, target, support, or handle in power relations. These points of resistance are present everywhere in the power network. Hence there is no single locus of great Refusal, no soul of revolt, source of all rebellions, or pure law of the revolutionary. Instead there is a plurality of resistances, each of them a special case: resistances that are possible, necessary, improbable; others that are spontaneous, savage, solitary, concerted, rampant, or violent; still others that are quick to compromise, interested, or sacrificial; by definition, they can only exist in the strategic field of power relations. But this does not mean that they are only a reaction or rebound, forming with respect to the basic domination an underside that is in the end always passive, doomed to perpetual defeat. Resistances do not derive from a few heterogeneous principles; but neither are they a lure or a promise that is of necessity betrayed. They are the odd term in relations of power; they are inscribed in the latter as an irreducible opposite. Hence they too are distributed in irregular fashion: the points, knots, or focuses of resistance are spread over time and space at varying densities, at times mobilizing groups or individuals in a definitive way, inflaming certain points of the body, certain moments in life, certain types of behavior. Are there no great radical ruptures, massive binary divisions, then? Occasionally, yes. But more often one is dealing with mobile and transitory points of resistance, producing cleavages in a society that shift about, fracturing unities and effecting regroupings, furrowing across individuals themselves, cutting them up and remolding them, marking off irreducible regions in them, in their bodies and minds. Just as the network of power relations ends by forming a dense web that passes through apparatuses and institutions, without being exactly localized in them, so too the swarm of points of resistance traverses social stratifications and individual unities. And it is doubtless the strategic codification of these points of resistance that makes a revolution possible, somewhat similar to the way in which the state relies on the institutional integration of power relationships.

It is in this sphere of force relations that we must try to analyze the mechanisms of power. In this way we will escape from the system of Law-and-Sovereign which has captivated political thought for such a long time. And if it is true that Machiavelli was among the few – and this no doubt was the scandal of his "cynicism" – who conceived the power of the Prince in terms of force relationships, perhaps we need to go one step further, do without the persona of the Prince, and decipher power mechanisms on the basis of a strategy that is immanent in force relationships.

To return to sex and the discourses of truth that have taken charge of it, the question that we must address, then, is not: Given a specific state structure, how and why is it that power needs to establish a knowledge of sex? Neither is the question: What over-all domination was served by the concern, evidenced since the eighteenth century, to produce true discourses on sex? Nor is it: What law presided over both the regularity of sexual behavior and the conformity of what was said about it? It is rather: In a specific type of discourse on sex, in a specific form of extortion of truth, appearing historically and in specific places (around the child's body, apropos of women's sex, in connection with practices restricting births, and so on), what were the most immediate, the most local power relations at work? How did they make possible these kinds of discourses, and conversely, how were these discourses used to support power relations? How was the action of these power relations modified by their very exercise, entailing a strengthening of some terms and a weakening of others, with effects of resistance and counterinvestments, so that there has never existed one type of stable subjugation, given once and for all? How were these power relations linked to one another according to the logic of a great strategy, which in retrospect takes on the aspect of a unitary and voluntarist politics of sex? In general terms: rather than referring all the infinitesimal violences that are exerted on sex, all the anxious gazes that are directed at it, and all the hiding places whose discovery is made into an impossible task, to the unique form of a great Power, we must immerse the expanding production of discourses on sex in the field of multiple and mobile power relations.

Which leads us to advance, in a preliminary way, four rules to follow. But these are not intended as methodological imperatives; at most they are cautionary prescriptions.

Rule of Immanence

One must not suppose that there exists a certain sphere of sexuality that would be the legitimate concern of a free and disinterested scientific inquiry were it not the object of mechanisms of prohibition brought to bear by the economic or ideological requirements of power. If sexuality was constituted as an area of investigation, this was only because relations of power had established it as a possible object; and conversely, if power was able to take it as a target, this was because techniques of knowledge and procedures of discourse were capable of investing it. Between techniques of knowledge and strategies of power, there is no exteriority, even if they have specific roles and are linked together on the basis of their difference. We will start, therefore, from what might be called "local centers" of power-knowledge: for example, the relations that obtain between penitents and confessors, or the faithful and their directors of conscience. Here, guided by the theme of the "flesh" that must be mastered, different forms of discourse – self-examination, questionings, admissions, interpretations, interviews – were the vehicle of a kind of incessant back-and-forth movement of forms of subjugation and schemas of knowledge. Similarly, the body of the child, under surveillance, surrounded in his cradle, his bed, or his room by an entire watch-crew of parents, nurses, servants, educators, and doctors, all attentive to the least manifestations of his sex, has constituted, particularly since the eighteenth century, another "local center" of power-knowledge.

Rules of Continual Variations

We must not look for who has the power in the order of sexuality (men, adults, parents, doctors) and who is deprived of it (women, adolescents, children, patients); nor for who has the right to know and who is forced to remain ignorant. We must seek rather the pattern of the modifications which the relationships of force imply by the very nature of their process. The "distributions of power" and the "appropriations of knowledge" never represent only instantaneous slices taken from processes involving, for example, a cumulative reinforcement of the strongest factor, or a reversal of relationship, or again, a simultaneous increase of two terms. Relations of power-knowledge are not static forms of distribution, they are "matrices of transformations." The nineteenth-century grouping made up of the father, the mother, the educator, and the doctor, around the child and his sex, was subjected to constant modifications, continual shifts. One of the more spectacular results of the latter was a strange reversal: whereas to begin with the child's sexuality had been problematized within the relationship established between doctor and parents (in the form of advice, or recommendations to keep the child under observation, or warnings of future dangers), ultimately it was in the relationship of the psychiatrist to the child that the sexuality of adults themselves was called into question.

Rule of Double Conditioning

No "local center," no "pattern of transformation" could function if, through a series of sequences, it did not eventually enter into an over-all strategy. And inversely, no strategy could achieve comprehensive effects if it did not gain support from precise and tenuous relations serving, not as its point of application or final outcome, but as its prop and anchor point. There is no discontinuity between them, as if one were dealing with two different levels (one microscopic and the other macroscopic); but neither is there homogeneity (as if the one were only the enlarged projection or the miniaturization of the other); rather, one must conceive of the double conditioning of a strategy by the specificity of possible tactics, and of tactics by the strategic envelope that makes them work. Thus the father in the family is not the "representative" of the sovereign or the state; and the latter are not projections of the father on a different scale. The family does not duplicate society, just as society does not imitate the family. But the family organization, precisely to the extent that it was insular and heteromorphous with respect to the other power mechanisms, was used to support the great "maneuvers" employed for the Malthusian control of the birthrate, for the populationist incitements, for the medicalization of sex and the psychiatrization of its nongenital forms.

Rule of the Tactical Polyvalence of Discourses

What is said about sex must not be analyzed simply as the surface of projection of these power mechanisms. Indeed, it is in discourse that power and knowledge are joined together. And for this very reason, we must conceive discourse as a series of discontinuous segments whose tactical function is neither uniform nor stable. To be more precise, we must not imagine a world of discourse divided between accepted discourse and excluded discourse, or between the dominant discourse and the dominated one; but as a multiplicity of discursive elements that can come into play in various strategies. It is this distribution that we must reconstruct, with the things said and those concealed, the enunciations required and those forbidden, that it comprises; with the variants and different effects – according to who is speaking, his position of power, the institutional context in which he happens to be situated – that it implies; and with the shifts and reutilizations of identical formulas for contrary objectives that it also includes. Discourses are not once and for all subservient to power or raised up against it, any more than silences are. We must make allowance for the complex and unstable process whereby discourse can be both an instrument and an effect of power, but also a hindrance, a stumbling-block, a point of resistance and a starting point for an opposing strategy. Discourse transmits and produces power; it reinforces it, but also undermines and exposes it, renders it fragile and makes it possible to thwart it. In like manner, silence and secrecy are a shelter for power, anchoring its prohibitions; but they also loosen its holds and provide for relatively obscure areas of tolerance. Consider for example the history of what was once "the" great sin against nature. The extreme discretion of the texts dealing with sodomy – that utterly confused category – and the nearly universal reticence in talking about it made possible a twofold operation: on the one hand, there was an extreme severity (punishment by fire was meted out well into the eighteenth century, without there being any substantial protest expressed before the middle of the century), and on the other hand, a tolerance that must have been widespread (which one can deduce indirectly from the infrequency of judicial sentences, and which one glimpses more directly through certain statements concerning societies of men that were thought to exist in the army or in the courts). There is no question that the appearance in nineteenth-century psychiatry, jurisprudence, and literature of a whole series of discourses on the species and subspecies of homosexuality, inversion, pederasty, and "psychic hermaphrodism" made possible a strong advance of social controls into this area of "perversity"; but it also made possible the formation of a "reverse" discourse: homosexuality began to speak in its own behalf, to demand that its legitimacy or "naturality" be acknowledged, often in the same vocabulary, using the same categories by which it was medically disqualified. There is not, on the one side, a discourse of power, and opposite it, another discourse that runs counter to it. Discourses are tactical elements or blocks operating in the field of force relations; there can exist different and even contradictory discourses within the same strategy; they can, on the contrary, circulate without changing their form from one strategy to another, opposing strategy. We must not expect the discourses on sex to tell us, above all, what strategy they derive from, or what moral divisions they accompany, or what ideology – dominant or dominated – they represent; rather we must question them on the two levels of their tactical productivity (what reciprocal effects of power and knowledge they ensure) and their strategical integration (what conjunction and what force relationship make their utilization necessary in a given episode of the various confrontations that occur).

In short, it is a question of orienting ourselves to a conception of power which replaces the privilege of the law with the viewpoint of the objective, the privilege of prohibition with the viewpoint of tactical efficacy, the privilege of sovereignty with the analysis of a multiple and mobile field of force relations, wherein far-reaching, but never completely stable, effects of domination are produced. The strategical model, rather than the model based on law. And this, not out of a speculative choice or theoretical preference, but because in fact it is one of the essential traits of Western societies that the force relationships which for a long time had found expression in war, in every form of warfare, gradually became invested in the order of political power.

DELEUZE

17

GILLES DELEUZE, DIFFERENCE, AND SCIENCE

Todd May

For some of the great Continental thinkers of the twentieth century, the engagement with science stands at or near the center of their thought. Merleau-Ponty's appropriation of Gestalt psychology, Heidegger's critique of science in the age of technology, and Foucault's analysis of the emergence of power in the human sciences stand as examples of this engagement. With Gilles Deleuze, the case is different. Deleuze has no extended discussion of science in any of his works before the collaboration with Félix Guattari. And even then, aside from a short chapter on science in *What is Philosophy?* (excerpted here) and a single plateau in *A Thousand Plateaus*, the scientific references tend to be clipped, oblique, and occasionally hidden in footnotes. A quick reading would leave one with the impression that for Deleuze science is largely a matter of indifference.

And yet a more sustained investigation brings out a greater concern with and reference to science that runs throughout his work, periodically breaking out into overt discussion, but lying often just beneath the text's surface. We might say of the role of science in Deleuze's work what he said of the scholia in Spinoza's *Ethics*: that it is a "subterranean book of fire"[1] that runs beneath the work and on occasion bursts through to scatter its effects across its landscape. To understand the role of science in Deleuze's work, then, we need to draw the disparate embers of that fire together, to more clearly see the light and warmth they are meant to cast.

If we draw those embers together, we will see that Deleuze's engagement with science happens on two levels. The first level involves the short, sustained discussion in *What is Philosophy?* The second level involves the references to science dispersed throughout his work and given sustained treatment in the third of the thousand plateaus. It would be tempting here to start with the discussion of the nature of science in *What is Philosophy?* and then to read the earlier references in light of the view of science developed there. This temptation would be a mistake. In the second level of engagement, Deleuze is not referring to science *as science*; he is not offering us a view of science that either conforms to or confirms his own philosophical project. He is, instead, appropriating and often reworking scientific themes for his own philosophical purposes. To understand what Deleuze's engagement in science amounts to, then, we will need to understand both what those philosophical purposes are and, more generally, what Deleuze thinks philosophy is all about.

Since many readers are likely to have a more passing understanding of Deleuze's philosophical project than that of many of the other thinkers in this book, I should perhaps start by isolating several important themes of that project, in particular that of the "philosophy of difference" which has become associated with Deleuze and is central to his engagement with science. Then I can turn to the engagement itself, focusing particularly on the biology that makes its appearance in the earlier works (with a nod to mathematics and calculus) and then on chemistry and physics that arise in the collaborative works with Guattari. From there, I will distinguish Deleuze and Guattari's account of philosophy from their account of science. In the light of that last discussion, I'll conclude by offering some remarks on the relation between philosophy and science in Deleuze's work.

Before embarking on this discussion, however, I would like to cite several works in the Deleuze literature that have offered a guiding orientation for what is to follow. Keith Ansell Pearson's treatment of biology in Deleuze in *Germinal Life: The Difference and Repetition of Deleuze*,[2] Brian Massumi's discussion of chemistry and physics in the second chapter of *A User's Guide to Capitalism and Schizophrenia*,[3] Manuel De Landa's treatment of both biology and chemistry in "Nonorganic Life"[4] and, recently, *Intensive Science and Virtual Philosophy*,[5] and Aden Evens's outstanding treatment of the calculus in "Math Anxiety,"[6] all provide more indepth discussions of the various aspects of Deleuze's appropriations of science in their specific areas than I can here. Although my own approach is distinct from each, I have drawn threads from those works in weaving this larger Deleuzian tapestry, and I recommend them to readers who would like to follow up on specific aspects of this discussion.

I

There are many ways into Deleuze's philosophy of difference. One might start with Deleuze's Nietzschean concept of the eternal return as the return of difference. One might begin at the end, contrasting Deleuze and Guattari's rhizomatic approach to philosophy with the arborescent approach they critique. Or one might plunge right into *Difference and Repetition*, seeking to understand what Deleuze means when he says that "difference is behind everything, but behind difference there is nothing."[7] Perhaps the easiest way in, however, is through Deleuze's treatment of Spinoza, whom Deleuze and Guattari once called "the Christ of philosophers," adding that "the greatest philosophers are hardly more than apostles who distance themselves from or draw near to this mystery."[8]

Spinoza's fame (and the infamy he incurred during his lifetime) result from his central ontological idea that there are not two substances in the universe, only one. Rather than distinguishing, with Descartes, a mental from a physical substance – a distinction that points the way toward, among other things, the immortality of the soul – Spinoza posits that there exists only a single substance that expresses itself in attributes and modes. Deleuze sees Spinoza adhering to the philosophical tradition of

Duns Scotus in holding to the univocity of Being. This heretical view consists in denying the existence of a god separate from the physical world, thus finding it immanent to the world we inhabit. The charge of pantheism leveled at Spinoza during his lifetime is not difficult to see here.

What is the significance of embracing the concept of a single substance and thus the univocity of Being? It lies in the abandonment of transcendence. Here we might recall Nietzsche's critique of transcendence, a critique with which Deleuze is in sympathy. The effect of positing any form of transcendence (of which the transcendence of the Judeo-Christian God would be the prime example) is to set up a tribunal, a judge that is not of this world but that nevertheless evaluates it and always finds it wanting. The transcendent is always the more nearly perfect (or the Perfect itself). It is always pictured as higher, above this world. It is the ideal toward which this world must strive through self-denial but which, because of some inherent flaw – be it the existence of the flesh or the finiteness of its creatures – it can never fully achieve. As Nietzsche puts the point in *The Genealogy of Morals*, "The idea at issue in this struggle is the *value* which the ascetic priests ascribe to our life: they juxtapose this life (along with what belongs to it, 'nature,' 'world,' the whole sphere of becoming and the ephemeral) to a completely different form of existence, which it opposes and ex-cludes, *unless* it somehow turns itself against itself, *denies itself*."[9] To reject transcend-ence, to return to immanence, is the first step in abandoning self-denial and *ressentiment*. It is the first step toward embracing the possibilities that this world, which is the only world, has to offer.

With the embrace of the univocity of Being, however, two questions arise. First, how is it that the perceived world exists as a manifold of differences in continuous evolution when there is only a single substance that comprises them? How can the univocity of Being be reconciled with the manifoldness of existence? This, of course, is the traditional philosophical question of the One and the Many. The second question, bound to the first one, is, What is the relation between the single substance and the manifold of existence? As Heidegger might put the question, what is the relation between Being and beings? The One and the Many; Being and beings. We are in the realm of some very traditional philosophical issues to which Deleuze gives a unique twist which constitutes his philosophy of difference.

The first question presents no insurmountable conceptual barrier if we jettison the idea that a single substance implies some kind of identity. For Deleuze, the single substance of Spinoza must be conceived not in terms of identity but in terms of difference. Substance, Being in its univocity, is difference itself. "Being is said in a single and same sense of everything of which it is said, but that of which it is said differs: it is said of difference itself."[10] *Difference is behind everything, but behind difference there is nothing.* If substance in some sense contains or comprises the differences that manifest themselves in the world, then there is no difficulty reconciling the One and the Many. The One is many; it is difference, difference itself, or, in the later term used in the collaborative works with Guattari, it is multiplicity.

Addressing the first question this way serves, however, to make the second ques-tion more urgent. For if Being is difference, doesn't it collapse into beings them-selves? If Being is as manifold as the beings that it comprises, doesn't Being just reduce itself to nothing more than the manifoldness of our particular world?

Deleuze denies this reduction, claiming instead that the kind of difference associated with substance or Being is distinct (different) from the kind of differences
associated with beings. One way of capturing this distinction emerges from a passage
discussing Bergson's attempt to distinguish the present and spatiality from his concept
of a pure duration. Deleuze writes:

> the decomposition of the composite reveals to us two types of multiplicity. One is
> represented by space ... It is a multiplicity of exteriority, of simultaneity, of juxtapos-
> ition, of order, of quantitative differentiation, of *difference in degree*; it is a numerical
> multiplicity, *discontinuous and actual*. The other type of multiplicity appears in pure
> duration: It is an internal multiplicity of succession, of fusion, of organization, of hetero-
> geneity, of qualitative discrimination, or of *difference in kind*; it is a *virtual and continuous*
> multiplicity that cannot be reduced to numbers.[11]

Being, then, is virtual difference, a qualitative multiplicity that is organized not
along the lines of spatial distinctions but along the lines of Bergson's *durée* or Spinoza's
single substance. By contrast, actual differences, which can be accounted for numeric-
ally, attach to spatial beings that appear in the phenomenal world.

It might seem as though, with this distinction, we have returned to transcendence,
where the world of difference in itself, separate and apart from the world of actual
differences, somehow produces or gives rise to the world of actual difference. However, this would be a mistaken view of things, one founded on seeing the relation
between that virtual and the actual as modeled on the relation of the possible and the
real. The relation between the virtual and the actual is, however, very different.[12] As
Deleuze uses these terms, the real is the mirror of the possible; it has the same structure
as the possible, with the sole but ontologically crucial exception that it is real and not
merely possible. So there are two ontological realms, a realm of the possible and a realm
of the real. By contrast, the virtual does not lack reality; it is part of the real. There is
only one reality, comprising aspects that are at once virtual and actual. The virtual
actualizes itself in order to become actual, but in actualizing itself it does not gain in any
reality it had lacked before, nor does it stand outside or behind the actuality that is
actualized. It is not part of the actual, but it remains real within the actual.

In his discussion of Spinoza, Deleuze utilizes the term "expression" to indicate the
relationship between the virtual (substance) and the actual (attributes and modes). In
contrast to medieval creationist or emanative theories of causality, in which God is
said to cause the beings of this world either by explicit authorship or by emanation,
Spinoza holds an expressive view of causality, in which that which is expressed is not
ontologically distinct from its expression. Attributes and modes may *explicate*, *involve*,
and *complicate* substance,[13] but they do not emerge from it on a distinct ontological
plane. A simple way of picturing this might be to recall how in Japanese origami, the
figures that are made are folds of the paper used to make them. The folds that
compose the figure are not distinct from the paper itself, they are "actualizations" of
the paper into a specific arrangement.[14] But the paper does not exist behind or
outside the folds or the figure. It is part of them. There are, of course, differences
between origami and the virtual/actual relationship, one of the most important being
that the paper in origami, unlike the virtual, remains phenomenologically accessible in

the folded figure. But the image, I hope, at least captures the general point. One can see that, by appeal to the concept of expression rather than emanation or some other form of causality, which require some sort of ontological transcendence (of creator to created), Spinoza retains the immanence of substance to its modes.

If the relation of virtual to actual is one of actualization, then two important implications follow from this. First, the virtual is not a mirror of the actual, as the real is of the possible. In *Difference and Repetition*, Deleuze marks this by saying: "We call the determination of the virtual content of an Idea differentiation; we call the actualization of that virtuality into species and distinguished parts differenciation."[15] In contrast to the possible/real distinction, the virtual/actual distinction involves three kinds of difference. First, there is the difference in itself of the virtual; second, there are the specific differences of the actualization of the virtual; finally, there is the difference *between* virtual and actual difference, between differentiation and differenciation.

The second important implication is that the ontological realm of the virtual is not distinct from that of the actual. The virtual and the actual must be mutually woven together in the real if they are not to be two separate ontological regions. The philosophy of this ontological inseparability is what Deleuze calls "transcendental empiricism." "Empiricism truly becomes transcendental ... only when we apprehend directly in the sensible that which can only be sensed, the very being *of* the sensible: difference, potential difference, and difference in intensity as the reason behind qualitative diversity."[16] The being *of* the sensible; part of it, not outside or above it.

II

With this admittedly quick overview of Deleuze's philosophy of difference in hand, we are ready to turn to the subterranean fire of science in his works. Based on what has been seen so far, it is not surprising that the aspects of science that most fascinate Deleuze are those that involve the ideas of internal difference, actualization, and fluidity. Among the sciences, it is biology that, in the works before (and, to an extent, within) the collaboration with Guattari, has pride of place.

We can begin to see the significance of biology if we move away, as biology has largely done, from seeing life primarily as a matter of living beings, toward more systemic or holistic treatments of life. In other words, the living being is not necessarily the primary unit of biological study; instead, life or the environment is. One of the biologists most influential on Deleuze's early work, particularly *Difference and Repetition*, Gilbert Simondon, has put it this way: "the individual is to be understood as having a relative reality, occupying only a certain phase of the whole being in question – a phase that therefore carries the implication of a preceding preindividual state, and that, even after individuation, does not exist in isolation, since individuation does not exhaust in the single act of its appearance all the potentials embedded in the preindividual state."[17] Deleuze states the influence of Simondon's work in a footnote to *The Logic of Sense*, noting that, "This entire book [from which the above quote is drawn], it seems to us, has special importance, since it presents the first thought-out theory of impersonal and pre-individual singularities."[18] He works out the implications of Simondon's ideas in an important section in *Difference and Repetition*, discussing the idea of intensities as preindividual singularities of difference that exist virtually in the organism and in the world.

Simondon's idea of a preindividual state is one that recognizes the significance of the virtual as a field of differences, of intensities. Regarding the term "intensity," Deleuze says: "The expression 'difference of intensity' is a tautology. Intensity is the form of difference in so far as this is the reason of the sensible."[19] We might think of intensities as contrasting with "extensities." Extensities exist in the actual realm, intensities in the virtual realm. The latter are relations of difference that give rise to the extensive world, the actual world, but are not phenomenologically accessible as such in the actual world. "Intensity is difference, but this difference tends to deny or cancel itself out in extensity and underneath quality."[20] What, then, does Deleuze see Simondon as proposing? That the biological individual is an actualization of a virtual intensive state, one that does not exhaust the potential of the virtual but that brings it into a specific state: a state of quantitative difference. Deleuze describes Simondon's contribution thus: "Individuation emerges like the act of solving a problem, or – what amounts to the same thing – like the actualization of a potential and the establishing of communication between disparates ... Individuation is the act by which intensity determines differential relations to become actualized, along the lines of differenciation and within the qualities and extensities it creates."[21]

Perhaps an example or two might help make this point more concrete. If we think of the gene not as a set of discrete bits of information but instead as part of a virtual field of intensities that actualizes into specific concrete beings, then we begin to see Deleuze's and Simondon's perspective. As biologists have argued, the gene is not a closed system of pregiven information that issues out directly into individual characteristics. Instead, the genetic code is in constant interaction with a field of variables that in their intensive interaction generate a specific living being. In fact, as Barry Commoner has argued recently, the failure to recognize this point is what leads to the dangers of genetic engineering, since genetic engineering as it is practiced assumes that the introduction of a gene into a foreign body will result in the passing on of that gene's information without alteration or remainder (what Commoner calls, following Watson, "the central dogma"). That assumption has been shown to be empirically false, with fatal, and, if practiced widely, potentially disastrous results.[22] We must conceive of genetic passage, then, not as the perpetuation of individuals by means of a closed genetic code, but rather as the unfolding of a genetic virtuality that has among its products the individuation of organisms. Or, as Simondon puts the point, "the being contains not only that which is identical to itself, with the result that being qua being – previous to any individuation – can be grasped as something more than a unity and more than an identity"; adding, in a turn of phrase whose implications form the basis of Deleuze's appropriation of his work, "This method presupposes a postulate of an ontological nature."[23]

A second, related example of biological differenciation from a field of prior intensities is that of the egg. Deleuze writes that

> Individuating difference must be understood first within its field of individuation – not as belated, but as in some sense in the egg. Since the work of Child and Weiss, we recognize axes or planes of symmetry within an egg. Here, too, however, the positive

element lies less in the elements of the given symmetry than in those which are missing. An intensity forming a wave of variation throughout the protoplasm distributes its difference along the axes and from one pole to another.... the individual in the egg is a genuine descent, going from the highest to the lowest and affirming the differences which comprise it and in which it falls.[24]

The egg, in other words, is a field of differential intensities, a preindividual field from within which an individual arises. The fact that the relation between egg and individual is one of descent indicates that the field of differentiation is not exhausted by the differenciation that is the actualization of the individual organism. It is worth noting as well that this example also shows how the field of difference in itself is not a transcendent field that creates or emanates out into an individual organism. It is rather a field of immanence that expresses itself in the individual.

As Ansell Pearson remarks, Deleuze's use of Simondon in *Difference and Repetition* implies that "it is the process itself that is to be regarded as primary. This means that ontogenesis is no longer treated as dealing with the genesis of the individual but rather designates the becoming of being."[25] Here the thought of both Bergson and Spinoza is apt. Bergson, with his concept of *durée*, offers the conceptual framework for a process of the unfolding of difference as the actualization of the virtual; Spinoza, with his concept of the univocity of being, sees this process as an immanent unfolding rather than a transcendent creation or emanation. In his turn, Deleuze takes Simondon's framework and generalizes it from ontogenesis to the unfolding of the world itself, going beyond the biological to the ontological. He writes:

> The world is an egg. Moreover, the egg, in effect, provides us with a model for the order of reasons: (organic and species related) differentiation–individuation–dramatization–differenciation. We think that difference in intensity, as this is implicated in the egg, expresses first the differential relations or virtual matter to be organized. This intensive field of individuation determines the relations that it expresses to be incarnated in spatio-temporal dynamisms (dramatization), in species which correspond to these relations (specific differenciation), and in organic parts which correspond to the distinctive points in these relations (organic differenciation).[26]

Thus, in the first major statement of his own philosophical ontology, Deleuze appropriates Simondon's view of individuation as a model for the ontological approach he had developed earlier in his works on Bergson and Spinoza. The works are conceptually continuous, although now Deleuze brings science to bear in articulating his philosophy of difference.

In his collaborative works with Félix Guattari, Deleuze does not abandon the references to Simondon.[27] He does, however, broaden the biological appropriation to include, among others, the work of Jacques Monod. *Anti-Oedipus*, published in 1972, two years after Monod's seminal *Chance and Necessity*,[28] adopts Monod's treatment of biological enzymes as having "cognitive" properties, allowing for the combination of disparate elements into new life-forms, and it articulates that idea in a way that converges with Simondon's earlier work on individuation.

Although *Chance and Necessity* is famous chiefly for its conclusion that human beings are the product of biological chance rather than any form of evolutionary necessity, it is not the conclusion but the biochemical analysis upon which it is based that draws the attention of Deleuze and Guattari. Monod analyzes particular enzymes, and especially a class of enzymes known as "allosteric" enzymes, which not only perform the function of binding chemical substrates, but can also regulate their own activity on the basis of the existence of other compounds. In Monod's words, allosteric enzymes[29] "have the further property of recognizing electively one or several *other* compounds, whose ... association with the protein has a modifying effect – that is, depending on the case, of *heightening or inhibiting its activity with respect to the substrate.*"[30] This property, which Monod calls a "cognitive" function, entails that there is a process of self-ordering at the molecular level that does not require the intervention of an already constituted highly ordered system. The formation of complex biological systems arises from a group of molecular interactions that require nothing more than the right opportunity provided by chance mixings in order to develop into those systems. "Order, structural differentiation, acquisition of functions – all these appear out of a random mixture of molecules individually devoid of any activity, any intrinsic functional capacity other than that of recognizing the partners with which they will build the structure."[31]

One can see how, given this approach to molecular biology, the rise of human beings would be more a matter of chance than necessity. The fortuitous meeting of certain enzymes under certain conditions facilitate the formation of specific types of biological compounds that give rise to certain life-forms, which are then reproduced or not depending on the capacity for those compounds to thrive in a given environment. That those life-forms would eventuate in human life is a random matter. Another, related lesson, however, and one more pertinent to the collaboration between Deleuze and Guattari, is that allosteric and other types of enzymes contain the capacity for all sorts of combinations at the preindividual level. Moreover, these combinations, because they are the product of chance, might have been and might become otherwise. The molecular level, in other words, is a virtual realm of intensities, a field of differentiation that, through chance, differenciates itself into specific biological arrangements. As Monod puts the matter, using a machinic metaphor of the type that lies at the heart of *Anti-Oedipus*, "With the globular protein we already have, at the molecular level, a veritable machine – a machine in its functional properties, but not, as we now see, in its fundamental structure, where nothing but the play of blind combinations can be discerned.... A *totally* blind process can by definition lead to anything; it can even lead to vision itself."[32]

The convergence of Monod's work with Simondon's in articulating at the biological level what Deleuze, and later Guattari, are articulating at the ontological level is straightforward. At different levels of biological complexity – Simondon at the level of the gene and its environment, Monod at the level of pregenetic molecular interaction that leads up to the gene – both of these biologists offer a picture of a complex preindividual field that allows for the generation of specific individual forms but also is not bound or reducible to those forms. Moreover, since that field is immanent to the physical realm, it requires no movement toward transcendence in order to conceive it. It is a realm of immanent difference that actualizes itself into particular differences while remaining purely and more generically differential itself.

I want to turn in a moment to a different scientific approach, that of the chemist Ilya Prigogine, whose work at the chemical level appears in Deleuze and Guattari's collaborative works, particularly in *A Thousand Plateaus* and *What is Philosophy?*, to provide a new touchstone for their thought. Before doing that it might be worth turning briefly to mathematics, which, although in the US is considered a distinct discipline from science, on the Continent is grouped alongside other sciences. One way in which Deleuze often refers to the virtual field is as a "problematic." In *Difference and Repetition*, for instance, he writes, "Problematic structure is part of the objects themselves ... More profoundly still, Being (what Plato calls the Idea) 'corresponds' to the essence of the problem or the question as such.... In this relation, being is difference. Being is also non-being, *but non-being is not the being of the negative*; rather, it is the being of the problematic, the being of the problem and question."[33] The problematic as a virtual field that does not dictate any particular solution but can actualize into a number of different ones converges with Deleuze's general approach to the virtual, and is rendered in *Difference and Repetition*,[34] among other ways, in an interpretation of the calculus, which, as Aden Evens points out, can form a point of intersection between mathematics and ontology.

Deleuze reads differential calculus not as a pragmatic matter of using differential equations to discover the slope of a particular function at a particular point. Rather, he sees in the differential an entire ontology of difference that can actualize itself into various functions and, consequently, specific curvilinear patterns. As Evens points out, "In calculus class we are presented with a function and told to differentiate it ... In Deleuze's rereading of the calculus, the primitive function does not precede the differential relation, but is only the ultimate result or byproduct of the progressive determination of that relation. The differential is a problem, and its solution leads to the primitive function."[35]

In differential calculus, the normal procedure is to start from a function, pick a point or set of points, and find the differential that corresponds to those points. In that way, differentials are derived from specific (or "primitive") functions. In Deleuze's terms, the virtual would seem to be derived from the actual. But suppose we invert the relationship, moving not from function to dy/dx, but the other way around. Suppose further that we give some sense (although not a specific meaning) to the dx and dy that compose the formula dy/dx. That is, although the generic formula dy/dx is founded on the idea that a given point is always indicated by both x and y coordinates, suppose we think of the domain of functional relationships among those points as *preceding* the points themselves. Then we will arrive at a virtual, differentiated field that differenciates itself in various ways according to various specific mathematical functions. As Deleuze writes, "a principle of determinability corresponds to the undetermined as such (dx, dy); a principle of reciprocal determination corresponds to the really determinable (dy/dx); a principle of complete determination corresponds to the effectively determined (values of dy/dx)."[36] If we think of the calculus this way, it corresponds to Deleuze's idea of Being as a virtual or problematic field of difference in itself that actualizes itself in specific forms. Mathematics becomes an ontological matter, not simply a matter of pragmatic calculations.

In the later collaboration between Deleuze and Guttari, the writings of Ilya Prigogine become increasingly important. Prigogine, whose book *La nouvelle alliance* (co-authored with Isabelle Stengers and partially translated as *Order Out of Chaos*) appeared in 1979, argues for a self-ordering of chemical components into patterns and relationships that cannot be read off from the previous state of chemical disarray. "The artificial," he writes, "may be deterministic and reversible. The natural contains essential elements of randomness and irreversibility. This leads to a new view of matter in which matter is no longer the passive substance described in the mechanistic world view but is associated with spontaneous activity."[37]

One of Prigogine's most arresting examples is that of the chemical clock. In conditions that move away from equilibrium (a move which, as Prigogine and Stengers argue, is a common condition in the natural world), a process they describe with the following image might occur:

> Suppose we have two molecules, "red" and "blue." Because of the chaotic motion of the molecules, we would expect that at a given moment we would have more red molecules, say, in the left part of a vessel. Then a bit later more blue molecules would appear, and so on. The vessel would appear to us as "violet," with occasional irregular flashes of red or blue. However, this is *not* what happens with a chemical clock; here the system is all blue, then it abruptly changes its color to red, then again to blue. Because all these changes occur at *regular* time intervals, we have a coherent process.[38]

One can see how unexpected this might be. It is not the introduction of some sort of ordering mechanism that makes the chemical clock appear. It is an inherent capability of the chemicals themselves for self-organization that gives rise to this phenomenon. It is as though there were virtual potentialities for communication or coordination contained in the chemicals themselves, or at least in their groupings, that are actualized under conditions that move away from equilibrium. As Manuel De Landa notes, in an echo of Deleuze's treatment of Spinoza, "Matter, it turns out, can 'express' itself in complex and creative ways, and our awareness of this must be incorporated into any future materialist philosophy."[39]

Another phenomenon that points to the self-organizing of matter concerns "bifurcations." These are situations, once again under far-from-equilibrium conditions, in which a chemical system will "choose" between two or more possible structures, but which structure it chooses cannot be predicted in advance. It is only a matter of probability, not natural law, which structure will result, and thus the resulting structure is not reversible into its initial conditions. Moreover, when bifurcation is about to occur, small changes in the surrounding environmental conditions might have large effects on the outcome of the bifurcation. This introduces another element of chance into the understanding of chemistry. "Self-organizing processes in far-from-equilibrium conditions correspond to a delicate interplay between chance and necessity, between fluctuations and deterministic laws. We expect that near a bifurcation, fluctuations or random elements would play an important role, while between bifurcations the deterministic aspects would become dominant."[40]

The lesson here, of course, is not simply that chemistry is a matter of chance, but rather that there is a self-ordering process within the chemical realm that cannot be reduced to strict laws because of the capacity of chemicals to combine with and react to disparate chemical elements or physical conditions in new and unpredictable ways.

This is a new understanding of the nature of matter, one that corresponds to Deleuze's positing of the virtual as a realm of differentiation out of which actualizations of diverse elements can appear. At the level of the virtual, the level of pure difference, there is no precluding what will combine with what or what will result. Disparate combinations and unexpected outcomes are the very possibilities of the virtual.

Our world, it turns out, has much more to offer than those who embrace transcendence would either admit or desire. Paraphrasing Monod's conclusion above concerning biology, in a chaotic and self-organizing chemical world, anything can happen.[41]

Thus, when Prigogine and Stengers conclude that *"Nonequilibrium brings 'order out of chaos',"*[42] we must be careful to interpret those words in a particular way. This does not mean, on the Deleuzian understanding, that when we move from a situation of chaos to one of order, chaos is left behind. Such an understanding would amount to a return to an emanative or creative view of causality. If matter expresses itself in particular organized or self-organized forms, this is not because we have jettisoned chaos or difference in itself for order or quantitative difference. Organized and self-organized matter brings its chaos along with it, not as actualized but as virtual. Deleuze and Guattari, in incorporating the work of Prigogine and Stengers, do not want to see organized matter as a resolution of the virtual but as an actualization of it. Thus, matter preserves its potentiality for disparate combinations and novel actualizations at every point. The virtual is immanent to matter as both chaotic and organized. It does not appear only in the former. As De Landa notes (discussing both bifurcations and attractors[43]), *"they are intrinsic features of the dynamics of physical systems, and they have no independent existence outside of those physical mechanisms."*[44]

III

In the last pages, we have ranged over biology, microbiology, mathematics, and chemistry. And yet a single ontological picture has emerged, one that is convergent with Deleuze's (and later Guattari's) overall ontological approach. In the third of the thousand plateaus Deleuze and Guattari construct, they attempt to weave together many of these aspects of science, often borrowing elements from the scientists we have discussed. Although a full treatment of this plateau is beyond the scope of the essay here,[45] a brief sketch might pull together several threads that have made their appearance.

The third of the thousand plateaus, "10,000 BC: The Geology of Morals (Who Does the Earth Think It Is?)," unfolds by means of an increasing proliferation of concepts offered in a lecture by Professor Challenger (a figure Deleuze and Guattari draw from the writings of Arthur Conan Doyle), each concept moving out from the previous ones in order to bring new conceptual territory within its reach. In this way the plateau is structured like a fractal of the kind that Deleuze and Guattari often discuss, or, for Marx Brothers fans, like the opera scene at the end of *A Night at the Opera* where an increasingly chaotic movement gradually envelops and brings together all of the film's players. If we seek to understand the underlying conceptual structure, however, we see that it starts with seeing matter as an absolute deterritor-ialization that lies on a (virtual) plane of consistency. "[I]f we consider the plane of consistency we note that the most disparate of things and signs move upon it: a semiotic fragment rubs shoulders with a chemical interaction, an electron crashes into language, a black hole captures a genetic message, a crystallization produces a passion, the wasp and the orchid cross a letter ... "[46]

Movements on the plane of consistency are organized by strata, which are "acts of capture"[47] that organize those movements into specific forms. This capture, however, requires a "double articulation" in which first (borrowing terms from the linguist Hjelmslev) content is established and then expression. The content/expression dis-tinction is not like the traditional substance/form distinction, since Hjelmslev distin-guishes *both* content and expression into substance and form. To illustrate this double articulation of content and expression, Challenger uses a geological example: "In a geological stratum, for instance, the first articulation is the process of 'sedimentation,' which deposits units of cyclic sediment according to a statistical order ... The second articulation is the 'folding' that sets up a stable functional structure and effects the passage from sediment to sedimentary rock."[48]

As the strata move outwards in their acts of capture, they do not simply meet an external environment, but also intermediate environments that Challenger calls "epi-strata." Epistrata can react back upon strata, forming new centers and introducing further complexity into the strata themselves. Moreover, any given stratum exists only within the intermediate milieu of its epistrata. (Think here of the relation, discussed above, between a gene and its environment.) "A stratum, considered from the standpoint of its unity of composition, therefore exists only in its substantial epistrata, which shatter its continuity, fragment its ring, and break it down into gradations."[49] In addition there are "parastrata," which are various milieus that can be associated with or annexed to a given strata in order to provide them with sources of energy.

Challenger associates parastrata with populations, (often organic) packs or multipli-cities that form various milieus that come in contact with other parastrata or other strata. These parastrata express codes (of which DNA would be an example) that are distributed throughout the parastrata's population. Epistrata, by contrast, are associated with "thresholds of deterritorialization" that "correspond to more or less stable inter-mediate states,"[50] rates of development or change that are inherent in the epistrata's own milieu. The understanding of the roles of epistrata and parastrata is ascribed to Darwin: "Darwin's two fundamental contributions move in the direction of a science of multiplicities: the substitution of populations for types and the substitution of rates or differential relations for degrees. These are nomadic contributions with shifting boundaries determined by populations or variations of multiplicities, and with differ-ential coefficients of variations of relations."[51]

The ascription to Darwin of the reference to populations and differential relations underlines what is at stake for Deleuze and Guattari in Challenger's proliferation of concepts. What is being constructed here is a view that rejects the stability of given forms or types of forms (thus the substitution of populations for species) that change by designated degrees in favor of a realm of virtual difference that expresses itself through variable codes across populations and that is always in flux and liable to mutation. The picture Challenger draws is meant to apply not only to a specific ontological region; it offers a more general ontology that brings together several of the scientific views discussed here (both Simondon and Monod are among the many scientists that appear in the plateau's footnotes) in an *ontological* framework that draws upon chemistry, biology, geology, linguistics, mechanics, and politics. The dominant motif, however, here as elsewhere, is biological. It is no surprise that the name of Darwin is invoked at the centerpiece of the "lecture" as the thinker who first jettisons terms implying stability and order for those that open up multiplicity and flux. What Deleuze first found in Simondon remains for his thought the touchstone of his embrace of science.

IV

At this point we are poised to consider a general question that has lain in the background of Deleuze's (and Guattari's) approach to science. It seems as though Deleuze, in constructing his ontology, casts about for scientific perspectives that match his and then appropriates them to his own thought. There is, of course, nothing objectionable in this. But it does raise an important matter of interpretation. Does Deleuze hold the scientific views he discusses as evidentiary? Does he invoke them to show that the ontology (or ontologies) he has created, however Dionysian, has resonance in the more sober realm of scientific inquiry? If so, then he would seem to think that there is enough overlap in the projects of science and of philosophy that the former could provide epistemic support for the latter. Alternatively, does he believe that science, rather than offering support for ontology, simply provides some illustrations of its concepts? On this latter view, the appeal to science is made because there are scientists whose concepts converge neatly with his. If this is true, then science is not a privileged realm of inquiry; any other realm would do just as well, so long as it exemplifies the ontology Deleuze constructs. What, then, is the relation between philosophy and science: is it one of evidentiary support or merely conceptual similarity?

I believe the relationship is neither. The relationship between philosophy and science, between Deleuze's ontology and the scientific approaches he has appropriated to it, is more complex than either of these alternatives. In order to understand how to think about that relationship, we need to turn to the final extended collaboration between Deleuze and Guattari, *What is Philosophy?* In that text the authors lay out their view of the respective projects of philosophy and science, and from that we can draw some conclusions about Deleuze's own relationship to the subterranean fire of science that burns in the interstices of his philosophical texts.

"Philosophy," Deleuze and Guattari write, "does not consist in knowing and is not inspired by truth. Rather, it is categories like Interesting, Remarkable, or Important that determine its success or failure."[52] (It should be emphasized, against some objections to this Deleuze, that Deleuze does not reject truth, but rather says that truth is not the primary object of philosophy.[53]) This view, although in marked contrast to traditional views of the philosophical project, accurately reflects Deleuze and Guattari's approach to philosophy. But how is one supposed to approach the Interesting, the Remarkable, and the Important? By creating concepts and inventing conceptual personae across a plane of immanence. What we must understand, then, is what concepts, planes of immanence, and conceptual personae are.

"The concept is defined by *the inseparability of a finite number of heterogeneous components traversed by a point of absolute survey at infinite speed.*"[54] A concept – which in Deleuze and Guattari's usage is always a *philosophical* concept – draws together intensive features on a plane of immanence into a term that linguistically surveys those intensive features. Recall that, for Deleuze, intensities are differences that occur in the virtual realm. Thus, if a concept surveys intensities, what it is surveying are those virtual differences. To survey intensive features, however, is not the same thing as to *capture* them, in the sense that strata engage in "acts of capture" on the plane of consistency. To capture is to give form to or to organize; to survey is to allow thought access to. What concepts do is allow thought access to the virtual realm of the plane of immanence, or at least to part of it. (In coordination with other concepts, each concept surveys certain intensive features of the virtual realm.) Thus, the virtual that concepts survey (those "finite number of heterogeneous components") retains its immanent chaos or difference or intensity (its "infinite speed").

The plane of immanence is the philosophical equivalent of the plane of consistency Deleuze and Guattari discuss in the third plateau. "It is the plane that secures conceptual linkages with ever increasing connections, and it is concepts that secure the populating of the plane on an always renewed and variable curve."[55] Concepts by themselves, because each surveys only certain intensive features, would spin in the void if they could not connect with one another across a plane that binds them together. The plane of immanence is what allows those connections to occur; "the plane is the abstract machine of which these assemblages [i.e. concepts] are the working parts."[56] In this sense, the plane is the virtual whole (*durée*, the plane of consistency, Being) that a philosophical position, comprised of concepts that link across it, articulates.

Conceptual personae are figures that play out the movements of a philosophy. Socrates for Plato; the dreamer and the madman and the Evil Demon for Descartes; the knight of faith for Kierkegaard; Zarathustra and Dionysus and the last man and many others for Nietzsche. "Conceptual personae constitute points of view according to which planes of immanence are distinguished from one another and brought together, but they also constitute the conditions under which each plane finds itself filled with concepts of the same group."[57] We might think of conceptual personae as actors in the dramatization of concepts. They take part in relating those concepts to one another and to the plane that those concepts express. In this sense, it is concepts that are primary to conceptual personae. The personae do not have a role to play outside the concepts they dramatize or the conceptual "points of view" they embody. When Deleuze and Guattari define philosophy as the art of creating concepts,[58] then, without reference to conceptual personae, they are emphasizing the dominance of concepts relative to the conceptual personae that dramatize them.

If philosophy works by means of concepts, planes of immanence, and conceptual personae, science works by means of functions, planes of reference, and partial observers. A function, in contrast to a concept, fixes the relationships among its variables rather than giving play to differentiated variations. It designates a specific relation or set of relationships that holds among its components a specific covariance that occurs among its variables. Contrasting concept and function, Deleuze and Guattari write, "In the one case we have a set of *inseparable variations* subject to a "contingent reason" that constitutes the concept from variations; and in the other case we have a set of *independent variables* subject to a "necessary reason" that constitutes the function from variables."[59] It is worth noting here that in contrast to traditional approaches to the relationship between philosophy and science, for Deleuze and Guattari it is philosophy that works with a contingent reason and science that works with a necessary one. This is because philosophy is an "art," an experimentation with concepts in attempting to express Being as difference. Thus there is no essential order of concepts, no essential or necessary reason. Science, by contrast, which does attempt to fix relationships, to engage in "acts of capture," does in fact construct a "necessary reason."

Rather than articulating a plane of immanence – or, as the authors sometimes put it, speak or express an event – science works on a plane of reference, articulating a state of affairs. It removes itself from the infinite speed and heterogeneity of the plane of immanence in order to give form and organization to the phenomena with which it deals. In this sense, philosophy is a matter of remaining within the virtual so that it does not lose any of its infinite quality, its differentiation. Science, by contrast, is a matter of actualizing the virtual by means of differenciation. "By retaining the infinite, philosophy gives consistency to the virtual through concepts; by relinquishing the infinite, science gives a reference to the virtual, which actualizes it through functions.... [Science] is a fantastic *slowing down*."[60] Science does not turn its back on the virtual; it still deals in chaos.[61] But it does not attempt to retain the infinite fluidity and movement, the speed, the difference in itself that characterizes that chaos. Rather, it actualizes it through functions, and thus works on a plane of reference rather than one of immanence.

Finally, partial observers are distinct from conceptual personae in that while the latter dramatize concepts, the former offer observations from specific viewpoints within a fixed framework. "Scientific observers ... are points of view in things themselves that presuppose a calibration of horizons and a succession of framings on the basis of slowing-downs and accelerations: affects here become energetic relationships, and perception itself becomes a quantity of information."[62] Information and finite points of view are opposed to dramatization and infinite speed; partial observers and conceptual personae reflect the related but divergent projects of science and philosophy.

In this contrast between the two, Deleuze and Guattari do not abandon the elements of science that they appropriate in their writings. They retain a viewpoint on science that retains the chaotic, differential, and preindividual character that we saw in the work of Simondon, Monod, and Prigogine as well as in Deleuze's interpretation of the calculus. It is not that philosophy works with the virtual or the chaotic and science with the actual or fixed forms. Rather, philosophy and science are two different ways of approaching the virtual: philosophy by means of concepts and science by means of functions.

This contrast leads us back to the question we raised earlier: what is the status of the scientific viewpoints that Deleuze incorporates into his own work? Given the contrast between science and philosophy that we have just seen, we cannot say that it is a matter of evidence or of illustration. Science cannot provide evidence for philosophy, since philosophy is not a matter of truth; it does not seek evidence. Nor is science merely an illustration of philosophical concepts, because functions work in too different a realm from concepts to stand as examples of them.

What I want to suggest here is that the incorporation of science into Deleuze's work is an attempt to "speed up" scientific viewpoints by offering them an onto-logical perspective that draws them "out of themselves" and brings them into contact with pure difference, difference in itself. By reinscribing scientific concepts (what Deleuze and Guattari call "functives") like preindividual state, allosteric enzyme, dy/dx, or bifurcation[63] into the realm of the virtual, Deleuze allows these functives to double as concepts, or better, he creates a point of intersection between science and philosophy, a point at which if we take the functive in one direction it becomes a component of a function, while if we take it in another it becomes a concept. Since the functive is originally part of scientific discourse before Deleuze borrows it, it is the borrowing that allows for the possibility of the "speeding up," of the appropri-ation into the heterogeneity of the pure difference of the virtual.

If we look at the relationship this way, it becomes easy to see why science appears mostly in Deleuze's work as a "subterranean book of fire." Philosophy is a project distinct from science. Its goal is neither to clarify science nor to offer it conceptual foundation. Science's role is neither to provide evidence for philosophy nor to illus-trate it. Philosophy's project is to create concepts while science's is to articulate functions. And yet, since both science and philosophy concern the virtual, they must inevitably come into contact; their trajectories must periodically intersect. When they do, Deleuze appropriates elements of that intersection for his own philosophical purposes, seeking to construct or to expand from those points of intersection philo-sophical concepts that open out into a perspective that is Interesting, Remarkable, or Important.[64]

Notes

1   *Essays Clinical and Critical*, trans. Daniel Smith and Michael Greco. Minneapolis: University of Minnesota Press, 1997 (originally pub. 1993), p. 151.
2   London and New York: Routledge, 1999.
3   Cambridge, MA: MIT Press, 1992.
4   In *Zone 6: Incorporations*, eds. S. Kwinter and J. McCrary. New York: Zone Books, 1992, pp. 129–61.
5   London: Continuum, 2002.
6   in *Angelaki* 5(3): 105–15.
7   *Difference and Repetition*, trans. Paul Patton. New York: Columbia University Press, 1994 (originally pub. 1968), p. 57.
8   *What is Philosophy?*, trans. Hugh Tomlinson and Graham Burchell. New York: Columbia University Press, 1994 (originally pub. 1991), p. 60. Deleuze's major treatments of Spinoza are to be found in *Expressionism in Philosophy: Spinoza*, trans. Martin Joughin. New York: Zone Books, 1990 (originally pub. 1968) and *Spinoza: Practical Philosophy*, trans. Robert Hurley. San Francisco: City Light Books, 1988 (originally pub. 1970, 1981).

9   *On the Genealogy of Morals*, trans. Douglas Smith. Oxford: Oxford University Press, 1996, p. 96.

10 *Difference and Repetition*, p. 36.

11 *Bergonism*, trans. Hugh Tomlinson and Barbara Habberjam. New York: Zone Books, 1988 (originally pub. 1966), p. 36.

12 Cf. e.g. *Difference and Repetition*, pp. 211–12.

13 These are the terms Deleuze uses in *Expressionism in Philosophy*.

14 In his books *Foucault* and *The Fold: Leibniz and the Baroque*, Deleuze uses the image of folding to capture this idea.

15 *Difference and Repetition*, p. 207.

16 Ibid., pp. 56–7. Note here that the "qualitative" diversity he refers to is the "quantitative" difference he cited in the Bergson passage above.

17 "The Genesis of the Individual," trans. Mark Cohen and Sanford Kwinter, in *Zone 6: Incorporations*, eds. S. Kwinter and J. McCrary. New York: Zone Books, 1992, p. 300. This text is the introduction to Simondon's *L'individu et sa génèse physico-biologique*. Paris: Presses Universitaires de France, 1964.

18 *The Logic of Sense*, trans. Mark Lester with Charles Stivale, ed. Constantin Boundas. New York: Columbia University Press, 1990 (originally pub. 1969), p. 344.

19 *Difference and Repetition*, p. 221.

20 Ibid., p. 222.

21 Ibid., p. 246.

22 Barry Commoner, "Unraveling the DNA Myth," in *Harper's*, Feb. 2002, pp. 39–47.

23 "The Genesis of the Individual," p. 312. The specific ontological – actually logical – postulate that Simondon has in mind is that the laws of identity and the excluded middle do not apply at the preindividual level, a point that Deleuze would certainly agree with. Deleuze goes much further, however, in his development of the ontological implications of Simondon's perspective.

24 *Difference and Repetition*, p. 250.

25 *Germinal Life*, p. 90.

26 *Difference and Repetition*, p. 251.

27 Cf. e.g. *A Thousand Plateaus*, trans. Brian Massumi. Minneapolis: University of Minnesota Press, 1987 (originally pub. 1980), pp. 408–10, where Deleuze and Guattari call upon Simondon's critique of the hylomorphic model of a fixed form and an amorphous matter to be molded in the form's image.

28 *Chance and Necessity: An Essay on the Natural Philosophy of Modern Biology*, trans. Austryn Wainhouse. New York: Knopf, 1971 (originally pub. 1970).

29 Deleuze and Guattari refer to Monod's discussion of the allosteric enzyme in *Anti-Oedipus: Capitalism and Schizophrenia*, trans. Robert Hurley, Mark Seem, and Helen R. Lane. New York: Viking Press, 1977 (originally pub. 1972). Cf. pp. 288–9.

30 *Chance and Necessity*, p. 63.

31 Ibid., p. 86.

32 Ibid., p. 98.

33 Pp. 63–4. Note also that Deleuze uses the idea of problems and solutions in his treatment of Simondon, discussed above. The idea of a problem as a virtual field, particularly as a virtual philosophical field, stretches back as far as Deleuze's earliest writings. Cf. e.g. *Empiricism and Subjectivity*, trans. Constantin Boundas. New York: Columbia University Press, 1991 (originally pub. 1953), p. 106.

34 Pp. 170–6.

35 "Math Anxiety," p. 111.

36 *Difference and Repetition*, p. 171.

37 *Order Out of Chaos: Man's New Dialogue with Nature*, by Ilya Prigogine and Isabelle Stengers. Boulder and London: New Science Library, 1984, p. 9. In this view of matter, the authors show some sympathy for Bergson when they write that *"durée*, Bergson's 'lived time,' refers to the basic dimensions of becoming, the irreversibility that Einstein was willing to admit only at the phenomenological level" (p. 294). I should note as well that in the original French edition of this book there are several complimentary references to Deleuze and borrowings from his work that do not appear in the English edition. For example, the following passage cites *Nietzsche and Philosophy* but could well stand as a summation of *Difference and Repetition*: "Science, which describes the transformations of energy under the sign of equivalence, must admit, however, that only *difference* can produce effects, which would in turn be differences themselves. The conversion of energy is nothing other than the *destruction* of one difference and the *creation* of another." *La Nouvelle Alliance: Métamorphose de la science*. Paris: Gallimard, 1979, p. 127. (My translation.)

38 *Order Out of Chaos*, pp. 147–8.

39 "Nonorganic Life," p. 133.

40 *Order Out of Chaos*, p. 176.

41 To say that anything can happen does not mean that anything can happen at any time, however. This is true for two reasons. First, the virtual unfolds within the actual, which means that the actual at a given time constrains the virtual. Second, as De Landa points out in *Intensive Science and Virtual Philosophy*, the chaos of the virtual is not completely random but can be ordered by various "singularities" that serve not as laws but as "attractors" for certain kinds of actualizations. We may read Prigogine and Stengers' bifurcations and chemical clocks as examples of such singularities.

42 *Order Out of Chaos*, p. 286.

43 Attractors are states of chemical systems to which they are "attracted" when they are in conditions of nonequilibrium.

44 "Nonorganic Life," p. 138.

45 The first two chapters of Massumi's *A User's Guide to Capitalism and Schizophrenia*, while not a summary of the third plateau, explicates a number of its concepts along the way.

46 *A Thousand Plateaus*, p. 69. This concept of the plane of consistency as absolute deterritorialized matter is in keeping with Deleuze's earlier treatments of the virtual as difference in itself, although Ansell Pearson argues that it transcends a residual humanism of the early works. Cf. *Germinal Life*, pp. 139–40.

47 *A Thousand Plateaus*, p. 40.

48 Ibid., p. 41.

49 Ibid., pp. 50–1.

50 Ibid., p. 53.

51 Ibid., pp. 48–9.

52 *What is Philosophy?*, p. 82.

53 There are at least two places where one might find truth in the traditional sense in his philosophical views. First, if one believes, for example, in scientific claims that are false, one is likely to be stymied in trying to develop anything that is Interesting, Remarkable, or Important. Believing one can walk through walls does not generally result in anything interesting. Second, as I argue elsewhere, Deleuze is also committed to such truths as that his philosophical approach will actually yield something more Interesting, Remarkable, and Important than more traditional approaches. (Cf. *Reconsidering Difference: Nancy, Derrida, Levinas, and Deleuze*. University Park: Penn State Press, 1997, esp. pp. 168–71.)

54 *What is Philosophy?*, p. 21.

55 Ibid., p. 37.

56   Ibid., p. 36.

57   Ibid., p. 75.

58   See e.g. ibid., pp. 2, 5.

59   Ibid., p. 126. See the extract below.

60   Ibid., p. 118. See the extract below.

61   When Deleuze and Guattari write that "chaos ... is a void that is not a nothingness but a *virtual*," *What is Philosophy?*, p. 118 (see the extract below), it is Prigogine and Stengers that they reference for that perspective.

62   *What is Philosophy?*, p. 132.

63   Referring to *Order Out of Chaos*, Deleuze says in an interview that, "One of the many concepts created in that book is that of a region of bifurcation ... it's a good example of a concept that's irreducibly philosophical, scientific, and artistic, too." "On *A Thousand Plateaus*," in *Negotiations, 1972–1990*, trans. Martin Joughin. New York: Columbia University Press, 1995 (originally pub. 1990), pp. 29–30.

64   I would like to thank Daniel Smith for his close reading and helpful suggestions on an earlier draft of this paper, as well as the participants at the conference on Continental philosophy and science at Notre Dame in September 2002.

18

From *WHAT IS PHILOSOPHY?*

Gilles Deleuze and Félix Guattari

5.   Functives and Concepts

The object of science is not concepts but rather functions that are presented as propositions in discursive systems. The elements of functions are called *functives*. A scientific notion is defined not by concepts but by functions or propositions. This is a very complex idea with many aspects, as can be seen already from the use to which it is put by mathematics and biology respectively. Nevertheless, it is this idea of the function which enables the sciences to reflect and communicate. Science does not need philosophy for these tasks. On the other hand, when an object – a geometrical space, for example – is scientifically constructed by functions, its philosophical concept, which is by no means given in the function, must still be discovered. Furthermore, a concept may take as its components the functives of any possible function without thereby having the least scientific value, but with the aim of marking the differences in kind between concepts and functions.

Under these conditions, the first difference between science and philosophy is their respective attitudes toward chaos. Chaos is defined not so much by its disorder as by the infinite speed with which every form taking shape in it vanishes. It is a void that is not a nothingness but a *virtual*, containing all possible particles and drawing out all possible forms, which spring up only to disappear immediately, without consistency or reference, without consequence.[1] Chaos is an infinite speed of birth and disappearance. Now philosophy wants to know how to retain infinite speeds while gaining consistency, by *giving the virtual a consistency specific to it*. The philosophical sieve, as plane of immanence that cuts through the chaos, selects infinite movements of thought and is filled with concepts formed like consistent particles going as fast as thought. Science approaches chaos in a completely different, almost opposite way: it relinquishes the infinite, infinite speed, in order to gain *a reference able to actualize the*

Gilles Deleuze and Félix Guattari, pp. 117–20 and 125–29 from *What Is Philosophy?*, trans. H. Tomlinson and G. Burchell. New York: Columbia University Press, 1994. © 1994 by Columbia University Press. Reprinted with the permission of the publisher.

virtual. By retaining the infinite, philosophy gives consistency to the virtual through concepts; by relinquishing the infinite, science gives a reference to the virtual, which actualizes it through functions. Philosophy proceeds with a plane of immanence or consistency; science with a plane of reference. In the case of science it is like a freeze-frame. It is a fantastic *slowing down,* and it is by slowing down that matter, as well as the scientific thought able to penetrate it with propositions, is actualized. A function is a Slow-motion. Of course, science constantly advances accelerations, not only in catalysis but in particle accelerators and expansions that move galaxies apart. However, the primordial slowing down is not for these phenomena a zero-instant with which they break but rather a condition coextensive with their whole development. To slow down is to set a limit in chaos to which all speeds are subject, so that they form a variable determined as abscissa, at the same time as the limit forms a universal constant that cannot be gone beyond (for example, a maximum degree of contraction). The first functives are therefore the limit and the variable, and reference is a relationship between values of the variable or, more profoundly, the relationship of the variable, as abscissa of speeds, with the limit.

Sometimes the constant-limit itself appears as a relationship in the whole of the universe to which all the parts are subject under a finite condition (quantity of movement, force, energy). Again, there must be systems of coordinates to which the terms of the relationship refer: this, then, is a second sense of limit, an external framing or exoreference. For these protolimits, outside all coordinates, initially generate speed abscissas on which axes will be set up that can be coordinated. A particle will have a position, an energy, a mass, and a spin value but on condition that it receives a physical existence or actuality, or that it "touches down" in trajectories that can be grasped by systems of coordinates. It is these first limits that constitute slowing down in the chaos or the threshold of suspension of the infinite, which serve as endoreference and carry out a counting: they are not relations but numbers, and the entire theory of functions depends on numbers. We refer to the speed of light, absolute zero, the quantum of action, the Big Bang: the absolute zero of temperature is minus 273.15 degrees Centigrade, the speed of light, 299,796 kilometers per second, where lengths contract to zero and clocks stop. Such limits do not apply through the empirical value that they take on solely within systems of coordinates, they act primarily as the condition of primordial slowing down that, in relation to infinity, extends over the whole scale of corresponding speeds, over their conditioned accelerations or slowing-downs. It is not only the diversity of these limits that entitles us to doubt the unitary vocation of science. In fact, each limit on its own account generates irreducible, heterogeneous systems of coordinates and imposes thresholds of discontinuity depending on the proximity or distance of the variable (for example, the distance of the galaxies). Science is haunted not by its own unity but by the plane of reference constituted by all the limits or borders through which it confronts chaos. It is these borders that give the plane its references. As for the systems of coordinates, they populate or fill out the plane of reference itself. [. . .]

The first difference between philosophy and science lies in the respective presuppositions of the concept and the function: in the one a plane of immanence or consistency, in the other a plane of reference. The plane of reference is both one and multiple, but in a different way from the plane of immanence. The second difference concerns the concept and the function more directly: the inseparability of variations is the distinctive characteristic of the unconditioned concept, while the independence of variables, in relationships that can be conditioned, is essential to the function. In one case we have a set of *inseparable variations* subject to "a contingent reason" that constitutes the concept from variations; and in the other case we have a set of *independent variables* subject to "a necessary reason" that constitutes the function from variables. That is why, from this point of view, the theory of functions presents two poles depending on whether, n variables being given, one can be considered as function of the $n-1$ independent variables, with $n-1$ partial derivatives and a differential total of the function, or, on the contrary, whether $n-1$ magnitudes are functions of a single independent variable, without differential total of the composite function. In the same way, the problem of tangents (differentiation) summons as many variables as there are curves in which the derivative for each is any tangent whatever at any point whatever. But the inverse problem of tangents (integration) deals with only a single variable, which is the curve itself tangent to all the curves of the same order, on condition of a change of coordinates.[2] An analogous duality concerns the dynamic description of a system of n independent particles: the instantaneous state can be represented by n points and n vectors of speed in a three-dimensional space but also by a single point in a phase space.

It could be said that science and philosophy take opposed paths, because philosophical concepts have events for consistency whereas scientific functions have states of affairs or mixtures for reference: through concepts, philosophy continually extracts a consistent event from the state of affairs — a smile without the cat, as it were — whereas through functions, science continually actualizes the event in a state of affairs, thing, or body that can be referred to. From this point of view, the pre-Socratics had already grasped the essential point for a determination of science, valid right up to our own time, when they made physics a theory of mixtures and their different types.[3] And the Stoics carried to its highest point the fundamental distinction between, on the one hand, states of affairs or mixtures of bodies in which the event is actualized and, on the other, incorporeal events that rise like a vapor from states of affairs themselves. It is, therefore, through two linked characteristics that philosophical concept and scientific function are distinguished: inseparable variations and independent variables; events on a plane of immanence and states of affairs in a system of reference (the different status of intensive ordinates in each case derives from this since they are internal components of the concept, but only coordinates of extensive abscissas in functions, when variation is no more than a state of variable). *Concepts and functions thus appear as two types of multiplicities or varieties whose natures are different.* Although scientific types of multiplicity are themselves extremely diverse, they do not include the properly philosophical multiplicities for which Bergson claimed a particular status defined by duration, "multiplicity of fusion," which expressed the inseparability of variations, in contrast to multiplicities of space, number, and time, which ordered mixtures and referred to the variable or to independent variables.[4] It is true that this very opposition, between scientific and philosophical, discursive and intuitive, and extensional and intensive multiplicities, is also appropriate for judging the correspondence between science and philosophy, their possible collaboration, and the inspiration of one by the other.

Finally, there is a third major difference, which no longer concerns the respective presuppositions or the element as concept or function but the *mode of enunciation*. To be sure, there is as much experimentation in the form of thought experiment in philosophy as there is in science, and, being close to chaos, the experience can be overwhelming in both. But there is also as much creation in science as there is in philosophy or the arts. There is no creation without experiment. Whatever the difference between scientific and philosophical languages and their relationship with so-called natural languages, functives (including axes of coordinates) do not preexist ready-made any more than concepts do. Granger has shown that in scientific systems "styles" associated with proper names have existed – not as an extrinsic determination but, at the least, as a dimension of their creation and in contact with an experience or a lived[5] [*un vécu*]. Coordinates, functions and equations, laws, phenomena or effects, remain attached to proper names, just as an illness is called by the name of the physician who succeeded in isolating, putting together, and clustering its variable signs. Seeing, seeing what happens, has always had a more essential importance than demonstrations, even in pure mathematics, which can be called visual, figural, inde-pendently of its applications: many mathematicians nowadays think that a computer is more precious than an axiomatic, and the study of nonlinear functions passes through slownesses and accelerations in series of observable numbers. The fact that science is discursive in no way means that it is deductive. On the contrary, in its bifurcations it undergoes many catastrophes, ruptures, and reconnections marked by proper names. If there is a difference between science and philosophy that is impossible to over-come, it is because proper names mark in one case a juxtaposition of reference and in the other a superimposition of layer: they are opposed to each other through all the characteristics of reference and consistency. But on both sides, philosophy and science (like art itself with its third side) include an *I do not know* that has become positive and creative, the condition of creation itself, and that consists in determining *by* what one does not know – as Galois said, "indicating the course of calculations and anticipating the results without ever being able to bring them about."[6]

We are referred back to another aspect of enunciation that applies no longer to proper names of scientists or philosophers but to their ideal intercessors internal to the domains under consideration. We saw earlier the philosophical role of *conceptual perso-nae* in relation to fragmentary concepts on a plane of immanence, but now science brings to light *partial observers* in relation to functions within systems of reference. The fact that there is no total observer that, like Laplace's "demon," is able to calculate the future and the past starting from a given state of affairs means only that God is no more a scientific observer than he is a philosophical persona. But "demon" is still excellent as a name for indicating, in philosophy as well as in science, not something that exceeds our possibilities but a common kind of these necessary intercessors as respective "subjects" of enunciation: the philosophical friend, the rival, the idiot, the overman are no less demons than Maxwell's demon or than Einstein's or Heisen-berg's observers. It is not a question of what they can or cannot do but of the way in which they are perfectly positive, from the point of view of concept or function, even in what they do not know and cannot do. In both cases there is immense variety, but not to the extent of forgetting the different natures of the two great types. [...]

Notes

1   Ilya Prigogine and Isabelle Stengers, *Entre le temps et l'éternité* (Paris: Fayard, 1988), pp. 162–3. The authors take the example of the crystallization of a superfused liquid, a liquid at a temperature below its crystallization temperature: "In such a liquid, small germs of crystals form, but these germs appear and then dissolve without involving any consequences."

2   G. W. Leibniz, "D'une ligne issue de lignes" and "Nouvelle application du calcul," both in *Oeuvre mathématique de Leibniz autre que le calcul infinitésimal*, trans. Jean Peyroux (Paris: Blanchard, 1986). These texts are considered to be the foundations of the theory of functions.

3   Having described the "intimate mixture" of different types of trajectory in every region of the phase space of a system with weak stability, Prigogine and Stengers conclude: "We may think of a familiar situation, that of the numbers on the axis where each rational number is surrounded by irrational numbers, and each irrational number is surrounded by rational numbers. Equally, we may think of the way in which Anaxagoras [shows how] every thing contains in all its parts, even the smallest, an infinite multiplicity of qualitatively different seeds intimately mixed together." Ilya Prigogine and Isabelle Stengers, *La nouvelle alliance* (Paris: Gallimard, 1979), p. 241. [*Translators' note:* the English version of this book, *Order out of Chaos* (London: HarperCollins, 1985), differs considerably from the original French, but see p. 264 of the English version.]

4   The theory of two kinds of "multiplicity" is present in Bergson from *Time and Free Will*, trans. F. L. Pogson (New York: Macmillan, 1910), chap. 2: multiplicities of consciousness are defined by "fusion" and by "penetration," terms that are equally found in Husserl from *The Philosophy of Arithmetic*. The resemblance between the two authors is, in this respect, extremely close. Bergson will always define the object of science by mixtures of space-time, and its principal action by the tendency to take time as an "independent variable," whereas, at the other pole, duration passes through every variation.

5   Gilles-Gaston Granger, *Essai d'une philosophie du style* (Paris: Odile Jacob, 1988), pp. 10–11, 102–5.

6   Cf. the great texts of Evariste Galois on mathematical enunciation: André Dalmas, *Evariste Galois, revolutionnaire et geometre* (Paris: Nouveau Commerce, 1982), pp. 117–32.

IRIGARAY

19

ON ASKING THE WRONG QUESTION ("IN SCIENCE, IS THE SUBJECT SEXED?")

Penelope Deutscher

In "In Science, Is the Subject Sexed?," linguist, feminist, psychoanalyst, and philosopher Luce Irigaray seems too easily impressed by the scientific community, convinced that they will not be too impressed by her. Her doctorates in linguistics, philosophy, and psychoanalysis do not admit her to a discussion with physicists, mathematicians, or biologists. She will be seen as belonging to the human and inexact sciences. Addressing a "Séminaire d'histoire et sociologie des idées et des faits scientifiques" (Seminar on the history and sociology of scientific facts) at the Université de Provence in Marseilles, she breaks into a confession: "Not for a long while have I experienced so much difficulty with the idea of speaking in public. Most of the time, I can anticipate to whom I will speak, how to speak, how to argue, to make myself understood, plead my cause, even please or displease my audience."[1] Anxieties hover in the air, arising, according to her diagnosis, from the pervasive authority of science's judgment. For, as she depicts them, scientific disciplines form closed, totalizing worlds, isolated by their vision of the universe, their conventions for its depiction, and their experimental methods and procedures. Outside her areas of expertise, Irigaray has offered close readings of texts from the domain of anthropology, economics, and film. But the broaching of dialogue with the sciences, particularly physics, mathematics, and biology seems to her to be of a different level of audacity.

Irigaray the feminist philosopher and Isabelle Stengers the historian of science[2] share their interest in opening scientific communities to new patterns of debate with nonspecialist communities. Stengers laments that "scientific priorities are discussed amongst those whose right to discuss them is recognized. If the 'public' is outside, observing the spectacle, it is not because they are incompetent. It's because their ability to intervene is not recognized."[3] Is it realistic to expect the public's intervention? If not, we should not ascribe this to a rigorous separation of competence between science and the humanities, or between experts and amateurs. She points out: "of course, the 'public' can't intervene directly in the question of whether the Big Bang is a fiction or a reality. But no more can non-specialist physicists."[4] The right of discussion must either be drawn extremely narrowly, or, if it is opened to the nonspecialist, why not draw those boundaries more widely? Stengers aligns herself with theorists (particularly Bruno Latour)[5] who challenge "the ideal of pure science" and put "in question any separation between the sciences and society." Science is "a social undertaking like any other, neither more detached from the cares of the world nor more universal or rational than any other practice."[6] Within a perspective that would so reorient our perception of the boundaries of science, nonspecialist engagement should not be more daunting than engagement with the humanities and human sciences.

Stengers hearkens to a time "when one discussed science in literary salons"[7] and science was seen as a source of reverie, amusement, and wit. Participants in an eighteenth-century literary salon expected to be fascinated by the inventions of contemporary science, and the scientist expected to stimulate their interest. Today, we are not surprised to find popular interest in scientific innovation reflected in fiction, comedy, television, and film. Certain areas of scientific inquiry will certainly be judged more off-putting than others to a popular audience. But scientific disciplines, programs, and problems will not be discredited because they fail to stimulate the imagination of the nonspecialist, or fail, as Stengers puts it, to make us laugh. Certainly the failure to amuse would not be taken to impugn their scientificity. By contrast, Stengers hopes to revalue interest as "an affective-intellectual foundation for criticism in the sciences."[8] For Stengers, "the question, 'Is this scientific?,' could be understood as 'Is this interesting?'"[9] How does it reinvent the world in ways that capture imagination, senses, and emotions?

Stengers's opposition between the interesting and the uninteresting seems more than arbitrary at times. She protests at the absence of levity in Stephen Hawking's *A Brief History of Time*. She speaks passionately of the "biology which had taught [Diderot] to see in the smallest drop of liquid thousands of animalculae and … that science which has taught me to think of 'grey matter' as a jungle of millions of interwoven neurons; of the continents in constant movement; and the moon as falling, interminably, like an apple, towards Earth."[10] To be sure, there are many readers whose imaginations may have been more captured by *A Brief History of Time*. Stengers's project, however, is not to promote a dogmatic account of what is interesting, but to question any use of the name of science to relinquish the responsibility to capture the imagination of colleagues and nonspecialists alike. Not because one can arbitrarily distinguish the interesting from the dull, but because the motives and practice of theorists are questionable when they excuse their own dullness with the alibi of science, Stengers questions those who put rats in boxes and students in statistics:

> Instead of critically assessing the "objectivity" of anyone who "in the name of science" puts rats into boxes or students into statistics, we could then ask how doing so could possibly interest anyone who doesn't similarly bow down, *a priori*, before the empty power of figures?[11]

Clearly, science is not being practiced as Stengers thinks it ought to be, when scientists are overly imbued with the spirit of seriousness or allow themselves to be isolated from everyday cares, or from disciplines they deem irrelevant or illegitimate. Among the scientists of whom Stengers speaks with interest in *The Invention of Modern Science* and elsewhere (Galileo, Pasteur, Perrin, Alfred North Whitehead, Stephen Jay Gould,

Joseph B. Rhine, Barbara McClintock) are Jean-Pierre Changeux and Daniel Cohen who, working respectively in neurology and genetics, are not afraid to express an interest in the unconscious or in mysticism, are not frightened from being curious by scientific authoritarianism.[12]

One can only anticipate the agitated response likely to meet Isabelle Stengers's expression of interest in interest. Would she favor a frivolous but aesthetically pleasing project over a new cure for a disease? And even if one could imagine such preferences, on what basis would one identify the interesting, and by what criteria? Are conventional values for pleasure, stimulation, and wit exercised by Stengers herself when this discussion is broached? Isn't an artist as likely to favor the abject, the confrontational, or everyday detritus, as that which is pleasing to the senses? Isn't the humorist as likely to favor the tasteless or gross joke as she is a form of wit? In other words, does Stengers sometimes adopt the least interesting characterizations of art, novelty, and humor when she seeks to install them at the heart of science?

How is her interest in science deployed when Stengers discusses feminism? She calls for science to be answerable to nonspecialists who admire wit and imagination. One might expect her to nudge science to listen to contemporary feminism, amongst other social movements. Because she does not,[13] one is led to the suspicion that one of Stengers's main reservations about feminism is that it is uninteresting. She might consider feminism insufficiently imaginative, though certainly she doesn't refer to the more innovative of the feminist writers on science.[14] But her real objection is that feminism tends to make science boring. It drowns out the diversity in the history of science. It listens only for the bores rather than the wits, and the most tedious rather than the most imaginative refrains. Worst of all, from her point of view, feminist critics of science give too much credit to scientists as truthful witnesses:

> In fact, the price paid by the radical character of the critique, whether technoscientific or feminist, is the respect it accords the scientist as a privileged interpreter of what science is able to do ... In this sense, the "radical" critique of science grants scientists all their pretensions. It recognizes the sociotechnical mutations that affect our world as the products of rationality – (techno)scientific or male – and tends to accept what scientists "say" at face value.[15]

Certainly, when we hear Luce Irigaray's comments about scientific imperialism and its dismissive, omnipresent voice of judgment, we feel sympathetic with a concern that feminism is displacing a very conservative, if not paternalist, characterization onto science. This would be feminism's least innovative gesture: its excessive willingness to find the father in science, philosophy, or psychoanalysis.

But there are some moments of proximity between the positions of Irigaray and Stengers, seen in their shared interest in assessing science in terms of its political, aesthetic, even its loving or lovable qualities. Irigaray writes of science that it operates according to *schizes* which separate the realms of pure science, politics, nature, art, and love. These *schizes* make impossible mutual verification, encounter, and responsibility between these realms (whereby one might assess the aesthetic value of science or the political value of love, for example). But it is Stengers who steps back to note also the risk run by this critique: that we either institute or consolidate a stereotype, or accept and abet too readily the most conservative self-image of science. In the mode of critique, one doesn't listen for the extent (be it fleeting) to which contemporary science and history of science may have been loving, lovable, aesthetic, the zone of free and wild creation of concepts,[16] amusing, ethical, empathizing, intuitive, politically responsible. Feminism doesn't know – and, worries Stengers, given our gusto for critique, we wouldn't want to know.

A related concern is raised by French feminist philosopher Michèle Le Doeuff, discussing the reiterated role played by Bacon amongst feminist epistemologists. In *The Sex of Knowing*[17] she criticizes feminists who, she argues, abet the convention that reason is masculine. Such feminists are overly inclined to attribute that view to the history of philosophy they deem to have propagated it. Though it can appear otherwise, Le Doeuff has not misunderstood feminist epistemologists, though some may be startled to find themselves apparently accused in *The Sex of Knowing* of believing reason to be masculine. Instead, Le Doeuff claims that unless feminist methodology changes, there is an insignificant difference between those who argue that reason has been associated with masculinity and those who argue that reason is intrinsically masculine. Rather than confusing the former for the latter, Le Doeuff maintains that the effect is much the same. Either way, a mystique about science's association with the masculine continues to proliferate, a mystique which is counter to feminist interests. Rather than splitting hairs over the difference between masculinity, historical association with masculinity, and conventional metaphors for masculinity, Le Doeuff argues for new ways of telling history of science that emphasize the interruptions, fissures, and failures of the maleness of reason (symbolic, metaphorical, associative, or otherwise).

What alternative does she propose to the practice of identifying the distancing of women from reason throughout history? She asks us to pay maximum attention to the high variability in the associations attributed to women.[18] We should pay attention to the qualifications, the exceptions and alternatives, the multiplicity of voices in given historical periods, and the most complex and contradictory aspects of even the most misogynist texts. She criticizes Evelyn Fox Keller's and other feminist interpretations of Bacon because such characterizations simply give too much fuel to their opponents. As she explains:

> This is the source of my disagreement with *feminist epistemology*. To critique Bacon as Keller does is in effect to grant that both he and the founders of the Academies were right to confine the sciences within a masculine enclosure: [as if] we are indeed indebted to virility for the sciences.[19]

Though Le Doeuff willingly deems Bacon sexist – to be sure, he excludes women from the House of Solomon and does not represent women as knowledgeable agents – she also concludes that Bacon's sexism is one of the more contradictory elements of his writing. It could be removed from his texts; and the sexism of his metaphorics is in any case not stable.[20]

Le Doeuff is not defending Bacon against feminism, but asking for a more nuanced reading of him better suited to feminist purposes. The recurring refrain of *The Sex of Knowing* is her rejection of the view that women have monotonously been distanced from the practices of reason and science, and that reason, science, and knowledge are in this sense sexed, whether literally or by association. We are asking the wrong question, she claims, when we interrogate the sexing of science, or of the subjects of science, in these terms. For one thing, we should be asking how science and the subjects of science have not been sexed or are not sex specific. It is more in feminist interests to do so. For another thing, when it comes to the maleness of reason, we are misidentifying the constant. The constant is women's association with devalued qualities. Depending on the context, those qualities might be intuition, practical competence, reason, or emotion. The relations between differentials of masculinity and femininity are sufficiently flexible to allow women to be associated with scientific competence or innovation. It is not inconceivable that the subject of science be deemed feminine. The constant is that, as feminine, it would very likely be devalued. According to Le Doeuff, feminist preoccupations with characterizations of science and knowledge as masculine have served to distract us from this point.

For this reason, Le Doeuff resists the characterization of reason as masculine whether it occurs in feminist or antifeminist contexts. She pits against a feminist reading of Bacon that stresses the feminine imagery attributed to a nature dominated by a reason whose imagery is male, an alternatively feminist reading of Bacon that asks how those associations are undermined. To this end, she notes that Bacon's examples of poor rational method come from venerated male thinkers: Plato, Aristotle, the Aristotelians, Cicero. He gives to the unreformed intellect the same characteristics that Fox Keller associates with a masculine practice of reason: dogmatism, manipulating experience to suit one's desired results and distancing oneself from inconvenient facts, making one's way only with facts that suit, along with all the violence towards one's audience that such practices imply.[21] Do men truly hunger for knowledge? Don't they shy away from "dry" knowledge? Doesn't Bacon consider men "soft and watery," easily irritated by the dry light of theoretical work? Accordingly, one of the pleasures Le Doeuff finds in Bacon's work is his refusal to flatter (male) reason. She sees him as frustrating the "pride or narcissism of men of science."[22]

Where many feminists have highlighted in Bacon his emphasis on controlling and dominating a feminized nature,[23] she isolates from his recommendations other values with which feminists might sympathize. Be humble, don't tyrannize nature, and submit oneself to what one studies. Do not trust one's conclusions: always go back and ask nature anew. Le Doeuff's Bacon rejects the view that there are significant differences between the abilities of human minds. What matters is the quality of one's methodology. Bacon deems the identity of the agent or subject of knowledge irrelevant. This refrain is also seen in a different kind of leveling of the human agent. The agent, he points out, can be animal as well as human. Nature is not just the passive object, but an active producer as seen when we consider the parallel between humans producing sugar, and bees producing honey. Le Doeuff suggests that one must be particularly motivated – too motivated – to lift out of Bacon's texts only the depiction of feminine nature and of women as excluded from the house of knowledge. This is not borne out by alternative refrains, and feminists should have the most interest in any such refrains.[24]

For her part, Stengers resists what she takes to be a simplistic opposition between approaches to science considered more masculine, or more feminine. She expresses strong reservations about the idea of a science practiced *au féminin*. When she considers the feminist affirmation of sexual difference, whom does she have in mind? A more feminine approach is sometimes associated – famously, by Fox Keller – with genetic specialist Barbara McClintock's intuition and empathy for chromosomal transposition in maize genes. McClintock emphasized about her methodology her attention to every exception, her refusal to consider exceptions aberrant or a contaminant, her interest in the specificity of every individual cell in her studies, her respect for difference, her pleasure in, affection and love for her object.[25] Stengers dryly asks if we should see in this intuition an example of nondominating, holistic science, "capable of offering an alternative to the reductionist violence that some feminine discourses have identified with 'male science'."[26] McClintock suggested that one should not force maize to divulge its secrets, but instead listen to it with the appropriate respect. Fox Keller describes her interviews with McClintock:

> Precisely because the complexity of nature exceeds our own imaginative possibilities, it becomes essential to "let the experiment tell you what to do." [McClintock's] major criticism of contemporary research is based on what she sees as inadequate humility. She feels that "much of the work done is done because one wants to impose an answer on it – they have the answer ready, and they [know what] they want the material to tell them, so anything it doesn't tell them they don't really recognize as there, or they think it's a mistake and throw it out ... If you'd only just let the material tell you."[27]

Stengers's concern is that Fox Keller's feminist reading of McClintock reinforces the worst clichés about masculine and feminine values. Instead, shouldn't we use our interest in McClintock (just as Le Doeuff uses her interest in Bacon) to complicate those attributions of femininity and masculinity? This is just one of the examples that can provide the context for locating what is conventionally thought of as femininity at the heart of the masculine, and the reverse. Don't we often think of the best hunter as having empathy for its prey? Many of the characteristics that are sometimes thought of as typically male (the stubborn, relentless pursuit; the desire to penetrate the object's secrets) could also be attributed to McClintock, who is nothing, Stengers suggests, if not a tireless hunter, a brilliant analyst and practitioner of analytic reason.

While Stengers does appreciate in McClintock qualities sometimes associated with femininity, such as the famous feeling for the organism, she gives these a new or modified metaphorics. Stengers describes the McClintock–maize relationship as both a reciprocal and resistant relationship between hunter and prey, where both are subtle, wily participants in the chase, differently responsive to each other.[28] This depiction of the wiliness of the participants also emphasizes the positive function of McClintock's independence and marginality. The most innovative hunter is the lone hunter, in Stengers's imaginary.

This depiction allows Stengers to question scientific collusion in place of questioning rationality, and Le Doeuff proposes the same substitution. Where McClintock is a lone and marginalized hunter, most scientists, argues Stengers, hunt in packs. Perhaps the problem is not the masculine or reductive character of scientific reason, but rationality practiced according to a pack mentality.[29] We should question the latter, not the former – indeed, there is nothing especially rational about the pack mentality given its basis in the domain of identification, the imaginary, and disavowed affect.

For Le Doeuff, the issue is not whether certain sciences were, historically, "founded by men alone," but this same question of the pack mentality. What should preoccupy us, she suggests, is whether sciences are "managed by men who refuse entry to anyone but men," by those who wish to remain in an enclave (which she also describes as the "closed world of masculine scientific sociability").[30] She agrees that the issue is not the sex of the scientist, but the pact-like modes of discussion fostered by a professional milieu. Minimal consensus is one thing. But excess attachment to consensus destructively takes the place of a healthy questioning of consensus, and inclusion becomes exclusion:

> To illustrate the problem briefly: in order to think and know with others, we must accept certain minimal assumptions, points such as 2 and 2 makes 4 or, most important of all, ways of discussing methods, which normally claim to be accessible to everyone. But when such minimal understandings evolve into rigid pacts, explicit in their exclusions, implicit in their effects, we must wonder what fictions knowledge is instituting.[31]

Le Doeuff acknowledges the presence of male enclaves in scientific contexts. Even so, she questions the association of reason, masculinity, and science. We should not allow the phenomenon of the male enclave to reinforce science's association with masculinity. Rather, we should use it to highlight the imaginary and affective nature of this association, and the fictional direction towards which it propels knowledge. Stengers asks how reason might be reconceptualized if we rethink the pack mentality as an irrationality of the sciences. It is not, of course, that Stengers is committed to condemnations on the grounds of irrationality, but that she wants to reinstall the distanced other at the heart of the privileged same. Locating irrationality at the heart of what we think of as rationality, or the associatively feminine at the heart of the associatively masculine, is co-extensive with Stengers's interest in scientists who manifest values departing from the conventional imaginary about the male scientist. Among her examples is the 1940s embryologist Albert Dalcq's depiction of deduction:

> In experimental biology ... deduction often demands a kind of art, where feeling may have a part to play ... The very object with which the embryologist works can respond, and our research could happily take on the aura of a conversation: the riposte may be as unexpected and as charming as that of an intelligent interlocutor.[32]

Le Doeuff and Stengers both favor a history of science that can be told in such a way as to thwart its depiction as a masculine domain of mastery over nature and the feminine. They also share a conviction that the leading figures in feminist epistemology have not been committed to bringing to the surface currents of intuition, empathy, passion, imagination, self-doubt, amusing conversation with nature, and the frustration of masculine pride, and have erred in this respect.

Offering a positive account of the best feminist stance to take to science, Stengers and Le Doeuff are haunted by a certain image of the leading figures in feminist

epistemology. Many female voices among contemporary French philosophy and history of science echo each other in their characterization of feminist epistemology,
deemed Anglo-American. This is seen in the comments of Stengers, Le Doeuff,
Françoise Balibar, and the contributors to a 1992 anthology of essays on "Le Sexe des
sciences," including its editor, Françoise Collin. There is a perception from some
French feminist writers that France has lagged behind the feminist epistemology
movement in America. According to Françoise Balibar, "the question of women and
science is more advanced in England and America than in France."[33] Not all would
agree, but Balibar wonders if this might be due to a more explicit feminist engagement by many women working in scientific contexts in America, professionals who
have overtly combated institutional sexism. Women scientists working in France, at
least since the 1960s, have associated themselves less with the feminist movement and,
suggests Balibar, have favored a quiet discretion. But despite these writers' perception
that an important intellectual movement is absent from France and present in England
and America, they also describe Anglo-American feminist epistemology as having
taken a fatal turn. They depict it in terms of this menacing reinforcement of the
convention that science is a masculine domain. According to the most extreme version, women are alienated by science and should avoid it. So much the better for
women if they have not been able to enter science at all levels in equal numbers.
Doing so involves adopting traditionally male values, standards, and qualities one
would do better to reinvent. One should either stay out of science or, in the words
of Isabelle Stengers, as she characterizes this position (from which, evidently, she
distances herself), one should practice "another science," somehow generated from a
radically exterior standpoint.[34]

These French depictions of Anglo-American feminist epistemology take a common
group of texts as the point of reference. The earliest work of Evelyn Fox Keller and
Sandra Harding is widely cited. Caroline Merchant's *The Death of Nature* (1980),
Susan Bordo's interpretations of Descartes, and feminist denunciations of Bacon are
ubiquitous representatives of concepts of masculinity and femininity deployed in
feminist epistemology that the authors find particularly reckless.[35] A position identified by Le Doeuff as differentialist is considered quintessentially Anglo-American
when it is associated with feminist epistemology. French feminists do not associate
themselves with feminist epistemology, deeming it an over-the-Atlantic (or Channel)
phenomenon.

Yet Luce Irigaray's work has been haunted by its own ghost, a specter of science,
throughout her work. In the 1970s, Irigaray began her career with empirical work in
the field of linguistics. Her investigation of the speech patterns of those suffering
dementia was published as *Le langage des déments* (1973), a scientific study document-
ing her extensive testing in this area.[36] In smaller studies of this period, Irigaray also
tested the speech patterns of schizophrenics. This work of Irigaray is not discussed by
such colleagues as Stengers and Le Doeuff, despite her relevance to the movement Le
Doeuff terms differentialism. The omission of Irigaray's name from those considered
important to questions of feminist epistemology might be due to a desire to distance
feminist epistemology as American, and perhaps is also due to Irigaray's early abandonment of the first formulation of her empirical work. Yet her exclusion from
French reflections on the possible hazards of feminist epistemology is striking because

at least one of her essays, "In Science, Is the Subject Sexed?," has been included in Anglo-American anthologies on feminist approaches to science, and because she is one of the most innovative contributors to this field. Most important, her work disrupts the opposition set up between science on the one hand and feminist critique on the other.

Irigaray's earliest empirical work was in several respects determining for her subsequent career. The methodology of *Le langage des déments* contains a type of omission that Irigaray would later deem a conventional and telling blind spot: the failure to consider the distinction between male and female participants of potential significance for her findings. Despite the reporting of all test responses in appendices, the presentation of *Le langage des déments* makes it difficult to distinguish the responses of men and women because they are identified throughout the work by abbreviated first names, only translated into the full first name on one page. At this stage of her career, Irigaray did not ask whether significant differences might be found between the speech patterns of men and women suffering dementia. She later used the term "sexual indifference" to describe the failure to suppose sexual difference from the outset as potentially significant.

In her subsequent work,[37] the phenomenon she terms "sexual indifference" provides the basis for a new definition Irigaray gives to the concept of sexual difference. Insofar as it is introduced as an Irigarayan concept, sexual difference is not an empirically demonstrable fact. She does not present a case for considering women and men as importantly different. For example, she does not make a case for considering women as more inclined to intuition, empathy, or emotion. Instead, she considers oppositions such as reason versus intuition, empathy, or emotion as indications of the absence in our culture of sexual difference. There is no sexual difference in our culture worthy of the name, for as long as we think of the sexes in terms of such oppositions. Irigarayan sexual difference is therefore defined as an excluded possibility. Fragments of culture – patterns of speech, image, and representation devoted to such oppositions as reason versus emotion – will be analyzed as symptoms of the absence of sexual difference.

To return to the methodology employed in *Le langage des déments*, the aptness of the question, "Might it be relevant to group and distinguish the responses of women and men suffering dementia," would not be demonstrated through the proof that doing so does reveal significant differences. It might be that significant differences in the speech of men and women suffering from dementia could be established by a return to the data. But that is not the route taken by Irigaray. Instead, the aptness of the question concerning possible difference could be established only by the critical analysis of the failure to ask it. The concept of sexual difference is founded on Irigaray's demonstration that it is a regularly excluded question or excluded social possibility. It might be said, then, that Irigaray introduces a fiction into science. She rhetorically asserts sexual difference as forgotten by science. But (by the terms of her own analysis) there is no sexual difference. It is the reiterated forgetting that it is an alternative possibility which gives content to the suggestion that something can be conceived as forgotten.

After the publication of her initial work in the area of linguistics, Irigaray turned to the publication of a series of works in the history of ideas, focusing on psychoanalysis and the history of philosophy from Plato to Sartre, including *Speculum of the Other Woman*, *The Ethics of Sexual Difference*, and *Marine Lover*. These works take a diagnostic approach to the representations of women and femininity they reveal. Irigaray argues that women and the feminine are repeatedly represented either as the opposite of, the complement to, or the same as the masculine. Again, she does not make empirical reference to women to demonstrate the aptness or otherwise of these historical representations. Instead, she identifies what she takes to be their parameters of exclusion: women are only represented insofar as they may be the same as, the opposite of, or a complement to the male. Methodologically, Irigaray uses this pattern to build up a hypothetical concept of women. She asks her readers to try to imagine ways in which women might be represented, or might be conceived, or might understand their lives, or indeed might speak, in ways that did not fit the reiterated patterns she presents.

Given her original work with the analysis of speech patterns seen in dementia and her training in quantificative methodology, one of the most curious aspects of the second phase of her career must be seen in her apparent decision not to take the route of empirical study of women and men to justify her concept of sexual difference. Her work can be usefully compared to that of Carol Gilligan in this respect.[38] Gilligan draws on the same object–relations psychology which informs the early work of Evelyn Fox Keller and Susan Bordo, to the dismay of figures such as Le Doeuff and Balibar. According to Nancy Chodorow, men and women may be understood in terms of their different relations to separation from the mother in childhood development, the boy seeking to demarcate himself radically from her (leading to the subsequent preference for dichotomous opposition, strong separations between subject and other, the preference for impartiality, objectivity, the dislike of blurred boundaries). Girls are said to have a more interconnected and identificatory relationship with their mothers, to suffer separation anxiety, and to be prone to relationships to the world which favor the intermingling of boundaries and empathy, rather than the tendency to maintain separation and distance.[39] This material has been extremely influential amongst some feminist writers. For example, it has been used by Susan Bordo to argue that Cartesianism might be considered masculine, by Evelyn Fox Keller to argue that some of the fundamental values of science (impartiality, neutrality, objective distance) might be considered masculine, and by Carol Gilligan in her discussions of why, in her empirical studies, men and women seemed to favor different approaches to moral reasoning: the women prone to empathy and identification, and the men favoring the establishment of abstract, general, and universalizable principles.

Though it is not uncommon to consider both Gilligan and Irigaray differentialists, it is a mistake to align them. Gilligan establishes differences among men and women. Irigaray's work is based on an analysis of contexts in which the possibility of differences, irrespective of whether they exist, are not taken into consideration. While the most obvious next step for Irigaray might have been empirically based demonstration of overlooked differences between men and women (for example, with reference to their psychology, speech patterns, reasoning, or behavior), this is not the path immediately taken by her. Gilligan's and Irigaray's very concepts of difference are different. Where Gilligan establishes differences between existing subjects, Irigaray establishes a pattern of exclusion of alternative possibilities for difference we could only hypothesize. Irigaray's concept of difference, it might be said, is an absence: Gilligan's is a presence. Finally, while Irigaray might not contest Gilligan's data that men and women manifest different approaches to moral reasoning, she would interpret the structure of difference revealed by Gilligan as an absence of sexual difference by her own reconceptualization of this term.

This gives some background to the questions asked by Irigaray in "In Science, Is the Subject Sexed?," in which she calls for a reflection on excluded categories and concepts. She refers several times in the essay to the absence and silence of women. She does not mean that women are literally silent. In a given situation they may speak as much as men. She conceptualizes the absence or "silence" of women (by her own definition) from discourse in a different sense, in women's status as opposite to, complementary to, or the same as the male. And just as she tries to interpret the exclusion of alternative possibilities for women, "In Science, Is the Subject Sexed?" tries to articulate questions not favored by science, although according to a much less structured template. The similarity is that she does not propose questions whose scientific or social pertinence is defended — except with the claim that they are not typically asked. In a manifestly fragmented ensemble, she claims that economics tends more to interrogate rarity and the possibility of survival than abundance and the possibilities for life; that linguistics tends not to suppose important differences between the linguistic habits of men and women; that psychoanalysis tends to attribute the same kinds of drives (eros, thanatos) to men and women, and that the biological sciences are not preoccupied with the structure of the placenta or the mother/child relationship. Also, mathematics is said to prefer well-demarcated categories to permeability, logic to prefer binary oppositions over the tri- or polyvalent, and physics to avoid reliance on perception. Less favored areas of scientific inquiry, as proposed, are: male contraception; the possibility of reproduction by ova without sperm; respects in which girls might have a more advanced development compared to boys; the role of "intuition" in scientific methodology; the appeal of symbolic notation; or of binary opposition, permeability of categories, and of the principle of noncontradiction. In this respect, Irigaray suggests that scientific practitioners prefer not to interrogate their own imaginary, including the very preference for categories, symbols, stark oppositions, and principles.

Evelyn Fox Keller suggests three ways of thinking about the claims of sex bias in science. It is seen in the absence of an equal number of women in science, in the way that underrepresentation "skews" the choices of problems, design of experiments, and interpretation of data; and in the very goal and value of objectivity.[40] Of course it is easy to riposte that Boyle's law is experimentally replicable and in no way subject to the sex of the inquirer. But, in the words of Fox Keller,

> it is crucial to recognize that it is a statement about a particular set of phenomena, prescribed to meet particular interests, and described in accordance with certain agreed upon criteria of both reliability and utility. Judgments about which phenomena are worth studying, which kinds of data are significant — as well as which descriptions (or theories) of those phenomena are most adequate, satisfying, useful, and even reliable — depend critically on the social, linguistic, and scientific practices of those making the judgments in question ... the successes of Boyle's law must be recognized as circumscribed and hence limited by the context in which it arises.[41]

Irigaray clearly has this kind of critique in mind when she comments – as do many Anglo-American feminist epistemologists, Balibar and others – on the room for improvement in investigation into matters of contraception, the placenta relationship, the possibilities of spermless reproduction. She asks her readers to reflect more flexibly on the kinds of questions science does not take to be worth studying, and to see in the areas which seem to us most preposterous and least plausible the expression of a cultural imaginary, rather than the neutrality or objectivity of the practitioners. And, in a whimsical tone, Irigaray gives her questions about scientific interests the air of a woman's magazine, as if to suggest that this tone of whimsy is appropriate anywhere but the domains of scientific investigation, and to ask why that should be. Perhaps some women's imaginations are fulfilled by speculation on the possibilities of anything from reliable male contraception to the possibilities of spermless reproduction. Her suggestion is that the male imagination is no less implicated in the aversion of these questions. Perhaps readers can also see the potential for satisfaction of at least some men's imaginations in either greater control of reproduction, or no involvement in reproduction. Irigaray moves too hastily in her references to what might satisfy the imagination. But she shares with all those involved in, and reflecting on, feminist epistemology the desire to locate desire, partiality, affect, and the imaginary at the heart of scientific practice, and shares the resistance to its depiction as neutral, through the interrogation of "what science sets, or does not set, as goals for its research."[42]

According to an Irigarayan perspective, we must add to the criteria for good science whether one frames one's investigations in terms of possible imaginings of alternative prospects for how to understand the human. Her project is less to demonstrate how men and women are, than to pinpoint exclusions of the possibility that they are otherwise. Introduced by Irigaray as an absence, the value of this "otherwise" is largely in prompting the imagination: we are prompted to speculate on the possibility of new ways of thinking about masculinity and femininity. But Irigaray argues that the imagination already operates no less in the domains into which she wishes to insert her speculations, insofar as certain possibilities are conceptualized and others never thought of.

So, in "In Science, Is the Subject Sexed?," Irigaray notes that the possibility of sexual difference is not expressed in logic: one doesn't contemplate the possibility of "male" and "female" logic. She does not argue that there is a "male" and a "female" logic, nor indeed that there should be. Rather, she offers a diagnosis, suggesting that we understand a great deal about our culture and its blind spots by reflecting on the extent to which we exclude the possibility of sex-relative logic as one of the most ridiculous prospects. Similarly, Irigaray asks whether science is able to imagine differences other than quantitative, whether it is able to imagine permeability and fluidity of categories. She does not make a case for permeable and fluid categories in relation to some particular empirical object. No more does she argue that women favor permeable and fluid categories in their approach to science. Even if such a case could be made (for example, if it could somehow be demonstrated that women scientists consistently favored the values discussed by McClintock, for example), she would take the phenomenon as an indication of an impoverishment of sexual difference in the culture in which it was manifest, rather than a demonstration of the fact of sexual difference.

Though Irigaray has sometimes been grouped amongst those feminists who consider that knowledge is masculine, it is evident in "In Science, Is the Subject Sexed?" that she does not evoke the ideal of a radically other science for women. Instead, she tries to reflect on those questions in which it seems we are least interested. She works to pinpoint what we want least to think. The point is not to argue that men and women reason or know differently, but to identify contexts most based on the conceptual sacrifice of the very possibility that they might or should.

Rejecting the possibility of male and female science, Isabelle Stengers attributes to little girls and boys alike the naughty desire to break a vase into a thousand pieces with a single whack,[43] a desire connected by her to curiosity about how things are made, what will happen, a desire to know and effect change. If girls ever lose this desire, we must blame the conventions which suppress or reroute it. Her comments in this regard help us to clarify the nature of Irigaray's concept of sexual difference, in the context of science, philosophy, or language. A desire to break a vase, shared by boys but not shared by girls, is not for Irigaray an example of sexual difference, but of its absence. Any such differentiation, even if true, would indicate that girls live an existence understood in terms of the absence or sameness of qualities or characteristics deemed male. Stengers rhetorically pits against the specter of a revalued sexual difference the claim that we are born to a sameness of drives: such as the childish desire to exercise omnipotence, to break, to see what will happen. The problem, she writes, is that girls learn to be ashamed and to suppress such desires. Boys give them free expression (and transform them into valorized activities). Irigaray would be less inclined towards these claims – whether speculative or empirical – about such possible differences, except to note the apparent alacrity with which Stengers, as she turns to the depiction of men, women, and matters of feminism, gives one kind of highly oppositional articulation to the possibility of difference (shame versus free expression), and then confronts it with what appears to be a modeling of sameness (a primordial drive to break, undo, transform and inquire).

For her part, Michèle Le Doeuff worries that figures like Marie Curie (we surmise she might be grouped with the vase-breakers) are simply neglected in the feminist critique of epistemology. Feminists interested to show that women somehow know differently to men, deemphasize those women who do not support this thesis, she claims:

> For instance, in Lorraine Code, *What Can She Know? Feminist Theory and the Construction of Knowledge* (Ithaca: Cornell University Press, 1991), there is a vaguely defined norm. Since "the sex of the knower is epistemologically significant," women may know differently from men. The simple result is that any intellectual woman whose work doesn't demonstrate a difference disappears. There is no mention of Marie or Irène Curie, in spite of their importance to women physicists as exemplary figures and to everyone because of their scientific achievements.[44]

The evocation of Marie Curie by Le Doeuff and McClintock by Stengers suggests a desire to refer feminist reflections on science back to scientific credentials: Curie and McClintock were the first two women to win the Nobel Prize unshared. Barbara McClintock's work in genetic mutation in maize arose from a research methodology she described in terms of having a "feeling for the organism." According to Evelyn Fox Keller, this methodology is "marked by difference at every turn"[45] (according to Fox Keller's definition of difference). Intuition and bonding with one's research object are unlikely to be valued in research contexts favoring objectivity and impartiality. Perhaps some kinds of data are less likely to be perceived under such circumstances (as borne out by the long period during which both McClintock and her research were discredited). But it is the acclaimed scientific validity of the results that generates greater sympathy with McClintock's methodology and values. Should her feeling for the organism not have produced Nobel Prize-winning research on chromosomal mutation, would such material be cited in the name of "another" approach to science? Certainly, its having pretended to the status of science would be unlikely to give it a comfortable seat in the house of fiction.

The apparent lack of true mutual impact in the domains of literature, wit, and science poses problems for Luce Irigaray's "In Science, Is the Subject Sexed?" As a feminist assessment of science, the pertinent question is whether it is to be classed with science, fiction, or humor. Her work departs from the categorization which pits feminist epistemologists on the one hand, such as Harding and Fox Keller, against women of science – long neglected, eventually honored – Marie Curie and Barbara McClintock, on the other. Rather than being easily deemed scientific or unscientific, her work occupies a more successfully uneasy place at the boundaries of science, fiction, humor, and the creation of concepts. It poses a question relevant to the refiguring of science as the domain of the potential creation of concepts. Could we build into empirical investigation an attention to concepts of the human excluded by the parameters of the investigation? Imagining a conversation between them, Irigaray would ask Carol Gilligan for a more imaginative reflection on her findings, with a view to asking what concepts of sexual difference are excluded when Gilligan finds that men favor abstract principles, and women intuition and empathy. For this reason, it is in relation to a hypothetical concept of sexual difference – a constitutively excluded, impossible concept – that Irigaray claims in "In Science, Is the Subject Sexed?" that women are "held back in a potential language, [constituting] a reservoir of energy."[46] Even when she makes claims about men's and women's empirical differences, such claims tend to be accompanied by the question of how we can imagine them otherwise, and a refiguring of the sexes as a reservoir of potential for such alternative possibilities.

As a series of concrete, methodological propositions for those undertaking empirical work, this is one of the least-developed aspects in Irigaray's corpus. But her intentions are clear: she encourages science's maximal attention to its own imaginative blind spots, so as to encourage its serving, as much as possible, as the locus of new metaphors for the human. In her own work, Irigaray eventually embodied this proposal for a closer relationship between science and poetic invention when she finally returned to her starting discipline – empirical studies of language use – this time addressing the issue of sex difference in language use in volumes (published throughout the 1990s) such as *Sexes et genres à travers les langues* (1990) and *I Love to You* (1996).[47]

It will be recalled that her earliest work, *Le langage des déments*, failed to consider the possibility that language use among men and women suffering dementia might be significantly different. But her concern, as her career developed, was that asking such questions (though important) forces our attention only to what is, rather than to what is not. Her interest is not to defend the view that men and women use language differently. Rather, if they do, she is interested in the subjective possibilities excluded for men and women by such variations. What is revealed about what is not, rather than what is?

What effect does her interest in this question have on her own methodology? Her revised approach to her empirical work has involved negotiation with research teams studying the linguistic habits of Anglophones, French, and Italians. Notice how according to her description in the introduction to *To Speak is Never Neutral* of her loss of confidence in her earlier pretension to scientificity, her irritation and mirth about those pretensions[48] did not eventually lead her to a project which parodies empirical method. Rather, her revised approach to linguistic analysis adopts anew the methodology of repeatable testing and quantitative study.

On the basis of extensive tests in which subjects were usually given three or four word elements to construct into short phrases, Irigaray argues that "men are more apt to be the subjects of their own speech."[49] This is not to say that women do not construct sentences based around a reference to "I." But men and women do not use the pronouns "I," "you," "he" and "she," and "they" in the same way. Women tend to construct more direct sentences of the form "I am cuddling you." Men tend to place "he" in the subject position[50] and transform the "I" into he. Nonetheless, they can be seen as remaining "in the logic of their I"[51] as opposed to women, who construct sentences whose subject less often coincides with themselves. If they refer to "she," the "she" (as it is for sentences constructed by men) is more likely to be placed as the object, and often as the object of a male subject ("She is putting on a dress to go see him tonight"[52]). Negative expressions are more frequent with men, interrogative with women.[53] Rare selection of the pronoun "elles" (referring in French to women amongst themselves) suggests it is the least common pronoun element. Similarly, women least commonly construct sentences in which a female subject is posited as having another female interlocutor. Male speech elements easily suppose a relation between men. (In French this means that "ils" is not uncommon, but "elles" is.)

However, in tandem with such claims, Irigaray stresses the sexual difference she considers absent from, rather than present in, such speech patterns. She argues that this "absence of women" is seen in the reiteration by both women and men of such speech patterns as the preferential use by both men and women of "il" over "elle" (as subject) and of "ils" over "elles," the avoidance of the feminine subject as subject of a grammatical sentence, the avoidance of reference to feminine pronouns altogether, or the preference for reference to the feminine as object rather than subject.[54] This analysis partly takes on the guise of science (references to numbers of respondents, percentages of different responses, control groups, and cue words).[55] But her work is premised on an initial hypothesis concerning the significance of the category whose absence cannot be empirically demonstrated, except negatively.

When Irigaray's tests suggest to her that

Women seek communication, especially dialogue, but they particularly address themselves to lui/il (him) [and that] men are interested in the concrete object if it is theirs [my car, my watch, my pipe, etc.] or in the abstract object insofar as it is proper to a man or sanctioned by the already existing community of men ... they rarely seek dialogue,[56]

one is likely to conclude – or to take her to conclude – that such sex-specific variations demonstrate the fact of sexual difference; that men and women are different. Women, she seems to conclude, seek communication, men are more interested in concrete and abstract objects.

This is the kind of differentialism Le Doeuff, Balibar, and Stengers resist strenuously. But the inventive work specific to Irigaray is overlooked if she is reduced to those who argue that, in some respect or other, men and women are different. The "other science" resisted by Stengers deploys such oppositions as rational versus emotional, neutral versus empathizing, autonomous versus communicative, oppositions abhorred in the work of such figures as Chodorow, Bordo, and Merchant. By contrast, and underscoring the specificity of Irigaray's place in this respect, is that for her, variations between the sexes (such as preference for abstraction versus empathy, and those linguistic variations revealed by her own work) are taken to indicate not the presence of sexual difference, but its absence (an absence located in the dominance of the model: same, opposite, or complement). Women and men, according to her argument, never have come to be otherwise different. We have not yet created the cultural conditions necessary for provoking and promoting difference. We are bogged down in such sexed oppositions as seeking communication, or, failing to do so, favoring rationality versus empathy, object versus subject, and so on. For Stengers, asking whether the subject of science is sexed would be the wrong question, because we should attend to the ways in which practices of science disturb conventions about sexed subjects, rather than the contrary. But for Irigaray, it is the wrong question for another reason. We learn nothing about sexual difference from differentials such as objectivity versus empathy. We learn only of its absence.

Notes

1   Luce Irigaray, "In Science, Is the Subject of Science Sexed?," in *To Speak Is Never Neutral*, trans. Gail Schwab (London: Athlone, 2002), p. 248.

2   Stengers was awarded the Grand Prix de Philosophie from the Académie Française in 1993 and is author of the following works translated into English: with Ilya Prigogine, *Order out of Chaos: Man's New Dialogue with Nature* (Boulder, CO : New Science Library, 1984); with Léon Chertok, *A Critique of Psychoanalytic Reason: Hypnosis as a Scientific Problem from Lavoisier to Lacan*, trans. Martha Noel Evans (Stanford, CA: Stanford University Press, 1992); with Bernadette Bensaude-Vincent Bensaude, *A History of Chemistry*, trans. Deborah van Dam (Cambridge, MA: Harvard University Press, 1996); in addition solely authored works such as *Power and Invention: Situating Science*, trans. Paul Bains (Minneapolis: University of Minnesota Press, 1997) and *The Invention of Modern Science*, trans. Daniel W. Smith (Minneapolis: University of Minnesota Press, 2000).

3   Isabelle Stengers, " Another Look: Relearning to Laugh," *Hypatia: A Journal of Feminist Philosophy* 15(4): 50.

4   Ibid.

5   See, for example, Bruno Latour, *Science in Action: How to Follow Scientists and Engineers Through Society* (Cambridge, MA: Harvard University Press, 1987) and Bruno Latour, *We Have Never Been Modern*, trans. Catherine Porter (Cambridge, MA: Harvard University Press, 1993).

6   Stengers, *The Invention of Modern Science*, p. 3.

7   Stengers, " Another Look: Relearning to Laugh," p. 41.

8   Elizabeth Wilson, "Scientific Interest: Introduction to Isabelle Stengers 'Another Look: Relearning to Laugh,'" *Hypatia: A Journal of Feminist Philosophy* 15(4): p. 38.

9   Stengers, "Another Look: Relearning to Laugh," p. 48.

10 Ibid., pp. 42–3.

11 Ibid., pp. 43–4.

12 Stengers, *The Invention of Modern Science*, p. 153.

13 One possible exception is her fairly favorable response to Evelyn Fox Keller's book on Barbara McClintock, discussed below.

14 Donna Haraway being the most obvious example.

15 Stengers, *The Invention of Modern Science*, p. 11.

16 The subtitle of Stengers's 2002 book on Whitehead – see Isabelle Stengers, *Penser avec Whitehead: Une libre et sauvage création de concepts* (Paris: Seuil, 2002).

17 Michèle Le Doeuff, *The Sex of Knowing*, trans. Kathryn Hamer and Lorraine Code (New York and London: Routledge, 2003).

18 The real issue, she argues, is the status and role of qualities associated with women in different contexts. She argues that no set of qualities have been systematically attributed to women. She gives several examples. It is often thought that intuition is associated with femininity, but in the Cartesian tradition intuition is a "primordial element and indispensable anchoring point." Le Doeuff cites her colleague Andrée Michel on this point: "rather than reasoning in qualitative terms, tying the supposedly feminine quality of a particular task to the female essence of the person performing it or the masculine quality of another task to the maleness of its performer, we should focus on the *variable* social value of tasks." Another example of the malleability of qualities associated with masculinity and femininity is to be seen in Schopenhauer's devaluation of reason. Although Schopenhauer associates women with reason, his *The World as Will and Representation* devalues reason as compared to intuition. In a further example, Le Doeuff points out that prior to the twentieth century, women were systematically associated with sex, temptation, and desire. Then in the twentieth century desire is revalued by certain French psychoanalysts and philosophers. Suddenly, women are associated with frigidity and impoverished *jouissance* (Le Doeuff, *The Sex of Knowing*, pp. 8, 12, 16, 17).

19 Le Doeuff, *The Sex of Knowing*, p. 155.

20 Ibid., pp. 167, 164.

21 Ibid., p. 151.

22 Ibid., p. 152.

23 Sometimes, argues Le Doeuff, through reliance on partial readings and poor translations – this accusation is thoroughly discussed in *The Sex of Knowing* by reference to Evelyn Fox Keller. Some good – and varied – examples of feminist interpretations of Bacon include Evelyn Fox Keller, "Feminism and Science," pp. 28–40, p. 36, Genevieve Lloyd, "Reason, Science and the Domination of Matter," p. 46, and Mary Tiles, "A Science of Mars of or Venus," p. 227, all in *Feminism and Science*, eds. Evelyn Fox Keller and Helen E. Longino (Oxford: Oxford University Press, 1996).

24 Le Doeuff, *The Sex of Knowing*, pp. 152, 154, 155.

25 In an interview with Evelyn Fox Keller, McClintock comments, "I actually felt as if I was right down there, and these were my friends ... As you look at these things, they

become part of you. And you forget yourself." Evelyn Fox Keller, *Reflections on Gender and Science* (New Haven and London: Yale University Press, 1985), p. 165, and see Evelyn Fox Keller, *A Feeling for the Organism* (New York: W. F. Freeman, 1983).

26   Isabelle Stengers, "Avant-propos," in Evelyn Fox Keller, *L'intuition du vivant: La vie et l'oeuvre de Barbara McClintock*, trans. Rose-Marie Vassallo-Villaneau (Paris: Tierce, 1988), p. 9.

27   Fox Keller, *Reflections on Gender and Science*, p. 162.

28   Stengers, "Avant-propos," p. 15.

29   Ibid.

30   Le Doeuff, *The Sex of Knowing*, p. 162.

31   Ibid.

32   Cited in Stengers, "Avant-propos," p. 8.

33   Françoise Balibar, "Y a-t-il une science feminine?," in *Le sexe des sciences: Les femmes en plus*, ed. Françoise Collin (Paris: Autrement [Série Sciences en société], 1992), pp. 166–7.

34   Stengers, *The Invention of Modern Science*, p. 11.

35   In addition to Evelyn Fox Keller's *Reflections on Gender and Science*, the works most discussed are Sandra Harding, *The Science Question in Feminism* (New York: Cornell University Press, 1986); Carolyn Merchant, *The Death of Nature: Women, Ecology, and the Scientific Revolution* (New York: Harper and Row, 1980); and Susan Bordo, *The Flight to Objectivity: Essays on Cartesianism and Culture* (Albany: State University of New York Press, 1987).

36   Luce Irigaray, *Le langage des déments* (The Hague: Editions de mouton, 1973).

37   In English as Luce Irigaray, *This Sex Which Is Not One*, trans. Catherine Porter (Ithaca, NY: Cornell University Press, 1985) and Luce Irigaray, *Speculum of the Other Woman*, trans. Gillian C. Gill (Ithaca, NY: Cornell University Press, 1985).

38   Carol Gilligan, *In A Different Voice: Psychological Theory and Women's Development* (Cambridge, MA: Harvard University Press, 1982).

39   See Nancy Chodorow, "Gender, Relation, and Difference in Psychoanalytic Perspective," in H. Eisenstein, ed., *The Future of Difference* (Boston: G. K. Hall & Co., 1980).

40   Fox Keller, *Reflections on Gender and Science*, p. 177.

41   Ibid., p. 11.

42   Irigaray, "In Science, Is the Subject Sexed?," p. 252.

43   Stengers, *The Invention of Modern Science*, p. 10.

44   Le Doeuff, *The Sex of Knowing*, p. 221n.

45   Fox Keller, *Reflections on Gender and Science*, p. 158

46   Irigaray, "In Science, Is the Subject Sexed?," p. 256.

47   Irigaray, *Sexes et genres à travers les langues: Elements de communication sexué* (Paris: Grasset, 1990); *I Love To You: Sketch of a Possible Felicity in History*, trans. Alison Martin (New York and London: Routledge, 1996).

48   Irigaray, *To Speak is Never Neutral*, p. 1.

49   Irigaray, *Why Different? A Culture of Two Subjects*, trans. Camille Collins (New York: Semiotext(e), 2000), p. 48.

50   Irigaray, *I Love to You*, p. 82.

51   Irigaray, *Why Different?*, p. 48.

52   Ibid.

53   Irigaray, *An Ethics of Sexual Difference*, trans. Carolyn Burke and Gillian C. Gill (Ithaca, NY: Cornell University Press, 1993), p. 136.

54   Irigaray, *I Love to You*, pp. 69–78.

55   Ibid., p. 80.

56   Ibid., p. 95.

20

IN SCIENCE, IS THE SUBJECT SEXED?

Luce Irigaray

How does one speak with scientists?[1] What is more, with scientists of different disciplines, each discipline a separate domain, and each system within each domain claiming, at one time or another, to be global? Since, at every moment, every one of these domains is totalized, closed off, how can the various fields be reopened in order to encounter and speak to each other? In what language? Using what type of discourse?

The problem has no evident solution. Each scientific field seems to have its own vision of the world, its own goals, its own experimental protocols, its own techniques, its own syntax. Each appears isolated, cut off from all the others. Can one take a bird's-eye view of all these different horizons in order to locate common ground, viable intersections, possible passages from one to the other? Does one have the right to take an outside point of view? How does one claim this right? Historically, there was God, transcendent to any *episteme*. But if, as Nietzsche said, 'when science is in power, God is dead,' then how can these different worlds be brought together? My hypothesis is that the place for collective questioning is *inside* and not *outside*, subjacent and not simply transcendent, 'underground' as well as 'in the sky,' deeply buried and not relegated to some absolute, unquestionable guarantee.

How can we discover this space for inquiry and make it perceptible? How can we speak of it? In the language of science, there is no *I*, no *you*, no *we*. The subjective is prohibited, except in the more or less secondary sciences, the human sciences, and we cannot seem to decide whether they are indeed sciences, or substitutes for science, or literature, or poetry?. . . . Or even whether are they true or false, able to be proved or disproved, formalizable or always ambiguous because expressed in natural languages, too empirical or too metaphysical, dependent on the axiomatization of the so-called exact sciences or resistant to such formalization, etc.? Old debates and old quarrels, potentially involving reversals of power, rises and falls of imperialism, that are still current.

Luce Irigaray, "In Science, Is the Subject Sexed?" pp. 247–58 from *To Speak Is Never Neutral*, trans. Gail Schwab. New York: Routledge, 2002. English translation © 2002 by Continuum and reproduced by kind permission of Continuum International Publishing Group. Originally published in 1985 as *Parler n'est jamais neutre* by Les Editions de Minuit.

These cycles can repeat themselves indefinitely. However, one could perhaps wonder if, in some subterranean underground, there might not be one common producer making science. But who? Is anyone there?? Can we see them? How do we question them? Not for a long while have I experienced so much difficulty with the idea of speaking in public. Most of the time, I can anticipate to whom I will speak, how to speak, how to argue, make myself understood, plead my cause, even please or displease my audience. This time, I know nothing, because I do not know whom I have before me. Is this the reverse side of scientific imperialism: not knowing to whom one speaks, or how to speak? Anxiety in the face of an absolute power that hovers in the air, in the face of judgment by an imperceptible but ever present authority, in the face of a tribunal without judge, lawyer or defendant! The judicial system is in place nonetheless. There is a truth to which one must submit without appeal, against which one can unintentionally and unknowingly transgress. This high court is in session against your own will. No one is responsible for this terror, or this terrorism. Nevertheless, they are in operation. In this very classroom or conference hall. For me, in any case. If I met individually with each one of you, male or female, it seems to me that I would be able to find a way to say *you, I, we*. But here? In the name of science?

My first question would be: what schiz does science impose on those who practice or convey it in one way or another? What desire is in play when men and women are making science, and what other desire when they are making love or creating love, individually or socially?

What schiz or what rupture: pure science on one side and politics on another, nature and art on a third or as conditions of possibilities, love on a fourth? Are not this schiz and this rupture, which you claim are above scientific imperialism, already *programmed by it* in the separation of the subject from itself and from its desires, as well as in its dispersion into multiple sectors, including those of science, among which encounters become impossible, verifications of responsibility impracticable. What remains is an imperialistic *there is/there are*, or a *one*, that the power-holders, the politicians, take advantage of as opportunities arise. By the time the scientists react, the game is already over: in the name of science? Imperialism without a subject.

*

• So, looking at things a little bit differently, and playing the game of those question-naires that flourish in women's magazines (replacing the crossword puzzles found in gender-neutral daily newspapers, which actually are all too often exclusively male?) let us make an effort:

If I tell you that two ova can engender a new life, does this discovery seem to you possible? Probable? True? Purely genetic? Or related also to the social, economic, cultural, political order? To be within the domain of the exact sciences? Check the appropriate box or boxes. Is this type of discovery going to be encouraged, and funded? Will it be discussed in the media? Yes? No? Why or why not?

Your answer? How do we interpret the answer? Through the importance of sperm in partriarchy, and its link to property and the symbolic? Through the importance of reproduction and its ambiguous correlation to pleasure and desire in sexual difference?

And, while we are dealing with reproduction and its hormonal components:

- Is male contraception hormonally possible? Yes? No? Why or why not? If it is, is this information disseminated, is the practice encouraged?
- Is the left hemisphere of the brain less developed in women than in men? Yes? No? Would this discovery be used to justify the social, cultural, and political inferiority of women? Yes? No? Would this affirmation concern innate or acquired characteristics? Give your own interpretation and your own hypothesis. Explain how you establish a parallel with the inhabitants of certain oriental countries who, as science tells us, share the same anatomical destiny as women. Do the types of mental and physical practices found in these Asian countries signify an unconscious (?) desire on the part of men to become women? Or a resistance to the liberation of women and an appropriation of all values, accompanied by lack of recognition of a symbolic sexed morphology?
- The girl-child, according to a certain number of observations, develops more precociously than the boy-child: she speaks earlier on and her social skills are precocious. Yes? No? Can it be proved? Disproved? Does she employ these early accomplishments to make herself into a desirable object for others? Resulting in regression? True? False? Justify your response.
- What percentage of the world's population is men and what percentage is women? What are the percentages of men and women in positions of political, social, and cultural leadership? Does that seem a foregone conclusion to you, does it correspond to a male or a female *nature*, and to men's and women's desire? Is it innate or acquired?
- Are women *naturally* more limited, more ignorant, more animalistic, or better at language than men are? Are they inept at political, economic, social, or cultural leadership? Innate? Acquired? Verifiable? Unverifiable?
- Is a woman scientist really just a man? A genetic aberration? A monster? A bisexed . individual? A submissive or a non-submissive woman? Or ...?
- Is there or is there not a dominant discourse that claims to be universal and neuter with respect to sexual difference? Do you agree that it should be perpetuated? For a year? Two years? One hundred years? Or forever?
- Who, according to our epistemological tradition, is the keystone of the order of discourse?
- Why has God always been, and why is He still, at least in the west, God the *father*? That is, the strictly masculine pole of sexual difference? Is that the way we designate the sex that is hidden within and beyond all discourse? Or ...?

In fact, what claims to be universal is actually the equivalent of a male idiolect, of a male imaginary, of a sexed world – and not neuter. There is nothing surprising in this, unless one is a passionate defender of idealism. Men have always been the ones to speak and especially to write: in the sciences, in philosophy, in religion, in politics.

However, nothing is said about scientific *intuition*. It is supposedly produced *ex nihilo*. Certain aspects or qualities of this intuition can nevertheless be distinguished. It is always a question of:

- positing *one* world that one confronts, constituting a world before oneself, as separate from oneself;
- imposing a model on the universe in order to appropriate it, an invisible, imperceptible model, projected over it like some piece of clothing. Is that not the same thing as clothing it blindly in one's own identity?
- claiming that one is rigorously exterior to the model, in order to prove that the model is purely and simply *objective*;
- demonstrating that the model is not dependent on the senses, even through it is always prescribed at least through privileging the visual, and through the absence and distancing of a subject who is nonetheless surreptitiously present;
- ensuring independence from the senses through the mediation of instruments, through the intervention of techniques that separate the subject from the object of investigation, and through processes that distance and delegate power to that which intervenes between the observed universe and the observing subject;
- constructing an ideational or ideal model, independent from the physical or psychical existence of the producer, according to ideally elaborated rules of induction and deduction;
- proving the universality of the model, at least for *x* amount of time, and its absolute power to constitute (independently of its producer) a unique and totalized world;
- backing up this universality with experimental protocols about which at least two (identical?) subjects agree;
- proving that the discovery is efficacious, productive, profitable, exploitable (exploitative? of a more or less inanimate nature?), all of which means that it is progress.

The above characteristics exhibit isomorphism with the male sexual imaginary, a fact that is supposed to remain rigorously concealed. 'Our subjective experiences and our feelings or convictions can never justify any statement,' affirms the epistemologist of the sciences.

It should be added that discoveries must be expressed in a formal language, a language that makes sense. And that means:

- expressing oneself in symbols or letters, substitutions for *proper names*, that refer only to intra-theoretical objects, and therefore never to any real persons or real objects. The scientist enters into a world of fiction incomprehensible to all who do not participate in it.

The signs forming terms and predicates are:
  +: or the definition of a new term
  =: which marks a property through equivalence and substitution (belonging to a set or a domain)
  $\in$: signifying belonging to a certain type of objects.

The quantifiers (not qualifiers) are:

><

the universal quantifier
the existential quantifier, subordinated, as its name indicates, to the quantitative.

In the semantics of incomplete entities (Frege), the functional symbols are variables taken from the limit cases of the forms of syntax, and the preponderant role is accorded the symbol of universality or the universal quantifier.

The *connectors* are:
* negation: P or not P
* conjunction: P and Q
* disjunction: P or Q
* implication: P results in Q
* equivalence: P equals Q.

There is therefore no sign:
* for *difference* other than quantitative difference;
* for *reciprocity* (other than within the same property or the same set);
* for *exchange*;
* for *permeability*;
* for *fluidity*.

Syntax is dominated by:
* *identity to*, expressed by properties and quantities;
* *non-contradiction*, or reduction of ambiguity, of ambivalence, or multi-valency;
* *binary oppositions*: nature/reason, subject/object, matter/energy, inertia/movement.

Undoubtedly, formal language is not simply a set of game rules. It serves to define the game so that all the participants play the same way, and so that a decision can be made in case of disagreement over a move. But who are the participants? Is it possible to intuit something outside the language utilized? How could such an intuition be translated for the participants?

*

The non-neutrality of the subject in science is expressed in many ways. It can be extrapolated from what is, or is not, being discovered at any given moment in history, and from what science sets, or does not set, as goals for its research. For example, in relative disorder and disrespect for the hierarchy of the sciences:

* Psychoanalysis is based on the two main principles of thermodynamics underlying the Freudian model of the libido. These two principles appear to be more iso-morphic with male sexuality than with female sexuality. The latter is less subject to alternations of tension and discharge, to conservation of required energy, to the maintenance of states of equilibrium, to functioning as a circuit that is closed and then reopened by saturation, to the reversibility of time, etc. Female sexuality may harmonize better, if we must evoke a scientific model, with what Prigogine calls

'dissipative' structures, which function through exchange with the outside world, which proceed in energy stages, and whose ordering is based not on seeking equilibrium, but on crossing thresholds corresponding to leaving disorder or entropy behind, without discharge.

- Economics (and the social sciences as well?) has emphasized scarcity and survival phenomena rather than those associated with life and abundance.
- Linguistics remains attached to models of the utterance, to synchronic structures of speech, to models of language that every normally constituted subject can intuit. It has not considered the question of the sexualization of discourse, and sometimes even refuses to do so. It accepted out of necessity – that certain terms of the lexicon have been added to the accepted stock, that new figures of style eventually impose themselves, but is unable to imagine that syntax and syntactic–semantic organization could be sexually determined, and neither neuter nor universal nor atemporal.
- Biology is beginning to approach certain issues rather late: for example, the consti-tution of placental tissue, or the permeability of membranes. Are these questions more directly correlated with the female and maternal sexual imaginary?
- Mathematics is interested in set theory, in closed and open spaces, in the infinitely large and the infinitely small. It shows little interest in the question of the partially open, of fluid sets, of analysis of the problem of boundaries, of passages between, of fluctuations taking place between thresholds of defined sets. Even these questions are raised by topology; it emphasizes that which closes back up, rather than that which remains outside circularity.
- Logic is more interested in bivalent theories than in trivalent or multivalent theories that still appear marginal.
- Physics conceives its object of study according to a nature it measures in ever more formalized, ever more abstract, ever more modeled, ways. Its techniques, expressed through increasingly sophisticated axioms, done with matter that does still exist, of course, but that is not perceptible to subjects conducting experiments, at least for the most part. Nature, the target of physics, risks being exploited and disintegrated by the physicist even without his or her knowledge. The Newtonian revolution ushered scientific practice into a universe where sense perception is almost non-existent, and where the matter (however it is predicated) of the universe and of the bodies that constitute it – the stakes and the object of physics itself – may be annihilated. Inside physics itself there are cleavages: quantum theory/field theory, mechanics of solids/dynamics of fluids, for example. In any case, the imperceptibility of the matter that is studied often leads to a paradoxical privileging of solidity in discoveries, and to a lag in, even an abandonment of, analysis of the unfinished in-finite of force fields. Could this be interpreted as a result of the refusal to take into account the dynamics of the researcher–subject?

*

In the face of these observations and questions, are we faced with an alternative: *either* be a scientist *or* be a 'militant'? Or even: continue to be a scientist *and* divide oneself up into different functions, into several different people or characters? Should the truth of science and the truth of life remain separate, at least for the majority of researchers? What kind of science and what kind of life are we dealing with then? The question is all the more pertinent since life in our times is largely dominated by science and its techniques.

What is the origin of this schiz that is both imposed by and inflicted upon scientists? Is it a non-analyzed model of the subject? A 'subjective' revolution that never took place: the splitting of the subject having been programmed by the *episteme* and the power structures put in place by it? Is it that the Copernican revolution has occurred, and that the epistemological subject has so far neither acted upon nor moved beyond it? Has it modified this subject's discourse about the world in such a way that it is even more disappropriating than the language that preceded it? Scientists now claim to be standing *before the world*: naming it, establishing its laws, axiomaticizing it. They manipulate nature, use it, exploit it, but forget that they are also *in* it, that they are still physical, and not simply confronting phenomena whose physical nature they sometimes fail to recognize. Progressing according to an objective method that shelters them from all instability, all moods, all feelings and affective fluctuations, all intuitions not already programmed in the name of science, all interference from their desires, notably sexual ones, that could affect discoveries, they settle down into the systematic – into what can be assimilated to the already dead? Fearing, sterilizing the destabilizations that are, nonetheless, necessary for the coming of a new horizon of discovery.

Inquiry into the subject of science, and its psychic and sexual implication in discourse, and in discoveries and their development, is one of the sites most capable of provoking a re-evaluation of the scientific horizon.

In order to ask oneself if the so-called universal language(s) and discourse(s) (including those of the sciences) are neuter with respect to the sex that produces them, we must pursue research in view of accomplishing two goals: the interpretation of the law-making discourse as subject to an unrecognized sexual dimension of the speaking subject, and the attempt to define the characteristics of what a differently sexed language would be.

In other words: is there, within the logical and syntactico-semantic-mechanisms of accepted discourse, an openness or a degree of liberty that would permit the expression of sexual difference? We must analyze, in order to interpret their position within a sexed logic, the laws (including those that are not explicit) that determine the acceptability of language and of discourse. This work can be pursued from different angles:

- The causal mode that currently dominates discourse considered normal, as well as the conditional, unreal, and restrictive modes, etc., that fix its 'practicable' framework, limiting the liberty of a subject of enunciation who does not necessarily obey certain criteria of normality, may be studied. While these causal and restrictive modes (the two are linked) permit the accumulation of information and a certain type of already coded communication, do they not inhibit intra- discursivity and prevent all possibility of any qualitatively different enunciation?
- The means or conjunctions of co-ordination also participate in the economy of the principle of causality dominating so-called asexual discourse: juxtaposition, including the summation of clauses and subjects (and ... and); alternative (either ... or);

exclusion, including the eventual elimination of the subject of enunciation (neither ... nor); co-ordination proceeding in the direction of the syllogistics regulating discourse (for, therefore, but).

What modes of subordination or co-ordination would authorize the discursive relationship between two sexually different subjects?

- The symmetry (notably right–left) in intersubjective relations and its impact on the production of language may be analyzed. Can the issues of symmetry and asymmetry result in criteria that would be able to determine a qualitative difference between the sexes? Is the 'blind spot in the old dream of symmetry' (cf. *Speculum*) situated in the same place in a relationship between individuals of the same sex, as it is in a relationship between individuals of different sexes? The dream itself, however, dream that may underlie the economy of the speaking subject, seems to be invalidated by cosmic laws in the face of which no observer of nature and language can remain indifferent, any more than a speaker or interlocutor from the outside.
- When women are held back in a potential language, they constitute a reservoir of energy that could be eliminated, or could explode for lack of possible forms of expression. When they represent only the underside or the reverse side (in specular symmetry?) of discourse, they close it off on itself. Forced into a mimetic defense or offense, women risk absorbing the meaning of discourse, by collapsing it through lack of any possible response. They may be intercepting the goal or the intentionality of discourse, and thus accelerating the destructuring process – which could be acceptable if a new language were to ensue. The question that must be asked is whether women's language would fulfill an as yet unrealized potential for meaning, while remaining within the same general discursive economy, or whether what women think and may be able to say would require a mutation of the horizon of language. That would explain the resistance to their entry into the networks of communication, and the even greater resistance to their entry into the spaces – theoretical and scientific – that determine the values and laws of exchange.

*

Certain questions should be asked regarding the access of women to language and discourse.

(1) Why is their potential energy for language always on the vanishing point, never able to get back to the subject of enunciation? Recent research in discourse theory, as well as in physics, may shed light on the site, in darkness until now, of women's lack of access to discursivity. We must come back to a study of temporalization and its relationship to the place from which the subject is either able, or unable, to position itself as producer of language. If the discourse of the hypothetical interlocutor intercepts the word, cutting it off from memory of the past and from anticipation of the future, all that is left for the subject are attempts to get back to that place from which she or he can be heard. We should emphasize in this context the importance of locality in the constructions

of women's language. The circumstances of place largely determine the pro-gramming of 'discourse.'

(2)   Do we not find, in this insistence on the question of place, an attempt to give form to a subject of enunciation, for lack of temporalization in a dynamics of communication? The utterance's potential for reversibility, or lack thereof, notably between speaker and interlocutor, should be approached from this per-spective, as well as its eventual repetition or reproduction. These conditions are absolutely essential for admissible discourse, since the other is placed in the position of a mirror that both inverts the received discourse, and responds to it after this retroaction.

(3)   The problem of the possible, or impossible, mirror in the other, dominates the enigma of the language and silence of women. Whatever the case may be, 'they' do not say nothing, and the fascination felt, by certain practitioners in particular, for what they do say certainly indicates that some kind of deciphering of the production of language is expressed through them.

These issues could also be approached from the following angle:

(1)   Does what we call the *mother tongue* establish a space for a specific production of language by the mother, and for exchange between mother and children? Is socially admissible language not always paternal? Does a fault-line open up at the entry into discourse? A fault-line that ceaselessly threatens discourse with total collapse, with madness, with sclerotic normalization.

(2)   The creation of language – in all forms – by the maternal has been barred since the origin of our culture. The maternal has been allocated to the procreation of children, and has never been a site for the functioning of a productive matrix. From this perspective, it is useful to reinvestigate and reinterpret the Freudian texts – notably *Totem and Taboo* – that define the foundation of the primitive horde as the murder of the father, and the sharing of his body by the sons. Deeper than the murder of the father, at the origin of our culture, can we not decipher (in Greek tragedy, mythology, and even philosophy) an even more archaic matricide? This murder of the mother in her cultural dimension as fecund lover, continues to govern the establishment of the symbolic and social order that is our own. What consequences does this matricide have for the production of language and the programming of discourses, including scientific discourses?

(3)   Since psychoanalytic 'science' is supposed to be the theory of the subject, Freud's hypothesis concerning the constitution of the relation of the subject to discourse calls for reconsideration and reinterpretation. Freud puts forth, as the scene of the introduction of the subject into language, the 'spool game.' The child – a boy, as it happens – tries to master the absence of his mother by using an instrument he throws away and then pulls back, first banishing it, and then bringing it in close to his space, into his space, alternating vowel sounds along with his gestures: o-o-o (far), a-a-a (near).

   This 'game,' the so-called *fort–da* game, complete with its alternating vowels, supposedly marks the entry of the child into the realm of symbolic distancing.

The boy-child (Freud does not provide any hypothesis as to how all of this might happen for a girl) is able to make this transition – while producing sounds, a kind of musical scale – by assimilating his mother to an object attached to a string that allows him to control, or even eliminate, the distance between her and himself. Does the *fort–da* scene still have a significant function in the constitution of the meaning of language? How are the vowels articulated with the consonants?

This scenario, as it is described by Freud, requires the absence of the mother as interlocutor, and the presence of the grandfather as observer and regulator of 'normal' language. What gestures, what other kinds of language, between child and mother, mother and child, are left out of acceptable discourse? Do the systematicity and the madness of so-called admissible discourse not result from this 'outside' of the spoken and the speakable, since a scenario for *exchange between* mother and son, mother and man-subject, has not been put into place in language? We had better make sure that this means of distancing does not become deadly.

(4)   Freud says nothing about the entry of the little girl into language, except that it takes place earlier than for the little boy. He does not describe her first scene of gestural and verbal symbolization, in particular in relation to her mother. On the other hand, he does affirm that the girl will have to leave her mother, turn away from her, in order to enter into the desire and the order of the father, of man. A whole economy of gestural and verbal relations between mother and daughter, between women, is thus eliminated, abolished, forgotten in so-called normal language, which is neither asexual nor neuter. Does discourse then consist only of partially theoretical exchanges between generations of men, concerning the mastery of the mother and of nature? What is lacking is the fecundity of the sexed word, and of a creation, beyond procreation, that is sexual.

Note

1   *Author's note.* The following questions were presented, in part, at the 'Seminar on the History and Sociology of Scientific Facts and Ideas,' at the University of Provence, in Marseilles.

HABERMAS

21

BISECTED RATIONALITY: THE FRANKFURT SCHOOL'S CRITIQUE OF SCIENCE

Axel Honneth[1]

When, in 1931, Max Horkheimer assumed the leadership of the Institute for Social Research in Frankfurt, thus founding, in a certain sense, the theoretical tradition we have since come to call the Frankfurt School, it seemed entirely natural for him to present the analysis and criticism of modern science's self-conception as forming one of the main tasks facing a critical program of social research.[2] The idea that in the methodological structure of the sciences the social conditions of daily life, situated in human history, too, are continuously reflected, belonged to the intellectual heritage of a generation grown up within the horizons of Hegel, Marx, and Lukács; for all the colleagues assembled by Horkheimer in the Institute, accordingly, there was no question but that one could gain just as much insight concerning the condition of social life through a critique of science as by the empirical means of social research. It was this shared point of departure which ensured that an ongoing confrontation with the transformations in the modern conception of science would take place within Frankfurt Critical Theory; the spectrum of the works owed to this interest, still strong today, extends from Horkheimer's early programmatic writings to Franz Borkenau's studies of the emergence of the world-picture of the bourgeoisie, from Theodor Adorno's inaugural lecture to Jürgen Habermas's treatise on *Knowledge and Human Interests*.[3] As disparate as these authors' approaches, encompassing both analyses of the history of science and those of a purely methodological character, may have been on particular points, they nonetheless strongly resemble one another in the critical accent always placed on the cardinal failure of the modern conception of science: the "positivism" of the theories of knowledge and of science prevalent in the modern age was taken to lie at the root of every methodological wrong turn, because it allows the particular sciences to forget the degree to which they are anchored in the frameworks of prescientific praxis.

The charge that praxis had been forgotten or ignored is the theoretical brace that has held the Frankfurt School's critique of science together, from its beginnings through to the present day. Whether it is Horkheimer, Adorno, or Habermas, each always criticizes the modern sciences for their denial, as a result of their positivistic self-conception, of the practical experiences with which they are bound up on account of both their genesis and the field within which they operate; however much the picture of contemporary science will be specified and rendered more complex in the course of Critical Theory's development, however differently the methodological particularities of the various sciences will be sketched out in any single instance, analysis of them nonetheless continues to be governed by the thesis that there is an internal connection between knowledge and action. Even though the concept was only introduced later on by Habermas and Karl-Otto Apel, it does not seem mistaken to apply the expression "anthropology of knowledge" to the theoretical framework of the Frankfurt School's critique of science: the analysis and critique of the sciences carried out by these authors is supported by anthropological assumptions concerning the particulars of the linkages among action, the constitution of reality, and scientific methodology.

Yet the claim of the existence of such an anthropological linkage between knowledge and praxis, between scientific methodology and social action, has repeatedly assumed new forms throughout the history of the Frankfurt School. In retrospect, it makes sense to distinguish three phases, in each of which the critique of positivism, taken as the key to a critique of the sciences, has been utilized in a different manner; and it may suffice, at least to begin with, simply to distinguish these various phases with the names of Horkheimer, Adorno, and Habermas. Each of these authors shall be addressed in turn in what follows.

I

In the first phase of the Institute's activities, which lasted up until the end of the 1930s, it was not only Max Horkheimer but Herbert Marcuse as well who worked out the methodological self-conception of Critical Theory. Although they came from the most different intellectual traditions imaginable – Horkheimer had developed an overwhelming interest, in his early years, in logical positivism, whereas Marcuse was a student of Heidegger's – they still were in accord in taking their bearings from the anthropology of knowledge of the young Marx; for this reason they both set out the specificity of Critical Theory in the course of reflections intended to expose the practical roots of every science. The starting-point of their program in the theory of science was provided by their diagnosis that the intellectual situation was marked by opposition between empirical research and philosophical thought: the unifying link that had still held both branches of knowledge together during the nineteenth century in the wake of Hegel, was torn asunder with the decline of idealist philosophy; as a result, by the beginning of the twentieth century the two constitutive parts of the Hegelian philosophy of history, now split up, stood opposed to one another, in the shape of contemporary positivism on the one hand and the new metaphysics of *Lebensphilosophie* on the other. In positivism, empirical knowledge of reality, because it had been separated from any attempt to arrive at philosophical self-certainty, was reduced to the level of the mere investigation of facts; while in the metaphysics of those years, in the philosophical sketches of Max Scheler and Nicolai Hartmann, reflection upon reason, since it was cut off from any contact with historical and empirical reality, shriveled up into mere speculation on essence.[4]

Horkheimer went on to take the real problem posed by this intellectual situation to be the fact that it had done away with any possibility of thinking along the lines of the philosophy of history; for, in the abstract division of labor between scientism and metaphysics to which developments in thought had led after Hegel's death, the idea of reason incarnate in history, which had inspired the classical philosophy of history from the very beginning, was no longer given credence. Along with this aspect of the philosophy of history, however, every social theory abandoned the possibility of mounting a critique that transcended existing society; social theory no longer attempted to find the conceptual means that would have been needed to measure the condition of a given society against an overarching idea of reason. Therefore, laying the foundations of a critical theory of society presupposed, to begin with, that this intellectual division between empirical research and philosophy had been overcome. This task was what both Horkheimer's and Marcuse's essays aimed to take on, first and foremost; in terms of epistemology they were based upon a systematic critique of positivism, while their objective was to arrive at an interdisciplinary conception of Critical Theory.

In the critique of positivism put forward by the Institute, the key is provided by the materialist anthropology of knowledge of the young Marx. Horkheimer took up this approach, which Marx's writings do no more than outline, in a way that drew on Lukács, while Marcuse adopted it in a version influenced by Heidegger.[5] Both proceeded from the premise that the empirical sciences, even as concerns their methodologies, are determined by the requirements of labor in society; reaching theoretically valid results is a function of the same interest in controlling physical nature that, prior to all science, already directs the activity of human labor. As soon as science's practical constitutive context had been investigated epistemologically, however, it became clear to what misunderstandings positivism must necessarily lead: by no longer providing anything but a merely methodological justification of the sciences, positivism did not sever them only from the awareness of their own roots in society but also from an understanding of their own choice of aims in practice. In this denial of the way in which scientific theories exist within a practical framework constituted by life's requirements, Horkheimer and Marcuse saw of course not merely a mistake committed by contemporary positivism but also a deficiency characteristic in general of the modern conception of theory's role – Horkheimer pursued the roots of this positivist conception, which allowed science to appear to be a pure profession entirely detached from any interest in practice, all the way back to Descartes. The name he gave to this tradition of scientism, coeval with modernity, was "traditional theory," to which, together with Marcuse, he proposed the alternative of "critical theory," a kind of science that would remain permanently aware both of the social context within which it had emerged and of the practical context in which it might be applied.

The conception of the anthropology of knowledge Horkheimer and Marcuse both followed would then require that for this other kind of science, too, one identify the roots in practical life upon which its claim to be a transcultural frame of reference was based; for if the empirical sciences owed their function and method solely to their satisfying a material interest that derived, anthropologically, from the conditions governing the reproduction of the human species, then Critical Theory would also have had to heed an imperative that constituted, in a comparable manner, a necessary precondition for the species' survival. Already at this early stage in the history of the Frankfurt School, accordingly, there arose the problem that Habermas, around 40 years later, would attempt to solve with his renowned concept of an emancipatory interest: once the approach of the anthropology of knowledge is taken and thus a particular type of science referred back to its basis of "natural" interests, one immediately stands in need of an anthropological justification of the higher-level vantage-point from which one hopes to investigate such a linkage between knowledge and interest in the first place.[6] The strategy adopted by the anthropology of knowledge also requires one to demonstrate the culturally invariant basis of interest in which a "critical theory," for its part, is supposed to be anchored. However, neither Horkheimer nor Marcuse concerned themselves sufficiently with this problem in their early critiques of positivism. Although they did charge positivism with ignoring the roots of the empirical sciences in practical life when it pursued the fiction of a pure methodology, nonetheless they themselves failed to draw, at least in a sufficiently decisive manner, the consequence of having to identify the invariant frameworks of human action to which their own kind of critical knowledge, too, remained methodologically as tightly bound as did technical know-how.

The reasons Horkheimer and Marcuse were unable to arrive at a satisfactory answer derive largely from their dependence on Marxism's traditional explanatory framework; an additional role may have been played by the fact that by their stark rejection of American pragmatism they robbed themselves of a philosophical resource which might well have been of great help in clarifying the epistemological problems. Their bond with Marxism is attested by the great difficulty each had in acknowledging that, besides the active process of labor, there was another form in which human experiences were constituted: like the young Marx they were compelled instead to contend that the reproduction of the species depended exclusively on success in dominating nature, such that no other modes of creating a world through praxis could be envisioned. For the question of how Critical Theory might for its part be anchored in terms of the anthropology of knowledge, such a reductive approach entailed the result that one was permitted to seek an answer only in the active sphere of labor. Accordingly, Horkheimer and Marcuse took care to give the process of social production an interpretation so thorough and complex that critical knowledge, too, could take its place there, epistemologically speaking, as a sort of higher-level reflection: either they accentuated the (concealed) cooperative character of labor, or they emphasized its function of sustaining human freedom, but in both cases they did so only in order to conclude that Critical Theory must conceive its task to be to articulate, in reflection, the normative potentials that are inherent in the process of production.[7] The problems called forth by this epistemological strategy first come fully into view, of course, when one attends more to the context in which theories are applied than to the one within which they first emerged; for it is far from persuasive to argue, on the one hand, that knowledge experimentally obtained is continuously plowed back into technology's control of natural processes, and then to claim, in the very next breath, that knowledge acquired in the same context of activity might also serve to foster critical insight into the realities of domination and repression.[8] The discrepancy between the function assigned to Critical Theory and the way in which the latter was situated with respect to an anthropology of knowledge was so great that Horkheimer, at least, did not manage to overlook it entirely; in his early essays there are passages in which he related Critical Theory not to the context of the activity involved in social labor, but to the quite distinct context of what he called "critical activity" or "attitude."[9] With this concept he pursued a very different path of justifying his own approach than the one that had been opened to him by the strategy of relating critical knowledge back to the functional sphere of cooperative labor, along the lines of the anthropology of knowledge: for the practical frame of reference within which the methodical generation of critical knowledge had to be located was no longer provided by the metabolic relationship between the challenges of nature and the instrumental solutions devised to meet them, but by the overarching tension between social domination and liberating insight.

However, Horkheimer does not seem to have possessed means adequate to actually have worked this first intuition up into an alternative explanatory framework. That would have required him to comprehend the field of emancipatory struggles and practices as being a constant variable in human history, in the same way as we might speak of social labor in its universality as being; and it would have been no less necessary for him to have sketched out the sort of practical prescientific knowledge that "critical activity" was to generate, such that this knowledge might form the basis for the methodical generalizations of Critical Theory. Both demands Horkheimer was quite unable to meet, however, because he failed to attain sufficient clarity regarding the implications of the anthropology of knowledge he had simply adopted in its entirety from Marx; neither he nor Marcuse really became aware of what it meant to conceive of the various branches of science as depending in each instance on a different objectification of reality, objectifications that only emerge in the first place on the basis of the requirements of human beings' prescientific praxis. In this connection, it may have been a great disadvantage that both authors were conscious of American pragmatism, in those years, only in the form of a caricature;[10] for they thus forfeited the possibility of gaining more precise insight, with the help of the works of Dewey, James, and Peirce, into the connection they believed to exist between prescientific practice and the methodically disciplined development of theory.

In the few places where Horkheimer and Marcuse deigned to mention pragmatism at all, they always summarily dismissed it. In doing so, it was quite clear to Horkheimer in particular that there was a certain affinity between his own position and that of the new philosophical currents in the USA, as both held that scientific understanding was derived from a methodical generalization of the knowledge acquired in actual practice: like James or Dewey, Critical Theory, too, assumed that "science is subject to the dynamisms of history" and that, therefore, "the separation of theory and action" or praxis represents a "historical phenomenon."[11] But at the same time Horkheimer was convinced that American pragmatism misunderstood this fact in the manner of utilitarianism when it claimed that the truth of scientific statements depended solely on the criterion that social goals be met; this reductive focus merely on satisfying the needs of life, according to Horkheimer, entails that one loses sight of the particularity of the conditions of validity of scientific knowledge, upon which scientific research as such is based independently of all connection to social interests. Granted, Horkheimer did not succeed in further indicating how the interplay between the relation to prescientific interests on the one hand and the rationality immanent in science on the other was to be conceived more precisely; from the tendency of his argument, most of the time it sounds as though he wished to see the influence of those social goals end precisely at the point beyond which science's particular conditions of validity take command, without even having considered how to mediate between these two domains. Here, a more intensive involvement with American pragmatism, in whose riper contributions to epistemology and the theory of truth the significance of imperatives immanent to science itself to justify its findings is by no means denied, might have been able to help Horkheimer advance further. For, on the contrary, Peirce or Dewey conceived of the vital interests which they believed influence the research process as being more akin to a transcendental framework which does indeed determine the direction to be taken by the efforts to obtain scientific knowledge while however not setting down conditions for its validity in detail. The room for maneuver that remains, under the assumption that knowledge is no more than directed by such interests, ought instead to be circumscribed by methodical rules in which the consensus of the researchers regarding the various criteria appropriate for the justification of scientific statements is reflected.[12]

However vague the works of pragmatism may have been, however susceptible they later showed themselves to be to a number of convincing objections, nonetheless, during the 1930s they would have been helpful to Horkheimer and Marcuse in these authors' attempts to claim an independent status for their own kind of theory while resituating the traditional sciences in terms of an anthropology of knowledge. The idea of thinking of anthropologically embedded interests as cognitive frameworks within which the conditions of the validity of scientific statements must repeatedly be specified, depending on the state of research, would have been relevant not only to the critique both authors had developed of the positivism of their times; here that idea could have led to a more precise description of the way in which our phylogenetic imperative to dominate nature finds expression in the methodological rules established by any nomological theory that addresses human convention. Above all, however, it is with respect to the aims both authors were pursuing in formulating the project of a "critical theory" that the proposal of the pragmatists would have been of the greatest conceivable significance; for then Horkheimer and Marcuse might have seen themselves forced to characterize the anthropological status of the question of action more precisely, since it is the guiding assumption of "the critical attitude" that human action constitutes social reality in such a way that the latter can become the object of science's methodically disciplined effort to develop a particular kind of knowledge. Yet because both authors were quick to push pragmatism to one side and did not even acknowledge its potential, their program of an anthropology of knowledge failed to clarify the matter of how far Critical Theory may claim a higher vantage-point for itself without in doing so abandoning its connection to a set of prescientific interests. To escape from the awkward situation in which Horkheimer and Marcuse thus found themselves, early in the 1940s, they thought for the most part only of invoking human reason through a recourse to the legacy of idealism; but with the change in the character of the philosophy of history that formed Critical Theory's conceptual framework, some years later under the influence of Theodor Adorno, this makeshift solution was no longer open to them – though in the *Dialectic of Enlightenment* an anthropology of knowledge still plays a dominant role, its framework is definitely too narrow to offer room for an epistemological justification of Critical Theory.

II

The *Dialectic of Enlightenment*, which Adorno and Horkheimer composed together early in the 1940s, represents the attempt to assimilate the historical experience of the catastrophe of National Socialism in the form of a negative philosophy of history. The very construction and style of the book, but especially its organizing ideas, reveal that behind the shared authorship there is primarily Adorno's signature to be found; he had already, towards the end of the 1920s, under the strong influence of Walter Benjamin, arrived at the conviction that philosophy, in view of existing circumstances, would have to be content with fulfilling the task, in the form of the essay, of decoding the irreducible limitations imposed by the progress of natural history upon human rationality.[13] Accordingly, the *Dialectic of Enlightenment*, too, unlike the constructive essays Horkheimer had earlier produced, has the character of an unsystematic anthology; it draws its primary material not from sources in social or economic history but from the oblique testimony of intellectual history. Through interpretations of literary and philosophical works – of Homer's *Odyssey*, Sade's novels and stories, the treatises of Kant and Nietzsche – Horkheimer and Adorno developed their thesis that, with the totalitarian domination of National Socialism, human history's character as a "regressive" process of civilization came to the fore; as the cause of this regression they identified the destructive effects emerging from the attitude of instrumental rationality in which Horkheimer had previously still seen the potential for bringing about the progress of the human species.

This reconsideration of instrumental rationality was a consequence of the fact that in the meantime they had ceased to emphasize the role of the labor process in preserving human freedom and focused instead on it exclusively as a repressive force of domination. In the *Dialectic of Enlightenment*, Horkheimer and Adorno described the way in which the human species, released from the particular security afforded by reliance on instinct, worked its way out of the dangers of nature, as being a process that involved the gradual replacement of mimetic modes of behavior by elementary forms of rational control: the human being raised himself far above the animal conditions of his existence by learning first to master the reflex motions with which prehuman life-forms attempted, when terrified, to imitate a threatening object physically, and then finally to replace them entirely by putting nature under preventive control. In the same process whereby the human species gradually subjected natural occurrences to its own power, it also began to abstract from the threat posed by the impressive forces of nature and to separate reality, since become an object to humanity, out into recurrent and utilizable events; hence, from the chaotic manifold of stimuli that was his natural environment the human being categorically cuts out only those constituent parts that have the function of marking reality for his instrumental interventions into it.[14] It is the activity of labor, in which the human being learns to shatter the ever-present threat of nature by forcing its sensuous manifold of impressions into a conceptual schema of comprehension, that shows him a world he can survey and master. With these first prehistorical acts whereby nature was brought under control, according to Horkheimer and Adorno, there commences the process they designate as a "regressive" course of increasing domination and reification: for the instrumental mastery of external nature – far removed as yet from possessing any potential to liberate human beings – soon gets reversed and becomes a means of disciplining inner nature, and finally culminates in the reification of social intercourse as well.[15]

If the epistemological implications of this genealogy of totalitarianism in Horkheimer and Adorno's philosophy of history are considered, one can easily understand why, with the shift that resulted in the *Dialectic of Enlightenment*, the Frankfurt School's critique of science would also have had to be radicalized. While in his early essays Horkheimer had still charged the nomological sciences, in their positivistic self-conception, with denying their own origin in the context of praxis, in the new approach the critique was extended, as it were, to encompass the very methods and procedures of those sciences: it was no longer merely the illusion of scientific theory to which they adhered, rather it was their methodical disposition in its entirety, the subsuming of the particular under general concepts which are instrumentally useful, that now began to be the object of the critique of positivism. This critique's core of argument, however, was formed by an anthropology of knowledge in which the sort of knowledge engendered in labor was viewed negatively: as long as our knowledge serves the aim of gaining control over natural processes, in a prescientific manner, it will have to operate with general concepts that allow us to recognize, in the reality around us, merely the sites of use to us in our instrumental operations; but the knowledge that results from these operations only does violence to the physical environment, because it abstracts from the variety of sensuous stimuli of the latter, such that there alone remains a surface to manipulate. Horkheimer and Adorno were not, of course, satisfied with this negative result, and extended their critique to cover other forms of knowledge as well, namely those that serve not to control nature but to subordinate one's own psychic processes or other human beings; for, in their view, even in these domains reality is constituted only with the help of the same instrumental conceptual schema that had originally been based solely in the primitive procedures of labor. Thus the Institute's critique of positivism was pushed further and became a critique of all the various disciplines, whose crucial lack was now thought to consist in their pressing a manifold of particularities into a system of general concepts suitable for use; not forgetfulness of the practical context in which they had originated, but this context itself, constituted Horkheimer and Adorno's objection to the disciplines.

The consequences that resulted for Critical Theory itself from this radical extension of the critique of science are extremely problematic in more than one respect. For one thing, the charge that all the particular disciplines participate in the system of domination, on account of their methodological frameworks, undercuts the program for interdisciplinary social research which Horkheimer had put forward a mere ten years earlier; if in the final analysis sociology or psychology also only pursue an intellectual interest that aims to gain instrumental control over social processes, then their results are not at all suited to provide the empirical foundations for a critical theory of society. On the contrary, it even seemed as though Critical Theory had to maintain the greatest possible distance from the particular disciplines, were it to avoid methodically obscuring its own interest, which was focused on the emancipation from domination and oppression, whereas the interest of the empirical disciplines pertained to the controlled manipulation of natural and social occurrences.

Of course, with this expansion of its critique of science, it was not only its relation to the particular disciplines that became problematic for Critical Theory but its own methodological status as well. Already during the first phase of the Institute's activities it had become clear that the epistemological framework formed by the anthropology of knowledge was basically insufficient to justify the emancipatory claims of Critical Theory in a reconstructive manner; neither Horkheimer nor Marcuse was successful in finding in the social reproduction of the species a durable interest that, when systematically followed up on, might have been able to legitimize the particular procedures of a critical science. With the transformation of the authors' philosophy of history that was carried out in the *Dialectic of Enlightenment*, however, this problem became even more pressing, for now one no longer finds, in terms of the anthropology, even the slightest effort to outline any practical course of action that could be thought of as having been a way of acting upon the human species' interest in its own emancipation; the reproduction of the species appeared instead to depend exclusively upon the controlled application of instrumental rationality, such that there no longer seemed to be an additional need for a form of knowledge to serve as the prescientific framework of a critical theory. Horkheimer and Adorno were no longer able to justify, on the basis of their own theory of knowledge, the type of science they themselves practiced in their book; for that would have required them to define, anthropologically, amidst the spectrum of the modes of human praxis, a special form of activity that could count as the pragmatic context within which Critical Theory itself emerged. As though they were aware of this difficulty, however, both authors tried to suggest at least the outlines of a possible solution. The passages where such suggestions may be found are concerned with the sole form of human conduct of which one could say, in conformity with the key thesis of the book, that they are not marked by the attempt to exert instrumental control; here, the authors were referring to the mimetic reactions that may still be found among human beings so long as they have not turned their attention to the practical domination of their environment. In what remains of such a non–instrumental attitude, for which their book gives examples in the various excurses, Horkheimer and Adorno now believed themselves able to discern the existence of a context within which scientific understanding might develop other than the one characterized by the metabolism of control and conceptual knowledge inherent in most human behavior: because reality, in mimetic conduct, is constituted not with a view to exerting control but rather in the interest of achieving the most exact imitation possible, here the kind of cognitive violation that otherwise pervades all forms of human praxis is no longer in force.[16] It is this sort of absence of violence that permits the two authors to discern in mimetic experience one anthropological root of all critical knowledge: the schema under which reality is comprehended, in such a mimetic attitude, is one of a synthesis without violence that does not subsume the manifold of the given under concepts, but rather permits it to come to expression in a constellation. However, in the *Dialectic of Enlightenment* it remains unclear whether Horkheimer and Adorno actually wished to introduce the strong thesis that Critical Theory ought to be based "transcendentally" in such a context of mimetic conduct; at the very least, the book fails to provide evidence which could support the assumption that behavior of this kind is constitutively built into the process of the reproduction of the human species.

But the turn Critical Theory took with the *Dialectic of Enlightenment* was aimed unambiguously at raising the status of the aesthetic in terms of their anthropology of knowledge. Even though the two authors did not deliver proof of their thesis that mimetic reactions are as fundamentally a part of social reproduction as are the practices of exerting instrumental control, nonetheless they did attempt to comprehend their own efforts as being a form of reflection on aesthetic experience opposed to a process of increasing violence and domination: it was no longer the kind of action represented by labor or the critical attitude, but that of a mimetic relation to the world, that was supposed to constitute the epistemological framework in which Critical Theory was anchored transcendentally and upon which it in turn ought to be able to have an effect in practice. Thus, the emancipatory interest, which Horkheimer, together with Marcuse, had still situated in society's labor process, was extricated from the central sphere of social reproduction and transferred to a peripheral domain; and the theoretical alternative Horkheimer had been toying with, when he spoke of the critical attitude's role in transcending history, no longer played a part in the *Dialectic of Enlightenment* either. Because they saw all these realms of action as being dominated by the exertion of instrumental control, to Horkheimer and Adorno there remained no option but to localize the interest in emancipation and within the sole domain of experience they assumed to be, by its structure, marked by the absence of violence.

Clearly, the consequences that came with this change in the self-conception of Critical Theory were far more weighty than the two authors themselves may perhaps have wanted at first to acknowledge. As the theory ought henceforth to have conceived of itself as being a form of reflection of and upon aesthetic experience, it could not for its part have continued to utilize the procedures and modes of presentation on which it had so self-confidently been based up to that point; rather, in its method, it had to appropriate for itself an element of that synthesis without violence characteristic in its view of the domain of aesthetic experience whose expression in reflection it was to suppose itself to be.[17] With this compulsion to render theory itself aesthetic, which was felt above all in Adorno's philosophical work, the distance from the various disciplines grew ever greater; what previously was supposed to represent the unique historical specificity of Critical Theory, namely its systematic crossing of philosophy and interdisciplinary research, was lost sight of more and more, for between these two branches of knowledge there was a fundamental difference in their respective ways of relating to reality. However constitutive the social sciences were to remain for the work of the Institute for Social Research, they were no longer lent legitimacy, on the plane of method, by the notion Critical Theory had of itself in Adorno's philosophy.[18]

III

Although from the very beginning Critical Theory's epistemological self-conception was based upon an anthropology of knowledge influenced initially by Marx and later by Nietzsche as well, this theoretical background remained nonetheless more or less entirely unexamined for a long period of time. Positivism was indeed charged early on with having rendered the practical motivations of the natural sciences illegible, behind the fiction of a pure methodology, but what it actually meant to say about an internal connection between actions and scientific insights was not ever a matter discussed in its own right. This first changed when, in the 1960s, Karl-Otto Apel and Jürgen Habermas initiated the attempt to think the methodological claims of Critical Theory through again from the start, in an altered scientific milieu; fundamentally more open with respect to the legacy of American philosophy, they enlisted the support of pragmatism in order to uncover the anthropology of knowledge that comprised the background of the theoretical tradition passed on by Horkheimer and Adorno.

In this enterprise, initially both authors employed once again only the example of positivism in their respective attempts to arrive at a more precise definition of the relationship between action and knowledge. In agreement with what Horkheimer had already established, thirty years before, they saw in positivism an approach to the theory of science that brought about the disappearance, behind the fiction of a pure methodology, of the natural sciences' practical interest; the genesis and the utilization of these sciences were also thought by Apel and Habermas to be situated in a context of action which compels human beings to strive to attain technical dominance over their natural environment. The productive reception of American pragmatism, above all of Charles Sanders Peirce's work, is reflected, however, in both authors in the fact that they were able to join this explanation of the genesis of the sciences to a presentation of the context within which the latter gain validity that was fundamentally more precise than Horkheimer's had been: whereas their predecessor had left the connection between prescientific interest and scientific rationality unclear, Apel and Habermas made use of Peirce's idea of the unending discourse of the community of researchers in order to clarify the conditions for the truth of social-scientific statements. Accordingly, though such statements are indeed obtained within the "transcendental framework" of a practical interest in gaining control over natural processes, their validity is not measured by their simply fulfilling these instrumental aims, but rather by their concordance with the conditions governing their own justification, conditions upon which, for the time being, the research community has agreed in its future-oriented discourse. The translation of interests directed by action into scientifically valid statements is thus accomplished by procedures of justification whose character always reflects aims given in advance and which yet also represent the result of a self-reflexive process of inquiry into the rationality of the research methods.

Up until this point, however, Apel and Habermas had only further illuminated an issue of concern in the anthropology of knowledge that Horkheimer's early essays, in principle, had already had in view, albeit in rough outline; with the help of Peirce's pragmatism the meaning of social-scientific statements was defined more precisely, but without contributing to a more differentiated view of science as a whole. A new stage in relation to the older generation of Critical Theory was first reached by these two authors when they proposed the thesis that the spectrum of scientific rationality is not exhausted by the single dimension of the instrumental interests involved in knowledge; alongside the natural sciences they granted to the historical-hermeneutic sciences as well an independent universal value by also referring the latter back to a practical interest they believed to be as deeply rooted anthropologically as the interest in gaining control over nature. The stimulus to extend their conceptions in this way, whereby an essential difference between the first and the second generation of Critical Theory is clearly delineated, came from these authors' working through the implications for the anthropology of knowledge of works from quite variegated traditions, among which Gadamer's hermeneutics belonged just as much as did Arendt's philosophy of action; however disparate the intellectual motivating forces may have been, they did provide Apel and Habermas with the opportunity to advance the thesis that the reproduction of the human species depends not only on technical control of natural processes but also, equally fundamentally, on reaching intersubjective agreement about shared horizons of meaning. Correspondingly, both authors set "interaction" beside "labor" as a second form of human action, for which the same connection between prescientific interest, the constitution of reality, and scientific rationality was to hold as existed in the case of the natural sciences: in communication among human beings, the social world is constituted with a view to reaching a possible understanding such that from this prescientific praxis there may arise the prospect of a methodical rationalization that will ultimately lead to the establishment, in the historical-hermeneutical disciplines, of an independent branch of knowledge. Extending their anthropology of knowledge in this manner, both Apel and Habermas closed a theoretical lacuna that had remained open, problematically, in the works of Adorno, Horkheimer, and Marcuse; for, in those authors of the first generation, on account of their unchanging concern with the model of exerting control over nature, no answer was ever provided to the question of where the distinctive forms of knowledge assembled today under the title of the human sciences [*Geisteswissenschaften*] ought to be situated anthropologically.

However, despite these great differences, Apel and Habermas still felt themselves obligated to the tradition of the Frankfurt School to the extent that they too made use of the critique of science as a key for diagnosing developments in society. Habermas, above all, undertook the project, in a sequence of essays in the 1960s, of interpreting the public preeminence of scientism as an indication of the fact that the communicative sphere of moral and political mutual understanding was threatened by the imperatives of technological rationality;[19] continuing to pursue themes he had already touched on in his study of *The Structural Transformation of the Public Sphere*, he described the expansion of a technocratic consciousness as a social process by which the historical achievement of the establishment of a sphere for the formation of the public will was once again being endangered in its very existence.[20] But these first attempts at producing a direct linkage between the critique of science and the theory of society still remained conceptually vague and underdeveloped; what is unclear about them, above all, is why social reproduction was thought to depend so essentially upon a mechanism for reaching mutual understanding through communication, that technical rationality's winning an independent institutional position could have been defined as posing a threat to the integration of society as a whole. Another twenty years had to go by before Habermas was able, in his *Theory of Communicative Action*, to bring into play the conceptual means that would allow him to present this original intuition in a fundamentally more precise version.[21]

Yet, in the 1960s, the challenge Habermas and Apel saw themselves primarily facing, with their anthropology of knowledge, was of quite a different kind; namely, they shared the problem with the representatives of the first generation of Critical Theory of having now for their part to justify, in terms of the anthropology of knowledge, the scientific vantage-point from which they believed themselves able to carry out their own sort of critique of science. The epistemological strategy of convicting established types of scientific knowledge of having a false conception of themselves by revealing their roots in contexts of prescientific praxis, required of Apel and Habermas that they too identify such a root in praxis for their own analyses, which they considered to be superior; thus they faced a recurrence of the difficulty Horkheimer had confronted when he had to name a prescientific source of knowledge for Critical Theory itself and could not finally decide between "labor" and the "critical attitude." In this regard, the situation in which Apel and Habermas found themselves was, of course, a great deal more favorable than Horkheimer's had been, because, after all, they had already reconstructed anthropologically two kinds of interest governing the development of knowledge; alongside the elementary interest in achieving instrumental control they had brought to the fore an equally deep-rooted interest in arriving at mutual understanding in language, with the result that the process of social reproduction offered them more room for identifying an emancipatory interest as well.

It was not entirely clear, however, what sort of proposals the two authors would devise in order to overcome the problem that thus arose. For in providing an answer they vacillated between two possibilities, alternately locating the emancipatory interest within the framework for action represented by the attempt to reach practical mutual understanding, or placing it in the critical attitude as a distinct and original mode of human conduct. In the first, what is at issue is an anticipation of the later idea that a transcending power always resides in the achievement of mutual understanding in language, because the latter requires those involved to embrace the normative assumption that the speech-situation ought to be free of domination: the emancipatory interest guiding Critical Theory is an expression of the implicit insight, shared by all reasonable individuals, that in their discursive practices they necessarily anticipate the condition of a just society. But in the second of these possibilities, alongside the two types of action previously distinguished by Apel and Habermas, a third kind of social action was brought into play, which they took to represent a "transcendental" framework for another basic mode of producing knowledge: in response to the political dependence and social oppression that have remained constitutive in human history up into the present, those affected have repeatedly had the inclination, stimulated by critical interpretation, to initiate a process of self-reflection that will liberate them from relations of domination of which they had previously been insufficiently aware.[22] As attractive as this second possibility may sound, of course, here the connection that was supposed to exist between action, interest, and knowledge remained quite unclear: what kind of interest, anthropologically deep-seated and guiding the course of human action, might it be whose satisfaction would require the production of a kind of knowledge uniquely involved in reflexive examination of the relations of power and domination in society? Apparently, a greater number of anthropological hypotheses than Habermas seemed willing to permit himself would have been required in order to back up the thesis that in social reproduction there is the compulsion to adopt a type of behavior that one cannot practice without having acquired critical knowledge.

But in principle matters are no better with the first option offered by Apel and Habermas for justifying the assumption of an emancipatory interest; for its tenability depends entirely on the persuasiveness of the premise of their discourse ethics that a transcendental requirement to anticipate conditions without domination is inherent in each and every attempt to reach mutual understanding in language. Over the last several years, Habermas at least has moved away from the strongest, the transcendental version of this thesis, and only Apel and his students presently still seem to defend the idea that human language has a moral telos.[23] Thus, to provide proof of the existence of an "emancipatory interest" remains the main challenge still facing Critical Theory today – so long as the framework of the anthropology of knowledge that was earlier its special concern is not completely abandoned, the methodological foundations of its own point of view will necessarily require Critical Theory to continue to seek out the practical roots of a universal interest in emancipation.

Notes

1   This chapter was translated by Jack Ben-Levi.
2   Max Horkheimer, "The Present Situation of Social Philosophy and the Tasks of an Institute of Social Research," trans. John Torpey, in Horkheimer, *Between Philosophy and Social Science: Selected Early Writings* (Cambridge, MA: MIT Press, 1993), pp. 1–14.
3   See: Max Horkheimer, "Notes on Science and the Crisis," trans. Matthew J. O'Connell, in Horkheimer, *Critical Theory: Selected Essays* (New York: Continuum, 1992), pp. 3–9; "The Present Situation of Social Philosophy and the Tasks of an Institute of Social Research"; and "Traditional and Critical Theory," trans. Matthew J. O'Connell, in *Critical Theory*, pp. 188–243. Franz Borkenau, *Der Übergang vom feudalen zum bürgerlichen Weltbild: Studien zur Geschichte der Philosophie der Manufakturperiode* (Paris: Librairie Félix Alcan, 1934). Theodor Adorno, "The Actuality of Philosophy," trans. Benjamin Snow, in *The Adorno Reader*, ed. Brian O'Connor (Oxford: Blackwell, 2000), pp. 23–39. And Jürgen Habermas, *Knowledge and Human Interests*, trans. Jeremy J. Shapiro (Boston: Beacon Press, 1971).
4   Horkheimer, "Traditional and Critical Theory" and "Notes on Science and the Crisis"; Herbert Marcuse, "Philosophy and Critical Theory," trans. Jeremy J. Shapiro, in Marcuse, *Negations: Essays in Critical Theory* (London: Free Association Books, 1988), pp. 134–58.
5   Regarding Horkheimer, see Martin Jay, *Marxism and Totality: The Adventures of a Concept from Lukács to Habermas* (Berkeley and Los Angeles: University of California Press, 1984), ch. 1; for Marcuse, see Stefan Breuer, *Die Krise der Revolutionstheorie: Negative Vergesellschaftung und Arbeitsmetaphysik bei Herbert Marcuse* (Frankfurt am Main: Syndikat, 1977).
6   Habermas, *Knowledge and Human Interests*, esp. ch. 3.
7   See Horkheimer, "Traditional and Critical Theory," and Herbert Marcuse, "Über die philosophischen Grundlagen des wirtschaftswissenschaftlichen Arbeitsbegriffs," in Marcuse, *Schriften*, vol. 1, ed. Peter-Erwin Jansen (Frankfurt am Main: Suhrkamp Verlag, 19xx), pp. 556–94.
8   Axel Honneth, *The Critique of Power: Reflective Stages in a Critical Social Theory*, trans. Kenneth Baynes (Cambridge, MA: MIT Press, 1993).
9   Horkheimer, "Traditional and Critical Theory," pp. 206 f.

10   On this point see Hans Joas, "An Underestimated Alternative: America and the Limits of 'Critical Theory'," in Joas, *Pragmatism and Social Theory* (Chicago: University of Chicago Press, 1993), pp. 79–93.

11   Horkheimer, "Notes on Science and the Crisis," pp. 3–4.

12   See John Dewey, *The Later Works, 1925–53*, vol. 4, *The Quest for Certainty*, ed. Jo Ann Boydston (Carbondale, IL: Southern Illinois University Press, 1984).

13   See Adorno, "The Actuality of Philosophy."

14   Max Horkheimer and Theodor Adorno, *Dialectic of Enlightenment: Philosophical Fragments*, ed. Gunzelin Schmid Noerr, trans. Edmund Jephcott (Stanford: Stanford University Press, 2002), p. 31.

15   See *The Critique of Power*, ch. 2.

16   See Josef Früchtl, *Mimesis: Konstellation eines Zentralbegriffs bei Adorno* (Würzburg: Verlag Königshausen & Neumann, 1986), esp. ch. 3.

17   See Jürgen Habermas, *The Philosophical Discourse of Modernity: Twelve Lectures*, trans. Frederick G. Lawrence (Cambridge, MA: MIT Press, 1987), ch. 5.

18   See Karl-Otto Apel, *Charles Peirce: From Pragmatism to Pragmaticism*, trans. John Michael Krois (Amherst: University of Massachusetts Press, 1981); and Habermas, *Knowledge and Human Interests*, chs. 5 and 6.

19   Jürgen Habermas, *Theory and Practice*, trans. John Viertel (Boston: Beacon Press, 1973).

20   Jürgen Habermas, *The Structural Transformation of the Public Sphere: An Inquiry into a Category of Bourgeois Society*, trans. Thomas Burger with the assistance of Frederick Lawrence (Cambridge, MA: MIT Press, 1989).

21   Jürgen Habermas, *The Theory of Communicative Action*, 2 vols., trans. Thomas McCarthy (Boston: Beacon Press, 1984, 1987).

22   See Jürgen Habermas, "Erkenntnis und Interesse," in Habermas, *Technik und Wissenschaft als "Ideologie"* (Frankfurt am Main: Suhrkamp Verlag, 1969), pp. 146–68.

23   See Karl-Otto Apel, *Auseinandersetzungen: In Erprobung des transzendentalpragmatischen Ansatzes* (Frankfurt am Main: Suhrkamp Verlag, 1998).

22

KNOWLEDGE AND HUMAN INTERESTS: A GENERAL PERSPECTIVE

Jürgen Habermas

I

In 1802, during the summer semester at Jena, Schelling gave his Lectures on the Method of Academic Study. In the language of German Idealism he emphatically renewed the concept of theory that has defined the tradition of great philosophy since its beginnings.

> The fear of speculation, the ostensible rush from the theoretical to the practical, brings about the same shallowness in action that it does in knowledge. It is by studying a strictly theoretical philosophy that we become most immediately acquainted with Ideas, and only Ideas provide action with energy and ethical significance.[1]

The *only* knowledge that can truly orient action is knowledge that frees itself from mere human interests and is based on Ideas – in other words, knowledge that has taken a theoretical attitude.

The word "theory" has religious origins. The *theoros* was the representative sent by Greek cities to public celebrations.[2] Through *theoria*, that is through looking on, he abandoned himself to the sacred events. In philosophical language, *theoria* was transferred to contemplation of the cosmos. In this form, theory already presupposed the demarcation between Being and time that is the foundation of ontology. This separation is first found in the poem of Parmenides and returns in Plato's *Timaeus*. It reserves to *logos* a realm of Being purged of inconstancy and uncertainty and leaves to *doxa* the realm of the mutable and perishable. When the philosopher views the immortal order, he cannot help bringing himself into accord with the proportions of the cosmos and reproducing them internally. He manifests these proportions, which he sees in the motions of nature and the harmonic series of music, within himself; he forms himself

Jürgen Habermas, "*Knowledge and Human Interests*: A General Perspective," pp. 301–17 from *Knowledge and Human Interests*, trans. Jeremy J. Shapiro. Boston: Beacon Press, 1971. German text © 1968 by Suhrkamp Verlag, Frankfurt am Main. English translation © 1971 by Beacon Press. Reprinted by kind permission of Beacon Press and Suhrkamp Verlag.

through mimesis. Through the soul's likening itself to the ordered motion of the cosmos, theory enters the conduct of life. In *ethos* theory molds life to its form and is reflected in the conduct of those who subject themselves to its discipline.

This concept of theory and of life in theory has defined philosophy since its beginnings. The distinction between theory in this traditional sense and theory in the sense of critique was the object of one of Max Horkheimer's most important studies.[3] Today, a generation later, I should like to reexamine this theme,[4] starting with Husserl's *The Crisis of the European Sciences*, which appeared at about the same time as Horkheimer's.[5] Husserl used as his frame of reference the very concept of theory that Horkheimer was countering with that of critical theory. Husserl was concerned with crisis: not with crises in the sciences, but with their crisis as science. For "in our vital state of need this science has nothing to say to us." Like almost all philosophers before him, Husserl, without second thought, took as the norm of his critique an idea of knowledge that preserves the Platonic connection of pure theory with the conduct of life. What ultimately produces a scientific culture is not the information content of theories but the formation among theorists themselves of a thoughtful and enlightened mode of life. The evolution of the European mind seemed to be aiming at the creation of a scientific culture of this sort. After 1933, however, Husserl saw this historical tendency endangered. He was convinced that the danger was threatening not from without but from within. He attributed the crisis to the circumstance that the most advanced disciplines, especially physics, had degenerated from the status of true theory.

II

Let us consider this thesis. There is a real connection between the positivistic self-understanding of the sciences and traditional ontology. The *empirical-analytic* sciences develop their theories in a self-understanding that automatically generates continuity with the beginnings of philosophical thought. For both are committed to a theoretical attitude that frees those who take it from dogmatic association with the natural interests of life and their irritating influence; and both share the cosmological intention of describing the universe theoretically in its lawlike order, just as it is. In contrast, the *historical-hermeneutic* sciences, which are concerned with the sphere of transitory things and mere opinion, cannot be linked up so smoothly with this tradition – they have nothing to do with cosmology. But they, too, comprise a *scientistic consciousness*, based on the model of science. For even the symbolic meanings of tradition seem capable of being brought together in a cosmos of facts in ideal simultaneity. Much as the cultural sciences may comprehend their facts through understanding and little though they may be concerned with discovering general laws, they nevertheless share with the empirical-analytic sciences the methodological consciousness of describing a structured reality within the horizon of the theoretical attitude. Historicism has become the positivism of the cultural and social sciences.

Positivism has also permeated the self-understanding of the *social sciences*, whether they obey the methodological demands of an empirical-analytic behavioral science or orient themselves to the pattern of normative-analytic sciences, based on presuppositions about maxims of action.[6] In this field of inquiry, which is so close to practice, the concept of value-freedom (or ethical neutrality) has simply reaffirmed the ethos that modern science owes to the beginnings of theoretical thought in Greek philosophy: psychologically an unconditional commitment to theory and epistemologically the severance of knowledge from interest. This is represented in logic by the distinction between descriptive and prescriptive statements, which makes grammatically obligatory the filtering out of merely emotive from cognitive contents.

Yet the very term "value freedom" reminds us that the postulates associated with it no longer correspond to the classical meaning of theory. To dissociate values from facts means counter-posing an abstract Ought to pure Being. Values are the nominalistic by-products of a centuries-long critique of the emphatic concept of Being to which theory was once exclusively oriented. The very term "values," which neo-Kantianism brought into philosophical currency, and in relation to which science is supposed to preserve neutrality, renounces the connection between the two that theory originally intended.

Thus, although the sciences share the concept of theory with the major tradition of philosophy, they destroy its classical claim. They borrow two elements from the philosophical heritage: the methodological meaning of the theoretical attitude and the basic ontological assumption of a structure of the world independent of the knower. On the other hand, however, they have abandoned the connection of *theoria* and *kosmos*, of *mimesis* and *bios theoretikos* that was assumed from Plato through Husserl. What was once supposed to comprise the practical efficacy of theory has now fallen prey to methodological prohibitions. The conception of theory as a process of cultivation of the person has become apocryphal. Today it appears to us that the mimetic conformity of the soul to the proportions of the universe, which seemed accessible to contemplation, had only taken theoretical knowledge into the service of the internalization of norms and thus estranged it from its legitimate task.

III

In fact the sciences had to lose the specific significance for life that Husserl would like to regenerate through the renovation of pure theory. I shall reconstruct his critique in three steps. It is directed in the first place against the objectivism of the sciences, for which the world appears objectively as a universe of facts whose lawlike connection can be grasped descriptively. In truth, however, knowledge of the apparently objective world of facts has its transcendental basis in the prescientific world. The possible objects of scientific analysis are constituted a priori in the self-evidence of our primary life-world. In this layer phenomenology discloses the products of a meaning-generative subjectivity. Second, Husserl would like to show that this productive subjectivity disappears under the cover of an objectivistic self-understanding, because the sciences have not radically freed themselves from interests rooted in the primary life-world. Only phenomenology breaks with the naive attitude in favor of a rigorously contemplative one and definitively frees knowledge from interest. Third, Husserl identifies transcendental self-reflection, to which he accords the name of phenomenological description, with theory in the traditional sense. The philosopher owes the theoretical attitude to a transposition that liberates him from the fabric of empirical interests. In this regard theory is "unpractical." But this does not cut it off from practical life. For, according to the traditional concept, it is precisely the consistent abstinence of theory that produces action–orienting culture. Once the theoretical attitude has been adopted, it is capable in turn of being mediated with the practical attitude:

> This occurs in the form of a novel practice ..., whose aim is to elevate mankind to all forms of veridical norms through universal scientific reason, to transform it into a fundamentally new humanity, capable of absolute self-responsibility on the basis of absolute theoretical insight.

If we recall the situation of thirty years ago, the prospect of rising barbarism, we can respect this invocation of the therapeutic power of phenomenological description; but it is unfounded. At best, phenomenology graps transcendental norms in accordance with which consciousness necessarily operates. It describes (in Kantian terms) laws of pure reason, but not norms of a universal legislation derived from practical reason, which a free will could obey. Why, then, does Husserl believe that he can claim practical efficacy for phenomenology as pure theory? He errs because he does not discern the connection of positivism, which he justifiably criticizes, with the ontology from which he unconsciously borrows the traditional concept of theory.

Husserl rightly criticizes the objectivist illusion that deludes the sciences with the image of a reality-in-itself consisting of facts structured in a lawlike manner; it conceals the constitution of these facts, and thereby prevents consciousness of the interlocking of knowledge with interests from the life-world. Because phenomenology brings this to consciousness, it is itself, in Husserl's view, free of such interests. It thus earns the title of pure theory unjustly claimed by the sciences. It is to this freeing of knowledge from interest that Husserl attaches the expectation of practical efficacy. But the error is clear. Theory in the sense of the classical tradition only had an impact on life because it was thought to have discovered in the cosmic order an ideal world structure, including the prototype for the order of the human world. Only as cosmology was *theoria* also capable of orienting human action. Thus Husserl cannot expect self-formative processes to originate in a phenomenology that, as transcendental philosophy, purifies the classical theory of its cosmological contents, conserving something like the theoretical attitude only in an abstract manner. Theory had educational and cultural implications not because it had freed knowledge from interest. To the contrary, it did so because it derived *pseudonormative power* from *the concealment of its actual interest*. While criticizing the objectivist self-understanding of the sciences, Husserl succumbs to another objectivism, which was always attached to the traditional concept of theory.

IV

In the Greek tradition, the same forces that philosophy reduces to powers of the soul still appeared as gods and super-human powers. Philosophy domesticated them and banished them to the realm of the soul as internalized demons. If from this point of view we regard the drives and affects that enmesh man in the empirical interests of his inconstant and contingent activity, then the attitude of pure theory, which promises *purification* from these very affects, takes on a new meaning: disinterested contemplation then obviously signifies emancipation. The release of knowledge from interest was not supposed to purify theory from the obfuscations of subjectivity but inversely to provide the subject with an ecstatic purification from the passions. What indicates the new stage of emancipation is that catharsis is now no longer attained through mystery cults but established in the will of individuals themselves by means of theory. In the communication structure of the polis, individuation has progressed to the point where the identity of the individual ego as a stable entity can only be developed through identification with abstract laws of cosmic order. Consciousness, emancipated from archaic powers, now anchors itself in the unity of a stable cosmos and the identity of immutable Being.

Thus it was only by means of ontological distinctions that theory originally could take cognizance of a self-subsistent world purged of demons. At the same time, the illusion of pure theory served as a protection against regression to an earlier stage that had been surpassed. Had it been possible to detect that the identity of pure Being was an objectivistic illusion, ego identity would not have been able to take shape on its basis. The repression of interest appertained to this interest itself.

If this interpretation is valid, then the two most influential aspects of the Greek tradition, the theoretical attitude and the basic ontological assumption of a structured, self-subsistent world, appear in a connection that they explicitly prohibit: the connection of knowledge with human interests. Hence we return to Husserl's critique of the objectivism of the sciences. But this connection turns *against* Husserl. Our reason for suspecting the presence of an unacknowledged connection between knowledge and interest is not that the sciences have abandoned the classical concept of theory, but that they have not completely abandoned it. The suspicion of objectivism exists because of the *ontological illusion of pure theory* that the sciences still deceptively share with the philosophical tradition *after casting off its practical content*.

With Husserl we shall designate as objectivistic an attitude that naively correlates theoretical propositions with matters of fact. This attitude presumes that the relations between empirical variables represented in theoretical propositions are self- existent. At the same time, it suppresses the transcendental framework that is the precondition of the meaning of the validity of such propositions. As soon as these statements are understood in relation to the prior frame of reference to which they are affixed, the objectivist illusion dissolves and makes visible a knowledge-constitutive interest.

There are three categories of processes of inquiry for which a specific connection between logical-methodological rules and knowledge-constitutive interests can be demonstrated. This demonstration is the task of a critical philosophy of science that escapes the snares of positivism.[7] The approach of the empirical-analytic sciences incorporates a *technical* cognitive interest; that of the historical-hermeneutic sciences incorporates a *practical* one; and the approach of critically oriented sciences incorporates the *emancipatory* cognitive interest that, as we saw, was at the root of traditional theories. I should like to clarify this thesis by means of a few examples.

V

In the *empirical-analytic sciences* the frame of reference that prejudges the meaning of possible statements establishes rules both for the construction of theories and for their critical testing.[8] Theories comprise hypothetico-deductive connections of propositions, which permit the deduction of lawlike hypotheses with empirical content. The latter can be interpreted as statements about the covariance of observable events; given a set of initial conditions, they make predictions possible. Empirical-analytic knowledge is thus possible predictive knowledge. However, the *meaning* of such predictions, that is their technical exploitability, is established only by the rules according to which we apply theories to reality.

In controlled observation, which often takes the form of an experiment, we generate initial conditions and measure the results of operations carried out under these conditions. Empiricism attempts to ground the objectivist illusion in observations expressed in basic statements. These observations are supposed to be reliable in providing immediate evidence without the admixture of subjectivity. In reality basic statements are not simple representations of facts in themselves, but express the success or failure of our operations. We can say that facts and the relations between them are apprehended descriptively. But this way of talking must not conceal that as such the facts relevant to the empirical sciences are first constituted through an a priori organization of our experience in the behavioral system of instrumental action.

Taken together, these two factors, that is the logical structure of admissible systems of propositions and the type of conditions for corroboration, suggest that theories of the empirical sciences disclose reality subject to the constitutive interest in the possible securing and expansion, through information, of feedback-monitored action. This is the cognitive interest in technical control over objectified processes.

The *historical-hermeneutic sciences* gain knowledge in a different methodological framework. Here the meaning of the validity of propositions is not constituted in the frame of reference of technical control. The levels of formalized language and objectified experience have not yet been divorced. For theories are not constructed deductively and experience is not organized with regard to the success of operations. Access to the facts is provided by the understanding of meaning, not observation. The verification of lawlike hypotheses in the empirical-analytic sciences has its counterpart here in the interpretation of texts. Thus the rules of hermeneutics determine the possible meaning of the validity of statements of the cultural sciences.[9]

Historicism has taken the understanding of meaning, in which mental facts are supposed to be given in direct evidence, and grafted onto it the objectivist illusion of pure theory. It appears as though the interpreter transposes himself into the horizon of the world or language from which a text derives its meaning. But here, too, the facts are first constituted in relation to the standards that establish them. Just as positivist self-understanding does not take into account explicitly the connection between measurement operations and feedback control, so it eliminates from consideration the interpreter's pre-understanding. Hermeneutic knowledge is always mediated through this pre-understanding, which is derived from the interpreter's initial situation. The world of traditional meaning discloses itself to the interpreter only to the extent that his own world becomes clarified at the same time. The subject of understanding establishes communication between both worlds. He comprehends the substantive content of tradition by *applying* tradition to himself and his situation.

If, however, methodological rules unite interpretation and application in this way, then this suggests that hermeneutic inquiry discloses reality subject to a constitutive interest in the preservation and expansion of the intersubjectivity of possible action-orienting mutual understanding. The understanding of meaning is directed in its very structure toward the attainment of possible consensus among actors in the framework of a self-understanding derived from tradition. This we shall call the *practical* cognitive interest, in contrast to the technical.

The systematic *sciences of social action*, that is economics, sociology, and political science, have the goal, as do the empirical-analytic sciences, of producing nomological knowledge.[10] A critical social science, however, will not remain satisfied with this. It is concerned with going beyond this goal to determine when theoretical statements grasp invariant regularities of social action as such and when they express ideologically frozen relations of dependence that can in principle be transformed. To the extent that this is the case, the *critique of ideology*, as well, moreover, as *psychoanalysis*, take into account that information about lawlike connections sets off a process of reflection in the consciousness of those whom the laws are about. Thus the level of unreflected consciousness, which is one of the initial conditions of such laws, can be transformed. Of course, to this end a critically mediated knowledge of laws cannot through reflection alone render a law itself inoperative, but it can render it inapplicable.

The methodological framework that determines the meaning of the validity of critical propositions of this category is established by the concept of *self-reflection*. The latter releases the subject from dependence on hypostatized powers. Self-reflection is determined by an emancipatory cognitive interest. Critically oriented sciences share this interest with philosophy.

However, as long as philosophy remains caught in ontology, it is itself subject to an objectivism that disguises the connection of its knowledge with the human interest in autonomy and responsibility (*Mündigkeit*). There is only one way in which it can acquire the power that it vainly claims for itself in virtue of its seeming freedom from presuppositions: by acknowledging its dependence on this interest and turning against its own illusion of pure theory the critique it directs at the objectivism of the sciences.[11]

VI

The concept of knowledge-constitutive human interests already conjoins the two elements whose relation still has to be explained: knowledge and interest. From everyday experience we know that ideas serve often enough to furnish our actions with justifying motives in place of the real ones. What is called rationalization at this level is called ideology at the level of collective action. In both cases the manifest content of statements is falsified by consciousness' unreflected tie to interests, despite its illusion of autonomy. The discipline of trained thought thus correctly aims at excluding such interests. In all the sciences routines have been developed that guard against the subjectivity of opinion, and a new discipline, the sociology of knowledge, has emerged to counter the uncontrolled influence of interests on a deeper level, which derive less from the individual than from the objective situation of social groups. But this accounts for only one side of the problem. Because science must secure the objectivity of its statements against the pressure and seduction of particular interests, it deludes itself about the fundamental interests to which it owes not only its impetus but the *conditions of possible objectivity* themselves.

Orientation toward technical control, toward mutual understanding in the conduct of life, and toward emancipation from seemingly "natural" constraint establish the specific viewpoints from which we can apprehend reality as such in any way whatsoever. By becoming aware of the impossibility of getting beyond these transcendental limits, a part of nature acquires, through us, autonomy in nature. If knowledge could ever outwit its innate human interest, it would be by comprehending that the mediation of subject and object that philosophical consciousness attributes exclusively to *its own* synthesis is produced originally by interests. The mind can become aware of this natural basis reflexively. Nevertheless, its power extends into the very logic of inquiry.

Representations and descriptions are never independent of standards. And the choice of these standards is based on attitudes that require critical consideration by means of arguments, because they cannot be either logically deduced or empirically demonstrated. Fundamental methodological decisions, for example such basic distinctions as those between categorial and noncategorial being, between analytic and synthetic statements, or between descriptive and emotive meaning, have the singular character of being neither arbitrary nor compelling.[12] They prove appropriate or inappropriate. For their criterion is the metalogical necessity of interests that we can neither prescribe nor represent, but with which we must instead *come to terms*. Therefore my *first thesis* is this: *The achievements of the transcendental subject have their basis in the natural history of the human species.*

Taken by itself this thesis could lead to the misunderstanding that reason is an organ of adaptation for men just as claws and teeth are for animals. True, it does serve this function. But the human interests that have emerged in man's natural history, to which we have traced back the three knowledge-constitutive interests, derive both from nature and *from the cultural break* with nature. Along with the tendency to realize natural drives they have incorporated the tendency toward release from the constraint of nature. Even the interest in self-preservation, natural as it seems, is represented by a social system that compensates for the lacks in man's organic equipment and secures his historical existence *against* the force of nature threatening from without. But society is not only a system of self-preservation. An enticing natural force, present in the individual as libido, has detached itself from the behavioral system of self-preservation and urges toward utopian fulfillment. These individual demands, which do not initially accord with the requirement of collective self-preservation, are also absorbed by the social system. That is why the cognitive processes to which social life is indissolubly linked function not only as means to the reproduction of life; for in equal measure they themselves determine the definitions of this life. What may appear as naked survival is always in its roots a historical phenomenon. For it is subject to the criterion of what a society intends for itself as *the good life*. My *second thesis* is thus that *knowledge equally serves as an instrument and transcends mere self-preservation.*

The specific viewpoints from which, with transcendental necessity, we apprehend reality ground three categories of possible knowledge: information that expands our power of technical control; interpretations that make possible the orientation of action within common traditions; and analyses that free consciousness from its dependence on hypostatized powers. These viewpoints originate in the interest structure of a species that is linked in its roots to definite means of social organization: work, language, and power. The human species secures its existence in systems of social labor and self-assertion through violence, through tradition-bound social life in ordinary-language communication, and with the aid of ego identities that at every level of individuation reconsolidate the consciousness of the individual in relation to the norms of the group. Accordingly the interests constitutive of knowledge are linked to the functions of an ego that adapts itself to its external conditions through learning processes, is initiated into the communication system of a social life-world by means of self-formative processes, and constructs an identity in the conflict between instinctual aims and social constraints. In turn these achievements become part of the productive forces accumulated by a society, the cultural tradition through which a society interprets itself, and the legitimations that a society accepts or criticizes. My *third thesis* is thus that *knowledge-constitutive interests take form in the medium of work, language, and power.*

However, the configuration of knowledge and interest is not the same in all categories. It is true that at this level it is always illusory to suppose an autonomy, free of presuppositions, in which knowing first grasps reality theoretically, only to be taken subsequently into the service of interests alien to it. But the mind can always reflect back upon the interest structure that joins subject and object a priori: this is reserved to self-reflection. If the latter cannot cancel out interest, it can to a certain extent make up for it.

It is no accident that the standards of self-reflection are exempted from the singular state of suspension in which those of all other cognitive processes require critical evaluation. They possess theoretical certainty. The human interest in autonomy and responsibility is not mere fancy, for it can be apprehended a priori. What raises us out of nature is the only thing whose nature we can know: *language*. Through its structure, autonomy and responsibility are posited for us. Our first sentence expresses unequivocally the intention of universal and unconstrained consensus. Taken together, autonomy and responsibility constitute the only Idea the we possess a priori in the sense of the philosophical tradition. Perhaps that is why the language of German Idealism, according to which "reason" contains both will and consciousness as its elements, is not quite obsolete. Reason also means the will to reason. In self-reflection knowledge for the sake of knowledge attains congruence with the interest in autonomy and responsibility. The emancipatory cognitive interest aims at the pursuit of reflection as such. My *fourth thesis* is thus that *in the power of self-reflection, knowledge and interest are one.*

However, only in an emancipated society, whose members' autonomy and responsibility had been realized, would communication have developed into the non-authoritarian and universally practiced dialogue from which both our model of reciprocally constituted ego identity and our idea of true consensus are always implicitly derived. To this extent the truth of statements is based on anticipating the realization of the good life. The ontological illusion of pure theory behind which knowledge-constitutive interests become invisible promotes the fiction that Socratic dialogue is possible everywhere and at any time. From the beginning philosophy has presumed that the autonomy and responsibility posited with the structure of language are not only anticipated but real. It is pure theory, wanting to derive everything from itself, that succumbs to unacknowledged external conditions and becomes ideological. Only when philosophy discovers in the dialectical course of history the traces of violence that deform repeated attempts at dialogue and recurrently close off the path to unconstrained communication does it further the process whose suspension it otherwise legitimates: mankind's evolution toward autonomy and responsibility. My *fifth thesis* is thus that *the unity of knowledge and interest proves itself in a dialectic that takes the historical traces of suppressed dialogue and reconstructs what has been suppressed.*

VII

The sciences have retained one characteristic of philosophy: the illusion of pure theory. This illusion does not determine the practice of scientific research but only its self-understanding. And to the extent that this self-understanding reacts back upon scientific practice, it even has its point.

The glory of the sciences is their unswerving application of their methods without reflecting on knowledge-constitutive interests. From knowing not what they do methodologically, they are that much surer of their discipline, that is of methodical progress within an unproblematic framework. False consciousness has a protective function. For the sciences lack the means of dealing with the risks that appear once the connection of knowledge and human interest has been comprehended on the level of self-reflection. It was possible for fascism to give birth to the freak of a national physics and Stalinism to that of a Soviet Marxist genetics (which deserves to be taken more seriously than the former) only because the illusion of objectivism was lacking. It would have been able to provide immunity against the more dangerous bewitchments of misguided reflection.

But the praise of objectivism has its limits. Husserl's critique was right to attack it, if not with the right means. As soon as the objectivist illusion is turned into an affirmative *Weltanschauung*, methodologically unconscious necessity is perverted to the dubious virtue of a scientistic profession of faith. Objectivism in no way prevents the sciences from intervening in the conduct of life, as Husserl thought it did. They are integrated into it in any case. But they do not of themselves develop their practical efficacy in the direction of a growing rationality of action.

Instead, the positivist self-understanding of the *nomological sciences* lends countenance to the substitution of technology for enlightened action. It directs the utilization of scientific information from an illusory viewpoint, namely that the practical mastery of history can be reduced to technical control of objectified processes. The objectivist self-understanding of the *hermeneutic sciences* is of no lesser consequence. It defends sterilized knowledge against the reflected appropriation of active traditions and locks up history in a museum. Guided by the objectivist attitude of theory as the image of facts, the nomological and hermeneutical sciences reinforce each other with regard to their practical consequences. The latter displace our connection with tradition into the realm of the arbitrary, while the former, on the levelled-off basis of the repression of history, squeeze the conduct of life into the behavioral system of instrumental action. The dimension in which acting subjects could arrive rationally at agreement about goals and purposes is surrendered to the obscure area of mere decision among reified value systems and irrational beliefs.[13] When this dimension, abandoned by all men of good will, is subjected to reflection that relates to history objectivistically, as did the philosophical tradition, then positivism triumphs at the highest level of thought, as with Comte. This happens when critique uncritically abdicates its own connection with the emancipatory knowledge-constitutive interest in favor of pure theory. This sort of high-flown critique projects the undecided process of the evolution of the human species onto the level of a philosophy of history that dogmatically issues instructions for action. *A delusive philosophy of history, however, is only the obverse of deluded decisionism.* Bureaucratically prescribed partisanship goes only too well with contemplatively misunderstood value freedom.

These practical consequences of a restricted, scientistic consciousness of the sciences[14] can be countered by a critique that destroys the illusion of objectivism. Contrary to Husserl's expectations, objectivism is eliminated not through the power of renewed *theoria* but through demonstrating what it conceals: the connection of knowledge and interest. Philosophy remains true to its classic tradition by renouncing it. The insight that the truth of statements is linked in the last analysis to the intention of the good and true life can be preserved today only on the ruins of ontology. However even this philosophy remains a specialty alongside of the sciences and outside public consciousness as long as the heritage that it has critically abandoned lives on in the positivistic self-understanding of the sciences.

Notes

1   Friedrich W. J. von Schelling, *Werke*, edited by Manfred Schröter (Munich: Beck, 1958–9), 3:299.

2   Bruno Snell, "Theorie und Praxis," in *Die Entdeckung des Geistes*, 3d ed. (Hamburg: Claassen, 1955), p. 401 ff.; Georg Picht, "Der Sinn der Unterscheidung von Theorie und Praxis in der griechischen Philosophie," in *Evangelische Ethik* (1964), 8:321 ff.

3   "Traditionelle und kritische Theorie," in *Zeitschrift für Sozialforschung*, 6:245 ff. Reprinted in Max Horkheimer, *Kritische Theorie*, edited by Alfred Schmidt (Frankfurt am Main: Fischer, 1968), pp. 137–91.

4   The appendix was the basis of my inaugural lecture at the University of Frankfurt am Main on June 28, 1965. Bibliographical notes are restricted to a few references.

5   *Die Krisis der europäischen Wissenschaften und die transzendentale Phänomenologie* in *Gesammelte Werke* (The Hague: Martinus Nijhoff, 1950), vol. 6.

6   See Gérard Gäfgen, *Theorie der wirtschaftlichen Entscheidung* (Tübingen: Mohr, 1963).

7   This path has been marked out by Karl-Otto Apel.

8   See Popper's *The Logic of Scientific Discovery*, and my paper "Analytische Wissenschaftstheorie," in *Zeugnisse* (Frankfurt am Main: Europäische Verlagsanstalt, 1963), p. 473 ff.

9   I concur with the analyses in Part II of Hans-Georg Gadamer, *Wahrheit und Methode*.

10   Ernst Topitsch, editor, *Logik der Sozialwissenschaften* (Cologne: 1965).

11   Theodor W. Adorno, *Zur Metakritik der Erkenntnistheorie*.

12   Morton White, *Toward Reunion in Philosophy* (Cambridge: Harvard University Press, 1956).

13   See my essay "Dogmatismus, Vernunft und Entscheidung" (Dogmatism, Reason, and Decision) in *Theorie und Praxis*.

14   In *One-Dimensional Man* (Boston: Beacon, 1964) Herbert Marcuse has analyzed the dangers of the reduction of reason to technical rationality and the reduction of society to

the dimension of technical control. In another context, Helmut Schelsky has made the same diagnosis:

With a scientific civilization that man himself creates according to plan, a new peril has entered the world: the danger that man will develop himself only in external actions of altering the environment, and keep and deal with everything, himself and other human beings, at this object level of constructive action. This new self-alienation of man, which can rob him of his own and others' identity ... is the danger of the creator losing himself in his work, the constructor in his construction. Man may recoil from completely transcending himself toward self-produced objectivity, toward constructed being; yet he works incessantly at extending this process of scientific self-objectification.

See Schelsky's *Einsamkeit und Freiheit* (Hamburg: 1963), p. 299.

INDEX